THE QUANTUM CLASSICAL THEORY

The Quantum Classical Theory

GERT D. BILLING

OXFORD
UNIVERSITY PRESS

2003

OXFORD
UNIVERSITY PRESS

Oxford New York
Auckland   Bangkok   Buenos Aires   Cape Town   Chennai
Dar es Salaam   Delhi   Hong Kong   Istanbul   Karachi   Kolkata
Kuala Lumpur   Madrid   Melbourne   Mexico City   Mumbai   Nairobi
São Paulo   Shanghai   Taipei   Tokyo   Toronto

Published by Oxford University Press, Inc.
198 Madison Avenue, New York, New York 10016

www.oup.com

Oxford is a registered trademark of Oxford University Press

Library of Congress Cataloging-in-Publication Data
Billing, Gert D.
The quantum classical theory / Gert D. Billing.
p. cm.
Includes bibliographical references and index.
ISBN 0-19-514619-0
1. Quantum chemistry. 2. Molecular dynamics. I. Title.

QD462 B55 2002
541.2'8—dc21   2002025772

9 8 7 6 5 4 3 2 1
Printed in the United States of America
on acid-free paper

To my mother Jørga and my grandchild Maria

Preface

When I started getting interested in quantum-classical or semi-classical methods, the theories available were of the type described in the beginning of chapter 3 of this book. These theories were later mainly called classical path theories, whereas the word semi-classical was reserved for the classical S-matrix theories developed in the seventies. These latter theories have been extensively covered in the two books by Professor M. S. Child, *Molecular Collision Theory* (1974) and *Semiclassical Mechanics with Molecular Applications* (1991). Since then the "old" classical path theory has been linked to the progress made in wave-packet theory. Today the theory can be derived from first principles and extended to cover a large range of molecular processes. The purpose of the present book is, therefore, to describe this development as well as other quantum-classical or semi-classical methods which have been not previously been reviewed in textbooks.

Often one does not distinguish between the two designations, semi-classical or quantum-classical. However, here we shall use the name "quantum-classical" and reserve "semi-classical" to methods as classical S-matrix theory, WKB theories, and others.

The possibility of blending classical and quantum mechanics provides the obvious choice if a flexible, as well as a general, theory for molecular processes is to be created. The method will and should be able to treat quantum effects when present and at the same time be able to take advantage of the ease with which classical treatments of even large systems can be formulated.

Some of the quantum-classical methods mentioned here have been successfully used to treat a large body of molecular events: Inelastic processes, energy transfer, collision induced dissociation, as well as photo-dissociation, chemical reactions, and non-adiabatic electronic processes.

During the preparation of the manuscript, the various models and methods have been used to study systems and processes, so as to obtain results not published previously. This is, for instance, the case for the calculations of vibrational relaxation of CO_2 colliding with oxygen atoms, the reaction path calculations on the $H_2+CN \rightarrow HCN+H$ reaction, and the quantum-classical results for the $HO+CO \rightarrow CO_2+H$ reaction, as well as for some of the model calculations in chapters 2 and 3. Hopefully, enough details are given for the practical implementation of the methods.

I would like to thank the many collaborators I have had in these areas over the last twenty to thirty years. I would also like to thank Professor John Avery, Department of Chemistry, University of Copenhagen, for proofreading the manuscript.

Contents

Abbreviations

CC	coupled channel
CMD	centroid molecular dynamics
CS	coupled states
CP	classical path
DOF	degree of freedom
DVR	discrete variable representation
FFT	fast Fourier transform
FMS	full multiple spawning
GH	Gauss-Hermite
GWP	Gaussian wave packet
IRP	intrinsic reaction path
IVR	initial value representation
LHA	local harmonic approximation
LZ	Landau-Zener
MCSCF	multi-configuration self-consistent field
MCTDH	multi-configuration time-dependent Hartree
MEM	minimum error method
ps	pico-second
RAM	random access memory
RP	reaction path
SCF	self-consistent field
SQ	second quantization
TDGH	time-dependent Gauss-Hermite
TDSCF	time-dependent self-consistent field
TDSE	time-dependent Schrödinger equation
VT	vibration-translation
VV	vibration-vibration
WKB	Wentzel, Kramer, Brillouin

THE QUANTUM CLASSICAL THEORY

Chapter 1

Introduction

Molecular dynamics deals with the motion of and the reaction between atoms and molecules. The fundamental theory for the description of essentially all aspects of the area has been known and defined through the non-relativistic Schrödinger equation since 1926. The "only" problem, therefore, is the solution of this fundamental equation. Unfortunately, this solution is not straightforward and, as early as 1929, prompted the following remark by Dirac (1929).

The underlying physical laws necessary for the mathematical theory of a large part of physics and the whole of chemistry are thus completely known, and the difficulty is only that the application of these laws leads to equations much too complicated to be soluble.

Dirac could, for that matter, have added the area of molecular biochemistry. But here the systems become even bigger and therefore the above statement is even more correct. What neither Dirac nor anybody else at that time could foresee was the invention of the computer. With that, a whole new area, namely that of computational chemistry, was created. The recent five-volume work *Encyclopedia of Computational Chemistry* (1998[1]), with several hundred entries, bears witness to the tremendous evolution in this particular area over the last fifty years or so. The success of computational chemistry has to do not only with computers and the increase in computational speed (see fig. 1.1) but also with the development of new methods. Here again it should be emphasized that the availability of computers makes the construction of approximate methods a very rich and diverse field with many possibilities. Thus, this combination of computer power and the invention of theoretical and computational methods has changed the pessimistic point of view into an optimistic one. To quote Clementi (1972), "We can calculate everything." Although this statement, at least in 1972, was somewhat optimistic, development since then has shown that the attitude should be quite optimistic.

The purpose of approximate methods should be, and always is, to try to circumvent the bad scaling relations of quantum mechanics (see fig. 1.2). The quantum mechanical scaling increases, roughly speaking, the computational effort by three orders of magnitude for each additional nucleus (three degrees of freedom) added to the system. Solution of the three-particle nuclear dynamical problem requires roughly giga-flop performance, whereas a four-particle problem requires tera-flop performance, and so on. Obviously, it is necessary to introduce

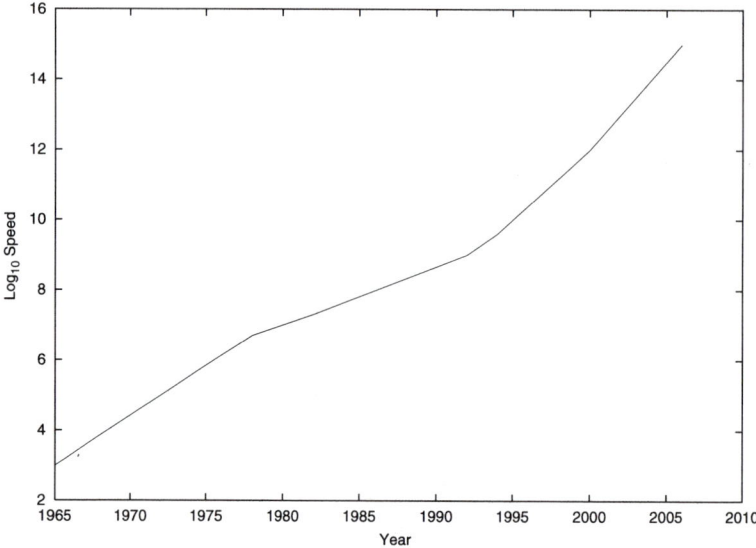

Fig. 1.1: Illustration of the increase in computer speed over the last 50 years.

approximations of some sort in order to improve on this scaling. In order to do so, scientists have customarily argued that the nuclear mass is large compared to that of the electrons. Hence, one might expect that a classical mechanical treatment of nuclear motion would be sufficiently accurate. Since classical mechanics involves the integration of equations of motion for the coordinates and

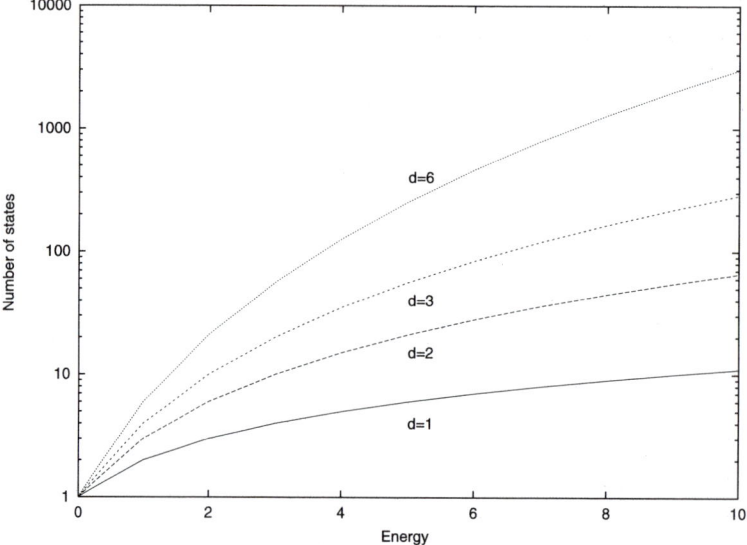

Fig. 1.2: Illustration of the increase in the number of quantum states with dimension and energy of the system.

momenta, the effort for a single trajectory in phase-space scales linearly with the number of particles. However, we do have to average over a number N_t of classical trajectories, that is, we must repeat the calculation many times. Since the classical phase-space has both coordinates and momenta, it is larger than the quantum space, which has one or the other. Thus, we have to sample over a larger space. However, the quantities we are interested in often converge quickly to a reliable number (as $\sqrt{N_t}$) but rather slowly to a very accurate number. Since it is inexpensive to run (integrate) one trajectory, it is possible to average over several hundreds of thousands of trajectories for small systems consisting of, say, less than 10 atoms for processes occurring on time-scales of a typical reaction (less than a pico-second). This large number of trajectories is necessary for the most detailed information, that is, state-to-state and angular resolved scattering cross sections. For larger systems and processes occurring on longer time-scales, the number of trajectories required is often much smaller because the information which is typically sought is not so dependent on details.

The obvious question is, then, Why do we not simply use classical mechanics for the nuclear motion? The reason is, of course, that many effects, also in nuclear dynamics, are really quantum mechanical and cannot be adequately described using classical mechanics, that is, by running classical trajectories. Tunneling through a potential barrier is a quantum phenomenon and it is important to include this aspect for some reactions, especially those involving small masses, low energies (temperatures), and narrow or small barriers for reaction. Classically forbidden events are in some sense also related to tunneling phenomena but are connected to state-resolved rates or cross sections. Thus, problems with using classical dynamics appear also for non-reactive scattering. Some transitions are even—though they are energetically allowed—not dynamically allowed in classical mechanics. This is a problem mainly in small molecules with large vibrational energy spacings. For polyatomic $(N > 3)$ molecules, another problem in classical mechanics arises: How do we assign quantum numbers to the trajectories describing molecular motion? This is especially problematic for polyatomic molecules or, in general, for what are called non-separable hamiltonians. For very large molecules, the "availability" in classical mechanics of a large amount of zero-point vibrational energy is also a problem. Also, for smaller systems the zero-point vibrational energy poses a problem since it is not automatically conserved in classical mechanics, that is, it may artificially be available to other degrees of freedom. Since the electronic motion is quantum mechanical, how do we cope with systems where, due to the breakdown of the Born-Oppenheimer approximation, we have coupling between electronic and nuclear motion? This coupling gives rise to various phenomena as geometric phase effect, Landau-Zener transitions, coupling between spin and orbital degrees of freedom, and Renner and Jahn-Teller couplings.

Since the various degrees of freedom are more or less classical (roughly according to their DeBroglie wavelength) (see table 1.2), the tempting answer to these problems would be to try to mix quantum and classical mechanics. This too, has been either attempted or done since the invention of quantum mechanics. It is, however, not clear how this mixing should be carried out, the reason being the different nature of the two mechanics—one being probabilistic

Table 1.1: Quantum effects of nuclear motion.

Tunneling
Classical forbidden events
Zero-point vibrational motion
Coupling of electronic and nuclear motion
Non-adiabatic transitions
Geometric phase effects
Spin-orbit coupling

in nature, the other being deterministic; one involving only coordinates or momenta, the other both; one involving time explicitly, the other involving, for time-independent hamiltonians, only time as a parameter; one obeying the Heisenberg uncertainty principle, the other not.

These differences in the mere nature of the two mechanics might give the impression that it is as difficult to mix quantum and classical mechanics as it is to mix oil and water. At some stage of development, this was also the opinion held by much of the scientific community. As a matter of fact, the consensus in the early seventies was that the quantum-classical approach was valid only under very limited conditions. But it turns out that it is, in fact, possible to derive the theory in a self-consistent manner and to extend its domain of validity considerably. The present book discusses how it should be done.

In the chapters 2 through 4, we describe a method which provides a quantum-classical approach derived from first principles (see flow diagram in fig. 1.3). The approach allows for an (in principle) exact treatment of dynamical processes through time-dependent multi-configuration self-consistent field (MCSCF) and the "quantum dressed" classical mechanics methods, but offers in addition a number of useful approximate schemes. In this manner, it is easy to evaluate the approximations involved in the various schemes, and it is possible to go from a completely quantum to a completely classical theory in a self-consistent manner. Since classical mechanics involves time, the natural starting point for a derivation of quantum-classical theories is the time-dependent rather than the time-independent Schrödinger equation, although early attempts have been made using the latter [2]. The book will, therefore, primarily deal with time-dependent theories such as path-integral methods, Gaussian wave-packet

Table 1.2: Degrees of freedom according to their classical behavior at a given energy.

Degree of freedom	Behavior
Translation	Classical
Rotation	Mainly classical
Vibration	Mainly quantum
Electronic	Quantum

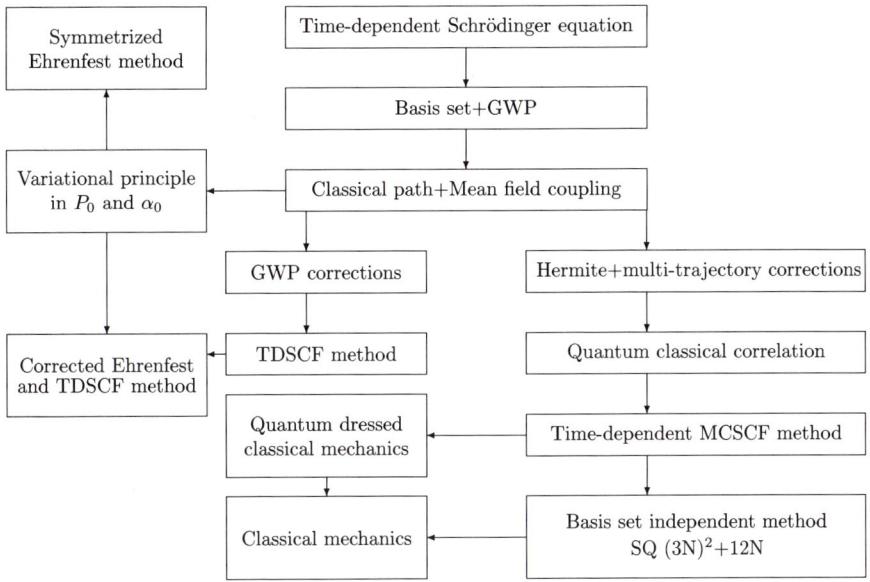

Fig. 1.3: Flow diagram for a quantum-classical approach to molecular dynamics. N is the number of atoms of the molecular system while P_0 and α_0 are momentum and width parameters of the GWP (see also chapter 3).

(GWP) propagation, the Gauss-Hermite expansion, time-dependent discrete variable representation (DVR) methods, that is, theories which are formally exact but which can be used to derive some very useful approximate schemes as time-dependent self-consistent field (TDSCF), the classical path theory, and trajectory hopping methods. The possibility of using a second quantization (SQ) instead of a first quantization picture furthermore constitutes the basis for efficient time-dependent methods for large systems. This method is reviewed in chapter 4. Chapter 5 gives examples where the SQ methodology has been used: Energy transfer to polyatomic molecules, surface chemistry, chemical reactions through the reaction path, and volume theories. For treating chemical processes in solution, special methods have been developed. Some of these are also mentioned in chapter 5.

Chapter 2

Rigorous theories

In this chapter we discuss theories which are rigorous in their formulation but which in order to be useful need to be modified by introducing approximations of some kind. The approximations we are interested in are those which involve introduction of classical mechanical concepts, that is, the classical picture and/or classical mechanical equations of motion in part of the system. At this point, we wish to distinguish between "the classical picture," which is obtained by taking the classical limit $\hbar \to 0$ and the appearance of "classical equations of motion." The latter may be extracted from the quantum mechanical formulation without taking the classical limit—but, as we shall see later by introducing a certain parametrization of quantum mechanics.

Thus there are two ways of introducing classical mechanical concepts in quantum mechanics. In the first method, the classical limit is defined by taking the limit $\hbar \to 0$ either in all degrees of freedom (complete classical limit) or in some degrees of freedom (semi-classical theories). We note in passing that the word *semi-classical* has been used to cover a wide variety of approaches which have also been referred to as classical S-matrix theories, quantum-classical theories, classical path theory, hemi-quantal theory, Wentzel Kramer-Brillouin (WKB) theories, and so on. It is not the purpose of this book to define precisely what is behind these various acronyms. We shall rather focus on methods which we think have been successful as far as practical applications are concerned and discuss the approximations and philosophy behind these.

In the other approach, the $\hbar$-limit is not taken—at least not explicitly—but here one introduces "classical" quantities, such as, trajectories and momenta as parameters, and derives equations of motion for these parameters. The latter method is therefore one particular way of parameterizing quantum mechanics. We discuss both of these approaches in this chapter.

2.1 Path-integral approach

The Feynman path-integral formulation is one way of formulating quantum mechanics such that the classical limit is immediately visible [3]. Formally, the approach involves the introduction of a quantity S, which has a definition resembling that of an action integral [101]. For now, we just consider a one-dimensional system, such as a particle with mass m moving along the x-axis in a potential $V(x,t)$. The quantity S is defined by

$$S(x_2 t_2; x_1 t_1) = \int_{t_1}^{t_2} L(\dot{x}, x, t) dt \tag{2.1}$$

where L is the Lagrangian, so that

$$L = \frac{m}{2}\dot{x}^2 - V(x, t) \tag{2.2}$$

and $\dot{x} = dx/dt$. The connection to the quantum mechanical formulation can be made through the equation

$$\psi(x_2, t_2) = \int_{-\infty}^{\infty} K(x_2, t_2; x_1, t_1)\psi(x_1, t_1) dx_1 \tag{2.3}$$

where $K(x_2, t_2; x_1, t_1)$ is the propagator, that is, the amplitude for a particle going from the position x_1 at time t_1 to x_2 at t_2. By summing or integrating over all such "transition" amplitudes multiplied by the amplitude for a given initial position—the wave function—we get the amplitude for a final position, namely $\psi(x_2, t_2)$. It is now possible to show that the propagator can be written as a sum or an integral over all "paths" leading from (x_1, t_1) to (x_2, t_2), each weighted by a phase factor [3]

$$\exp\left[\frac{i}{\hbar}S(x_2, t_2; x_1, t_1)\right] \tag{2.4}$$

Thus, one postulates that

$$\psi(x_2, t_2) = \frac{1}{N}\int_{-\infty}^{\infty} \exp\left(\frac{i}{\hbar}S(x_2, t_2; x_1, t_1)\right)\psi(x_1, t_1) dx_1 \tag{2.5}$$

where N is a normalization factor. The justification of the expression and the normalization factor can easily be found by considering a small time increment $t_2 = t_1 + \Delta t$ and showing that the above expression is (in this limit) the time-dependent Schrödinger equation. Thus the left-hand side is expanded as

$$\psi(x, t) + \Delta t\frac{\partial \psi(x, t)}{\partial t} \tag{2.6}$$

where x_2 has been replaced by x and t_1 by t. We note that the action given by eq. (2.1) in this limit can be approximated as

$$\frac{m}{2}\frac{(x - x_1)^2}{\Delta t} - \Delta t V(\bar{x}, t) \tag{2.7}$$

where $\bar{x}$ is an average distance $\bar{x} = \frac{1}{2}(x + x_1)$. Introducing the variable $\eta = x_1 - x$ and expanding in η and Δt we get

$$\psi(x,t) + \Delta t \frac{\partial \psi}{\partial t}$$

$$= \int_{-\infty}^{\infty} d\eta \frac{1}{N} \exp(im\eta^2/2\hbar\Delta t) \left(1 - \frac{i\Delta t}{\hbar} V(x,t)\right) \left(\psi(x,t) + \eta \frac{\partial \psi}{\partial x} + \frac{1}{2}\eta^2 \frac{\partial^2 \psi}{\partial x^2}\right)$$

$$(2.8)$$

where we have used that $\exp(-i\Delta t V(\bar{x},t)/\hbar) \sim 1 - i\Delta t V(\bar{x},t)/\hbar$.

The normalization constant is found by requiring that

$$\frac{1}{N} \int_{-\infty}^{\infty} d\eta \exp(im\eta^2/2\hbar\Delta t) = 1 \qquad (2.9)$$

which gives

$$N = \sqrt{\frac{2i\pi\hbar\Delta t}{m}} \qquad (2.10)$$

By equating terms of order Δt and performing the integrals over the variable η, we obtain

$$i\hbar \frac{\partial \psi}{\partial t} = -\frac{\hbar^2}{2m} \frac{\partial^2 \psi}{\partial x^2} + V(x,t)\psi \qquad (2.11)$$

that is, the time-dependent Schrödinger equation. These considerations not only determine the normalization constant N but also justify the expression (2.5).

The above result can now be generalized to a finite time interval—simply by repetition—such that the propagator becomes

$$K(x_2, t_2; x_1, t_1) = \lim_{\Delta t \to 0} \frac{1}{N^{(n-1)}}$$

$$\times \int\int \ldots \int \exp\left(\frac{i}{\hbar} S(x_2 t_2; x_1 t_1)\right) dy_1 dy_2 \ldots dy_{n-1} \quad (2.12)$$

The integration (see fig. 2.1) is carried out over the $n - 1$ time segments between the end points x_1 and x_2, that is, y_i is integrated from $-\infty$ to $+\infty$ for $i = 1, \ldots n - 1$. This way of evaluating the path integral is, however, prohibitive for many dimensional systems [5]. The exception is in case of complex time propagation, where the operator decays exponentially for paths away from the shortest between x_1 and x_2. Complex time propagation with $(t = -i\hbar\beta)$ is used in quantum statistical mechanics to evaluate for instance time-correlation functions involving a trace over the Boltzmann operator $\exp(-\beta\hat{H})$ $(\beta = 1/k_B T)$ (see, for instance, ref. [6] and section 5.7). Efficient Monte Carlo sampling techniques can be attempted in order to evaluate the path integral [7]. These techniques are, however, not sufficiently effective if the integrand is highly oscillatory, as is the case for real time propagation. Therefore, for real time propagation one has to consider other alternatives, for instance, the short time

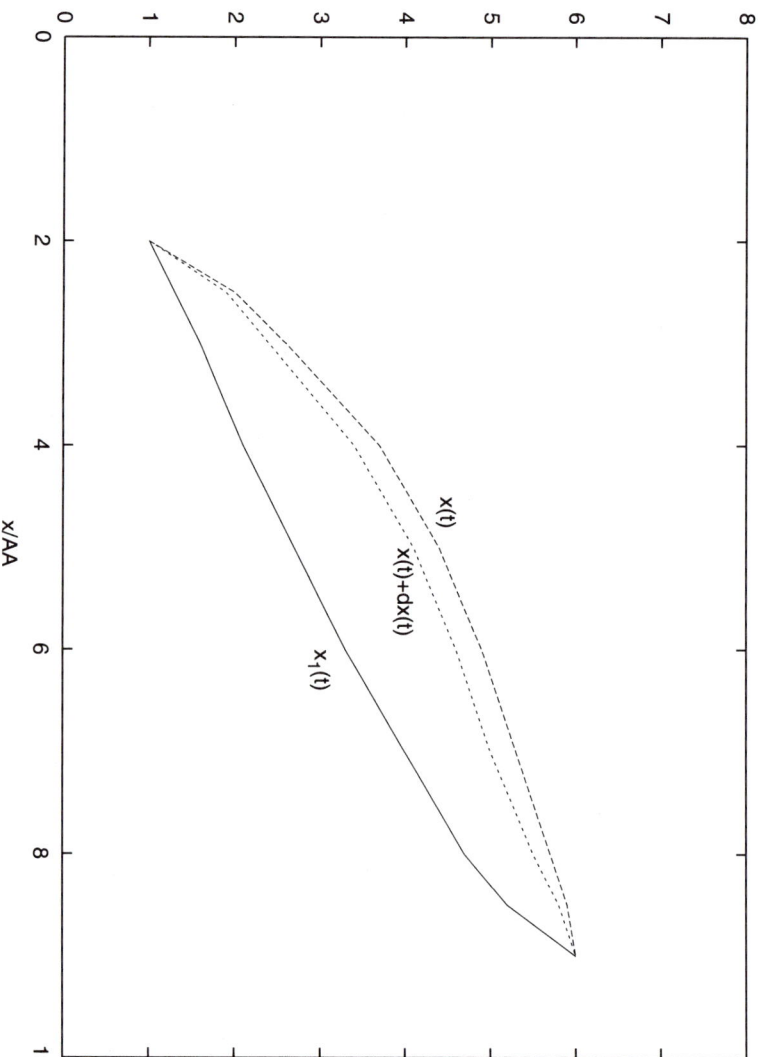

Fig. 2.1: The classical path is defined through a variation of the action integral, such that $\delta S(t_1, t_2) = 0$. The region around the classical path $x(t)$ contributes the most to the path integral. In the classical limit we can, therefore, neglect contributions from such paths as path $x_1(t)$.

propagator (see below). However, in some cases the path integration can be carried-out analytically—specifically, if the force is at most linear in the coordinate x, and in the general case for hamiltonians which can be written in a quadratic form. Consider, for example, a free particle, where the action can be written as

$$S(x_2, t_2; x_1, t_1) = \frac{im}{2\Delta t} \left[\sum_{i=2}^{n-1} (y_i - y_{i-1})^2 + (y_1 - x_1)^2 + (x_2 - y_{n-1})^2 \right] \quad (2.13)$$

Here, integration over the y_i variables results in the propagator for the free particle [3]

$$K(x_2, t_2; x_1, t_1) = \sqrt{\frac{m}{2\pi i\hbar(t_2 - t_1)}} \exp\left(\frac{im(x_2 - x_1)^2}{2\hbar(t_2 - t_1)} \right) \quad (2.14)$$

and for the harmonic oscillator, we may similarly obtain

$$K(x_2, t_2; x_1, t_1) = \sqrt{\frac{m\omega}{2\pi i\hbar \sin(\omega T)}} \exp\left(\frac{im\omega}{2\hbar \sin(\omega T)} [(x_2^2 + x_1^2)cos(\omega T) - 2x_1 x_2] \right) \quad (2.15)$$

where $T = t_2 - t_1$. But, in general, it is not possible to evaluate the propagator. We must therefore, as suggested, resort to considering the evaluation in the short time limit $(t_2 \to t_1)$.

2.1.1   The short time propagator

In order to apply the Feynman path concept in actual numerical calculations, we find it convenient to introduce the short time propagator. Various approximations in the limit of a small Δt have been used. Among those is the so-called Trotter formula [14]

$$K(x_2, t_1 + \Delta t; x_1, t) = \frac{1}{N} \exp\left[\frac{im(x_2 - x_1)^2}{2\hbar\Delta t} - \frac{i\Delta t}{\hbar} \frac{1}{2} (V(x_2) + V(x_1)) \right] \quad (2.16)$$

However, by considering the short time propagator for the harmonic oscillator (HO), one can see that for this particular case one gets from eq. (2.15) (to order Δt) an exponent

$$\frac{im(x_2 - x_1)^2}{2\hbar\Delta t} - \frac{1}{6} \frac{i\Delta t m\omega^2}{\hbar} (x_2^2 + x_1^2 + x_1 x_2) \quad (2.17)$$

Comparing with the Trotter formula, we see that the first term is identical but the latter is replaced by

$$-\frac{1}{4} \frac{i\Delta t m\omega^2}{\hbar} (x_2^2 + x_1^2) \quad (2.18)$$

noting that for the harmonic oscillator, $V(x) = \frac{1}{2}m\omega^2 x^2$. It has been demonstrated by Makri and Miller [15] that the proper short time propagator for small but finite Δt is

$$K(x_2, t_1 + \Delta t; x_1, t) = \frac{1}{N}\exp\left(\frac{im(x_2 - x_1)^2}{2\hbar\Delta t} - \frac{i\Delta t\langle V\rangle}{\hbar}\right) \quad (2.19)$$

where

$$\langle V\rangle = \frac{1}{\Delta x}\int_{x_1}^{x_2} dx V(x) \quad (2.20)$$

For higher order terms of order Δt^3, the reader is referred to ref. [15]. It can easily be demonstrated that expression (2.19) for the HO gives the correct Δt term. For finite time intervals, the above propagator can be used if it is divided into a number of time segments (N_t) such that $\Delta t = t/N_t$. The generalization of the propagator to multi-dimensional systems (with dimension d) is straightforward. But the evaluation of the multi-dimensional integral by, for example, Monte Carlo methods is prohibitively expensive. One reason for this, aside from the $N_t \times d$ dimensions of the integral, is that the integrand is oscillatory. Hence, efficient cancellation of the integral away from the stationary phase region occurs. However, this cancellation is, in fact, difficult to achieve by numerical methods.

There is, however, a possibility of introducing an effective propagator for which part of these oscillations are averaged out. The effective propagator can be introduced as [26]

$$\langle x_2|\exp(-iH\Delta t/\hbar)|x_1\rangle_{eff}$$
$$= \int_{-p_{max}}^{p_{max}} dp\langle x_2|\exp(-iH\Delta t/2\hbar)|p\rangle\langle p|\exp(-iH\Delta t/2\hbar)|x_1\rangle \quad (2.21)$$

where we have said that $\sum_p |p\rangle\langle p| = 1$. Thus the expression is strictly valid only for $p_{max} \to \infty$. But it can be used approximately for finite p_{max}. Using that now

$$\langle x|\exp(-iH\Delta t/2\hbar)|p\rangle = \frac{1}{\sqrt{2\pi\hbar}}\exp\left(\frac{i}{\hbar}(px - H\Delta t/2)\right) \quad (2.22)$$

we obtain [26]

$$K_{eff}(x_2, t_2; x_1, t_1) = K(x_2, t_2; x_1, t_1)f_{smooth}(\Delta x; p_{max}, \Delta t) \quad (2.23)$$

where the first factor is given by eq. (2.16), $\Delta x = x_2 - x_1$, $\Delta t = t_2 - t_1$, and the last factor is unity for $p_{max} \to \infty$ but for finite values is given as

$$f_{smooth}(\Delta x; p_{max}, \Delta t) = \frac{1}{2}[\mathrm{erf}(p_+) + \mathrm{erf}(p_-)] \quad (2.24)$$

where erf(x) is the error function [71] and

$$p_\pm = \sqrt{\frac{i\Delta t}{2m\hbar}}p_{max} \pm \sqrt{\frac{im}{2\hbar\Delta t}}\Delta x \quad (2.25)$$

Figure 2.2 shows the effective propagator together with the Trotter propagator for a particle in an Eckart potential $A \mathrm{sech}^2(ax)$. We notice that the oscillations die off much faster with the effective propagator and, therefore, it facilitates the evaluation of the path integral. Figure 2.3 shows the propagator for a higher value of the maximum momentum. This latter value is what is typical for molecular systems, and it is used to define the maximum energy in other time-dependent propagator methods as the Split operator [140], Lanczos [139], Chebyshev [141], and Hermite expansions of the evolution operator $\exp(-iHt/\hbar)$ [138, 25].

We are often interested in the formulation of a theory for the coupling of a large system, a "heat bath," to a smaller system. If the heat bath is approximated by a harmonic expression, that is, that only linear and quadratic terms appear in the hamiltonian, we may perform the evaluation of the short time propagator for the bath modes since it can be carried out analytically [26, 27]. However, it is also true that in many other approaches, the harmonic case can be formulated and solved efficiently.

2.1.2   The classical limit

The action S appearing in the Feynman path-integral formulation can be used to define a trajectory $\bar{x}(t)$ by introducing a least action principle, that is, the trajectory is defined by

$$\delta S = S[\bar{x} + \delta x] - S[\bar{x}] = 0 \tag{2.26}$$

By using the definition of the action integral, we can show that the least action principle leads to the Lagrangian equation of motion:

$$\frac{d}{dt}\frac{\partial L}{\partial \dot{x}} - \frac{\partial L}{\partial x} = 0 \tag{2.27}$$

In the classical limit, one expects that most of the contribution to the path integral comes from the region close to the classical path. In the limit $\hbar \to 0$, contributions from paths farther away from the one defined by $\delta S = 0$ are small due to the rapid oscillations of the integrand. It is then convenient to expand the action around this path, so that

$$S[x(t)] = S[\bar{x}(t) + y(t)] = \int_{t_1}^{t_2} dt \left[\frac{m}{2}(\dot{\bar{x}} + \dot{y})^2 - V(\bar{x}(t) + y(t)) \right] \tag{2.28}$$

Since the classical trajectory is defined by requiring that the linear term in the displacement from the path vanishes, this action integral can be written as

$$S[x(t)] = S_{cl}[\bar{x}(t)] + \int_{t_1}^{t_2} dt \left(\frac{m}{2}\dot{y}^2 - \frac{1}{2}\frac{\partial^2 V}{\partial x^2}|_{\bar{x}(t)}y^2 \right) \tag{2.29}$$

and the propagator can in this approximation be expressed as

$$K(x_2, t_2; x_1, t_1) = \exp\left(\frac{i}{\hbar} S_{cl}[\bar{x}(t)] \right) F(t_2, t_1) \tag{2.30}$$

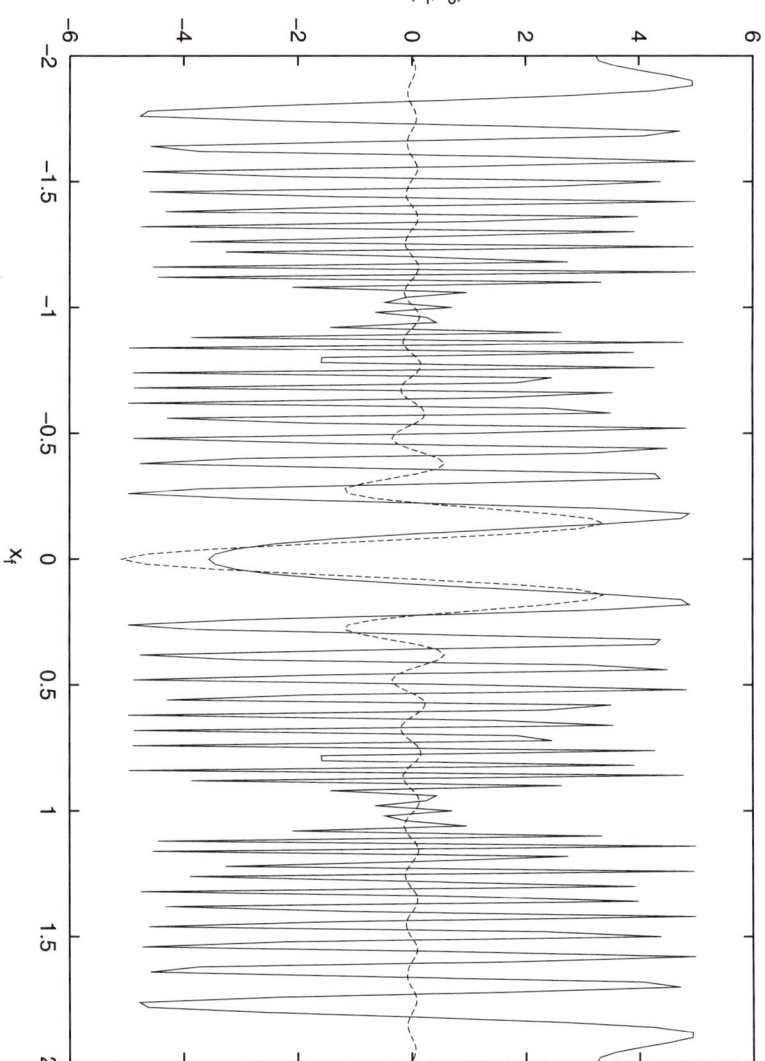

Fig. 2.2: The real part of the propagator for a particle with mass 1 amu in an Eckart potential $A/\cosh(ax)^2$. The values of A and a are 100 kJ/mol and 1 Å^{-1}, respectively. For the evaluation of the propagator we have used $x_0 = 0, \Delta t = 0.01\,\tau\,(1\tau = 10^{-14}\,\text{sec})$ and $p_{max} = 2\,\text{amu}\,\text{Å}/\tau$. The dashed line shows the effective propagator.

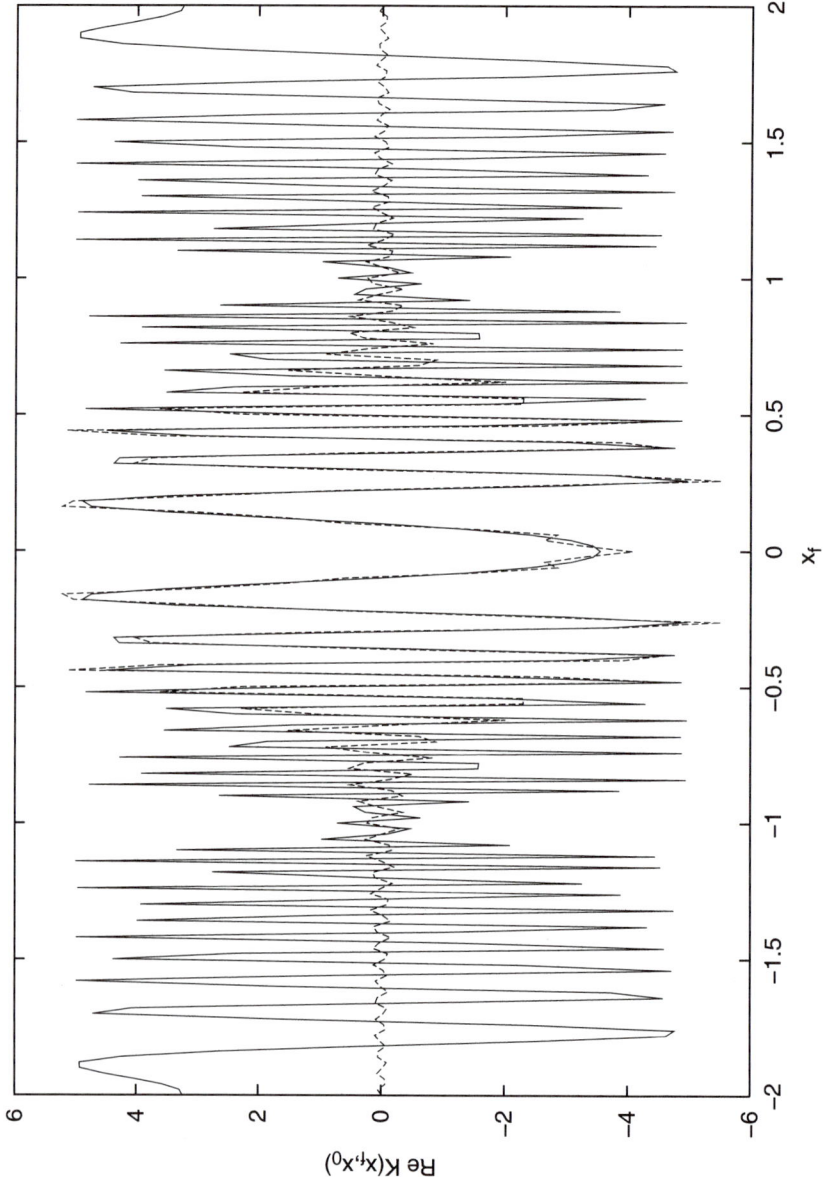

Fig. 2.3: Same as fig. 2.2 but with $p_{max} = 6\,\mathrm{amu}\,\text{Å}/\tau$.

where we have made use of the fact that the classical trajectory is assumed to start at x_1 and end at x_2. Hence, the integration over the deviation y must be carried out subject to the condition that $y(t_1) = y(t_2) = 0$. Then the path integral over y can only depend on time. Thus, in this "semi-classical" limit, the propagator consists of the classical action times a time-dependent normalization function $F(t_2, t_1)$. One can furthermore demonstrate that the function $F(t_2, t_1)$ can be obtained from the classical action as [16]

$$F(t_2, t_1) = \left(\frac{1}{2\pi i\hbar} \left| \frac{\partial^2 S_{cl}}{\partial x_1 \partial x_2} \right| \right)^{\frac{1}{2}} \tag{2.31}$$

Introducing the hamiltonian $H(x, p_x) = \frac{1}{2m} p_x^2 + V(x)$, we can rewrite the classical action as

$$S_{cl}(x_2, x_1, \Delta t) = \int_{x_1}^{x_2} dx p_x - H(x_1, p_1)\Delta t \tag{2.32}$$

where $\Delta t = t_2 - t_1$. From the above expression we obtain

$$\frac{\partial S_{cl}}{\partial x_1} = -p_1 = -p_x(t_1) \tag{2.33}$$

$$\frac{\partial S_{cl}}{\partial x_2} = p_2 = p_x(t_2) \tag{2.34}$$

The semi-classical propagator is exact if the action is a quadratic function of the coordinates x_1 and x_2. In general, it will be a locally correct (in t) or a globally approximate propagator.

If the hamiltonian is harmonic we can, as mentioned, evaluate the exact propagator but the Feynman path formulation contains another interesting possibility, namely introducing the so-called reduced propagator.

$$K_{mn}(R_2 t_2; R_1 t_1) = \int D(R(t)) \exp\left(\frac{i}{\hbar} \int_{t_1}^{t_2} L_0(R(t), \dot{R}(t), t) \right) \alpha_{mn}([R(t)]) \tag{2.35}$$

This propagator assumes a traditional quantum mechanical treatment of some degrees (r) of freedom and introduces a wave function for these $\phi_n(r)$. Thus, $\alpha_{mn}([R(t)])$ is the amplitude for a transition from quantum state n to state m in the r system, associated with or induced by the "trajectory" $R(t)$ for the degree of freedom R. The Lagrangian L_0 involves the R system only. For this degree of freedom, the path integral formulation can be used and the classical limit taken as shown above. By expressing α_{mn} as

$$\alpha_{mn} = \exp(\ln \alpha_{mn}) \tag{2.36}$$

and defining the action as the imaginary part of the exponent, such that, as $S = \int L_0 dt + \hbar \text{Im} \ln \alpha_{mn}$, the stationary phase approximation defines the trajectory $R(t)$ under a certain quantum transition $n \to m$.

This approach was first formulated by Pechukas in 1969 [108] and has mainly been used for treating non-adiabatic electronic transitions, see, for instance, Laing and Freed ref. [32]. Non-adiabatic processes will be discussed further in chapter 3. But the Pechukas theory is completely general and may, therefore, also be used to treat, for instance, inelastic vibration/rotation excitation processes quantally using the path description for the relative translational motion [4, 8].

We notice that in this formulation the strict classical limit $\hbar \to 0$ is taken in order to define $R(t)$. This limit causes some problems as far as the potential defined in this manner is concerned. It becomes non-local in time (see eq. (3.189)). Thus, the system of equations, that is, the equations for the amplitudes $\alpha_{nm}(t)$ for the states of the r-system and the equations of motion for $R(t)$ will have to be solved iteratively (see also section 3.3.1). Problems with this iterative procedure have been noticed [4], [9].

It may, therefore, be more advantageous to formulate theories in which one avoids taking the strict classical limit but retains some uncertainty or spread in the wave function for the degrees of freedom, which are near the classical limit.

Thus, the problem with the formulation given by the Pechukas theory is not that it is wrong. The problem is actually that it is right, meaning that it is the right solution but to the wrong problem. It is, in fact, the solution to the problem *"How do you mix quantum and classical mechanics?"*

The other correct and probably more fruitful answer to that question is: You do not! The fact that you are forced to take the limit $\hbar \to 0$ creates problems if that limit is taken only in one or a few degrees of freedom. By doing so, we create a situation that violates the uncertainty relation in some degrees of freedom while keeping it in others. There is a penalty (non-locality of the classical mechanics) to be paid for this and it is, therefore, better to avoid it.

However, if the reduced propagator is only used in a short time interval Δt, the problems with non-locality can be diminished.

There is, however, an alternative approach, namely to invoke some uncertainty also in the "classical" degrees of freedom. In practice, this can be achieved by introducing a wave-packet formulation in these degrees of freedom. But before we consider wave packets, another semi-classical propagator should be mentioned—the Van Vleck propagator.

2.1.3   The Van Vleck propagator

In classical mechanics, canonical transformations play an important role [101]. The transformation is used to change the coordinates and momenta such that the new coordinates and momenta are canonically conjugate variables. These transformations can be carried out through the generating functions, as, for instance, the $F_2(q, P, t)$ function [101], which depends on the old coordinates q and the new momenta P in such a way that the old momenta are given as

$$p_i = \frac{\partial F_2}{\partial q_i} \tag{2.37}$$

and the new coordinates as

$$Q_i = \frac{\partial F_2}{\partial P_i} \tag{2.38}$$

If the transformed hamiltonian depends on variables Q, P which are constant in time, then the generating function F_2 must satisfy the equation [101]

$$H(q, p, t) + \frac{\partial F_2}{\partial t} = 0 \tag{2.39}$$

Under these conditions, this equation (usually with $p = \frac{\partial F_2}{\partial q}$ inserted) is known as the Hamilton-Jacobi equation, and the solutions are denoted by S rather than F_2. The equation has the variables q_i and t, and the new momenta are, as mentioned, constants usually denoted by α_i, that is, $P_i = \alpha_i$. Likewise for the new coordinates, we use the notation $\beta_i = Q_i$. The generating function is then $S(q_i; \alpha_i, t)$ and the Hamilton-Jacobi equation

$$H\left(\frac{\partial S}{\partial q_i}; q_i, t\right) + \frac{\partial S}{\partial t} = 0 \tag{2.40}$$

The close resemblence of this equation to the TDSE was noted as early as 1926 by Wentzel [17].

It was shown by Van Vleck [18] that the wave function

$$\Psi(q, t) = A\Delta^{1/2} \exp(S(q; \alpha, t)/i\hbar) \tag{2.41}$$

where A is a constant, satisfies the equation

$$F\left(i\hbar\frac{\partial}{\partial q}; q\right)\Psi(q, t) = \Psi(q, t)F\left(\frac{\partial S}{\partial q}; q\right) + \text{terms of } \hbar \tag{2.42}$$

that is, a function F of the quantum operators. For example, the Hamiltonian function works on this wave function, giving the wave function back, times a function in which the operator $i\hbar\frac{\partial}{\partial q}$ is replaced by $\partial S/\partial q$.

$S(q_i; \alpha_i, t)$ is, for a conservative system, $S(q_i; \alpha_i) - Et$ where $-\frac{\partial S}{\partial t} = E$ is the system energy and the determinant Δ in eq. (2.41) is defined as

$$\Delta = \left|\frac{\partial^2 S}{\partial q_k \alpha_l}\right| \tag{2.43}$$

The action S satisfies the classical Hamilton-Jacobi equation (2.40) and can be determined by "running trajectories." Similarly, the value of Δ is available through its dependence on the variables q_k and α_l. Thus, the approximate wave function can be propagated using the classical action $S(q, \alpha, t)$ and information carried by the trajectories.

A similar procedure can be used for the asymptotic ($\hbar \to 0$) solution of the time-dependent Schrödinger equation for the propagator $K(xt; x_0t_0)$, that is,

$$\left[\hat{H}(x) - i\hbar\frac{\partial}{\partial t}\right]K(xt; x_0t_0) = 0 \tag{2.44}$$

where

$$\lim_{t \to t_0} K(x, t; x_0 t_0) = \delta(x - x_0) \qquad (2.45)$$

Thus, the approximate propagator, the Van Vleck propagator, is

$$K(x, t, x_0, t_0) = (2\pi i\hbar)^{-1/2} \sqrt{-\left|\frac{\partial^2 S}{\partial x \partial x_0}\right|} \exp(iS(x, t; x_0 t_0)/\hbar) \qquad (2.46)$$

where [101]

$$S(x, t; x_0, t_0) = \int_{t_0}^{t} dt' [p_x(t')\dot{x}(t') - H(p, x)] \qquad (2.47)$$

is the classical action. We now notice that

$$\frac{\partial S(x, t, x_0, t_0)}{\partial x_0} = -p(t_0) \qquad (2.48)$$

and, hence,

$$\frac{\partial^2 S}{\partial x \partial x_0} = -\frac{\partial p(t_0)}{\partial x} = -\left(\frac{\partial x}{\partial p(t_0)}\right)^{-1} \qquad (2.49)$$

Thus, an alternative expression for the semi-classical propagator is obtained as

$$K(x, t, x_0, t_0) = \frac{1}{\sqrt{2\pi i\hbar}} \frac{\exp(iS(x, t, x_0, t_0)/\hbar)}{\sqrt{\left|\frac{\partial x(t)}{\partial p(t_0)}\right|}} \qquad (2.50)$$

This expression can now be extended to n dimensions and more than one trajectory as

$$K(\mathbf{x}, t; \mathbf{x}_0, t_0) = (2\pi i\hbar)^{-n/2} \sum_{paths} \frac{\exp(iS(\mathbf{x}, \mathbf{x}_0, t)/\hbar - i\pi\nu/2)}{\sqrt{\det\left|\frac{\partial \mathbf{x}(t)}{\partial \mathbf{p}(t_0)}\right|}} \qquad (2.51)$$

where ν is an integer counting the number of caustics (zeros of the determinant) and where the sum runs over all trajectories starting at $\mathbf{x}(t) = \mathbf{x}_0$ at $t = t_0$ and ending at $\mathbf{x}(t)$.

Although the above propagator is apparently straightforward to evaluate, the practical application of it runs into a number of problems. But, the problems with the propagator are well known and summarized, for example, by Kay in ref. [42]:

1) Root search problem: We have to search for all the trajectories satisfying the boundary conditions.
2) The expression is invalid at caustics.
3) Numerical problems will occur for chaotic trajectories.
4) Many trajectories are needed to obtain an accurate wave function.

Various techniques—smoothing, filtering etc. are devised to circumvent some of these difficulties.

Below we will mention one of the successful methods. But before we do so, it is necessary to introduce the concept: *Gaussian wave packets*.

2.2 Gaussians at work

The problem that arises by letting $\hbar \to 0$ in some degrees of freedom is as mentioned previously, that the uncertainty principle is violated in one part of the system while being kept in other parts.

It turns out that a more fruitful avenue is one in which the full classical limit is not taken, that is, where the uncertainty relation is retained. Such an approach can be obtained by introducing a description based on Gaussian wave packets (GWP). Pioneering work along these lines was carried out by Lebedeff and Heller already in the sixties and seventies [48], [49].

We will express the GWP in the notation due to Heller

$$\Phi(R,t) = \exp\left(\frac{i}{\hbar}(\gamma(t) + P_R(t)(R - R(t)) + A(t)(R - R(t))^2)\right) \tag{2.52}$$

where $\gamma(t)$ and $A(t)$ have imaginary, as well as real, parts. Since the wave packet is normalized over space ($\int dR\Phi^*\Phi = 1$), we have a connection between $\mathrm{Im}A$ and $\mathrm{Im}\gamma$. It is

$$\mathrm{Im}\gamma(t) = -\frac{\hbar}{4}\ln(2\mathrm{Im}A(t)/(\pi\hbar)) \tag{2.53}$$

or

$$\exp(-\mathrm{Im}\gamma(t)/\hbar) = (2\mathrm{Im}A(t)/\pi\hbar)^{1/4} \tag{2.54}$$

The GWP has, furthermore, a momentum parameter $P_R(t)$ and is centered around a path $R(t)$ being also the average value of the momentum and position operator, that is,

$$P_R(t) = \int dR\Phi^*(R,t)\frac{\hbar}{i}\frac{\partial}{\partial R}\Phi(R,t) \tag{2.55}$$

and

$$R(t) = \int dR\Phi^*(R,t)R\Phi(R,t) \tag{2.56}$$

The packet obeys the uncertainty relation and has a spread in momentum, as well as in coordinate space. Thus, the width in momentum space is

$$\Delta P = |A(t)|\sqrt{\frac{\hbar}{\mathrm{Im}A(t)}} \tag{2.57}$$

and in coordinate space

$$\Delta R = \frac{1}{2}\sqrt{\hbar/\mathrm{Im}A(t)} \tag{2.58}$$

giving (for $\mathrm{Re}A(t_0)=0$) $\Delta R\Delta P \geq \hbar/2$.

The momentum distribution is centered around k_0, that is,

$$P(k, k_0) = \Delta R \sqrt{2/\pi} \exp(-2(\Delta R(k - k_0))^2) \tag{2.59}$$

where $k = P/\hbar$ and $k_0 = P_R(t_0)/\hbar$.

By inserting the expression (2.52) in the TDSE and expanding the potential to the second order around the path $R(t)$, we obtain by equating terms with equal powers of $R - R(t)$ the following equations of motion

$$\dot{R}(t) = P_R(t)/\mu \tag{2.60}$$

$$\dot{P}_R(t) = -\left.\frac{\partial V}{\partial R}\right|_{R=R(t)} \tag{2.61}$$

$$\dot{A}(t) = -\frac{2A(t)^2}{\mu} - \frac{1}{2}\left.\frac{\partial^2 V}{\partial R^2}\right|_{R=R(t)} \tag{2.62}$$

$$\dot{\gamma}(t) = \frac{i\hbar A(t)}{\mu} + \frac{P_R(t)^2}{\mu} - V(R(t)) \tag{2.63}$$

Strictly speaking, the first equation (2.60) has to be postulated. But this does not pose any fundamental problems. Thus, we see that the Gaussian wave packet is an exact solution to the TDSE if the potential is of the second order in the coordinate R. The center of the GWP follows a classical trajectory and, hence, it gives in this way a classical picture of the motion. The wave packet spreads or contracts as a function of time according to the potential, but the form remains the same. It is possible to use the GWP as a basis set, that is, to expand the wave function in a number of functions initially located at certain positions $R_i(t_0)$. Each of these wave packets then has the parameters $P_i(t)$ and $A_i(t)$. The initial values of these parameters can be determined from the wave function, that is, from the expansion

$$\psi(R, t_0) \sim \sum_i \Phi_i(R, R_i(t_0), P_i(t_0), A_i(t_0))c_i \tag{2.64}$$

where the c_is are expansion coefficients. If the local harmonic approximation is made, the GWPs Φ_i are propagated independent of each other using the above equations (2.60–2.63). This method is known as the LHA method.

The equations of motion for the parameters can be more accurately obtained from the time-dependent variational principle, that is, by minimizing the integral.

$$I = \int_{-\infty}^{\infty} dR \left(-i\hbar\frac{\partial \Psi(R, t)^*}{\partial t} - H\Psi(R, t)^*\right)\left(i\hbar\frac{\partial \Psi(R, t)}{\partial t} - H\Psi(R, t)\right) \tag{2.65}$$

where H is the hamiltonian operator. The time-dependence of the wave function comes about through the parameters $R_i(t)$, $P_i(t)$, $A_i(t)$, and $\gamma_i(t)$, that is, we have

$$\frac{\partial \Phi(R, t)}{\partial t} = \sum_i \frac{\partial \Phi(R, t)}{\partial x_i(t)}\dot{x}_i(t) \tag{2.66}$$

where $x_i(t) = R_i(t), P_i(t), A_i(t), \gamma_i(t)$.

This minimization leads to the so-called minimum error method (MEM), that is, we obtain with a hamiltonian of the type $H = -\frac{\hbar^2}{2\mu}\frac{\partial^2}{\partial R^2} + V(R)$, the equations [11, 19]

$$\dot{R}_i = P_i(t)/\mu \tag{2.67}$$

$$\dot{P}_i(t) = \frac{V(i,1;i,0)}{C(i,1;i,1)} \tag{2.68}$$

$$\dot{A}_i(t) = -\frac{2A_i^2}{\mu} + \frac{C(i,2;i,0)V(i,0;i,0) - C(i,0;i,0)V(i,2;i,0)}{C(i,0;i,0)C(i,2;i,2) - C(i,2;i,0)C(i,2;i,0)} \tag{2.69}$$

$$\dot{\gamma}_i(t) = \frac{i\hbar A_i(t)}{\mu} + \frac{P_i(t)^2}{2\mu}$$
$$+ \frac{C(i,2;i,0)V(i,2;i,0) - C(i,2;i,2)V(i,0;i,0)}{C(i,0;i,0)C(i,2;i,2) - C(i,2;i,0)C(i,2;i,0)} \tag{2.70}$$

where we have adopted the notation from ref. [19], that is,

$$C(i,n;j,m) = \int dR(R - R_i(t))^n (R - R_j(t))^m \Phi_i^* \Phi_j \tag{2.71}$$

$$V(i,n;j,m) = \int dR(R - R_i(t))^n (R - R_j(t))^m \Phi_i^* V \Phi_j \tag{2.72}$$

We notice that an approximation concerning the overlap between two wavepackets i and j has been made in the above equations of motion—it has been set to zero. Thus, each GWP is propagated independent of the others. The reason for introducing this so-called independent Gaussian wave packet approximation is to reduce the complexity of the scheme and to avoid the singularities in the matrix needed to invert, in order to get the equations of motion for the center of the Gaussians. We will in a later chapter avoid this inversion problem by introducing an orthorgonal basis set, the Gauss-Hermite basis set. Problems with singularities in the matrix, which has to be inverted in each time step, have been discussed in refs. [19], [24]. The MEM method is more accurate than the LHA method mentioned above. For a comparison, see, for example, ref. [20].

Letting the expansion coefficients c_i in eq. (2.64), as well as the parameters of the wave packet, depend upon time, we obtain aside from the equations of motion for parameters $R(t)$, $P(t)$, and $\gamma(t)$, also, a set of coupled equations for the expansion coefficients $c_i(t)$ [33]. The Gaussian basis functions are then moved in space according to classical mechanics, with one trajectory for each basis function. The solution of the time-dependent equations for the expansion coefficients gives an in-principle exact formulation if enough basis functions are included. This approach is the formal basis of the so-called spawning method for non-adiabatic processes. This method is described in more detail in chapter 3.

2.2.1 Cellular dynamics

The advantage of using classical mechanics for molecular dynamics problems is that each trajectory is propagated independent of the others. This makes methods based on trajectories advantageous for parallel and PC-based computing.

Several attempts have, therefore, been made to formulate quantum mechanics in such a manner that it, too, would involve trajectory propagation. One such method is the cellular dynamics method of Heller [29]. The method is based on the semi-classical Van Vleck propagator and introduces, in order to circumvent some of the problems mentioned previously with this propagator, an initial value representation (IVR), as well as a Gaussian smoothing of the integrand. Thus, we have as the first step (the IVR step), that

$$K(x, t, x_0, t_0) = (2\pi i\hbar)^{-1/2} \int dp_0 \left| \left(\frac{\partial x(t)}{\partial p_0} \right) \right|^{-1/2}$$

$$\times \, \delta(x - x(t; x_0, p_0)) \exp\left(\frac{i}{\hbar} S(x_0, p_0) - i\nu\pi/2 \right) \qquad (2.73)$$

where $x(t) = x(t; x_0, p_0)$ is the trajectory evolving from the initial condition $x = x_0$ and $p = p_0$. Thus, assuming that

$$\Psi(x, t) = \int dx_0 K(x, t, x_0, t_0) \Psi(x_0, t_0) \qquad (2.74)$$

and a Gaussian smoothing is introduced; by using that we can express the identity operator as

$$1 \sim A \sum_n \exp(-\alpha(x_0 - x_n)^2) \qquad (2.75)$$

$$1 \sim B \sum_m \exp(-\beta(p_0 - p_m)^2) \qquad (2.76)$$

which finally gives

$$\Psi(x, t) = (2\pi i\hbar)^{-1/2} AB \sum_{nm} \int dx_0 \int dp_0 \left| \frac{\partial x(t)}{\partial p_0} \right|^{-1/2} \delta(x - x(t))$$

$$\times \exp(-\alpha(x_n - x_0)^2 - \beta(p_n - p_0)^2 + iS(x_0, p_0)/\hbar - i\nu\pi/2) \Psi(x_0, t_0)$$

$$(2.77)$$

If we now expand the action $S(x_0, p_0)$ and the trajectory $x(t; x_0, p_0)$ as

$$S(x_0, p_0) = S(x_n, p_n) + \left. \frac{\partial S}{\partial x_0} \right|_{p_0} (x_0 - x_m) + \left. \frac{\partial S}{\partial p_0} \right|_{x_0} (p_0 - p_m)$$

$$+ \frac{1}{2} \left. \frac{\partial^2 S}{\partial x_0^2} \right|_{p_0} (x_0 - x_m)^2 + \left. \frac{\partial^2 S}{\partial p_0^2} \right|_{x_0} (p_0 - p_m)^2$$

$$+ \frac{\partial^2 S}{\partial x_0 \partial p_0} (x_0 - x_n)(p_0 - p_m) \qquad (2.78)$$

and

$$x(t; x_0, p_0) = x(t; x_n, p_m) + \left(\frac{\partial x(t)}{\partial x_0} \right)_{p_0} (x_0 - x_n) + \left(\frac{\partial x(t)}{\partial p_0} \right)_{x_0} (p_0 - p_m)$$

$$(2.79)$$

then the integral over p_0 can be carried out analytically [29]. The final result is

$$\Psi(x,t) = \sum_{nm} \int \left| \frac{\partial x(t; x_0, p_0)}{\partial p_0} \right|^{-1/2} g_{nm}(x, x_0)\Psi(x_0, t_0)dx_0 \qquad (2.80)$$

where the g_{nm} functions are Gaussians [29]. If the initial wave function $\Psi(x_0, t_0)$ is represented as a sum of Gaussians in x_0, then the integral over x_0 can be evaluated analytically. The cellular method as outlined here is similar in spirit to the considerations involved in the formulation of the Herman-Kluk propagator described in the next section.

2.2.2 A semi-classical IVR propagator

The problems with the Van Vleck propagator are mentioned above. One of these, the root search problem, can be removed by suitable transformation to an IVR (initial value, representation). Such transformations can be carried out by using the fact that the Dirac delta function can be expressed as

$$\delta(x) = \frac{1}{\sqrt{2\pi\hbar}} \int dp \exp(ipx/\hbar) \qquad (2.81)$$

Hence, a propagator in coordinate representation can be transformed to momentum representation as

$$K(p, p_0) = \frac{1}{2\pi\hbar} \int dx \int dx_0 \exp(-ipx/\hbar)K(x, x_0)\exp(ip_0 x_0/\hbar) \qquad (2.82)$$

The integrals can be carried out in the stationary phase approximation using the semi-classical expression for the propagator.

A very convenient transformation was introduced by Herman and Kluk [36] using the fact that the delta function can also be represented as

$$\delta(x - x') = \frac{1}{2\pi} \int dy \int dp_y \Phi(x; y, p_y)\Phi^*(x'; y, p_y) \qquad (2.83)$$

where Φ is a GWP

$$\Phi(x; y, p_y) = \exp\left(\frac{i}{\hbar}(\gamma(t) + p_y(x - y(t)) + A(t)(x - y(t))^2) \right) \qquad (2.84)$$

We now replace the propagator by its semi-classical limit (eq.(2.30)) as obtained either in the path integral or in the Van Vleck method, that is, as

$$K(x_2 t_2; x_1 t_1) = \sqrt{\frac{1}{2\pi i\hbar} \left| -\frac{\partial^2 S}{\partial x_1 \partial x_2} \right|} \exp(iS(x_2 t_2; x_1 t_1)/\hbar) \qquad (2.85)$$

By introducing the delta function twice and performing integrals by stationary phase methods, an IVR representation of the semi-classical propagator is obtained [36]. For a system with N dimensions, the propagated wave function can be expressed as

$$\Psi(\mathbf{x}, t) = \frac{1}{(2\pi\hbar)^N} \int d^N\mathbf{x}_0 d^N\mathbf{p}_0 \langle \mathbf{x}|\mathbf{x}_t\mathbf{p}_t\rangle C(\mathbf{x}_0, \mathbf{p}_0, t)$$
$$\times \exp\left(\frac{i}{\hbar}S(\mathbf{x}_0, \mathbf{p}_0, t)\right) \langle \mathbf{x}_0, \mathbf{p}_0|\Psi(\mathbf{x}_0)\rangle \tag{2.86}$$

where $S(\mathbf{x}_0, \mathbf{p}_0, t)$ is the classical action

$$S(\mathbf{x}_0, \mathbf{p}_0, t) = \int_{t_0}^{d} dt' \left(\mathbf{p}(t') \cdot \dot{\mathbf{x}}(t') - H(\mathbf{p}(t'), \mathbf{x}(t'))\right) \tag{2.87}$$

for a trajectory started at t_0 with position and momenta given by $\mathbf{x}_0$ and $\mathbf{p}_0$. The initial wave function $\Psi(\mathbf{x}_0)$ is projected on the Gauss basis set $|\mathbf{x}_0, \mathbf{p}_0\rangle$ and the trajectory is propagated to time t, defining the final GWP

$$\langle \mathbf{x}|\mathbf{x}(t), \mathbf{p}(t)\rangle = (2\text{Im}A/\pi\hbar)^{N/4} \exp\left(-\frac{\text{Im}A}{\hbar}(\mathbf{x} - \mathbf{x}(t))^2 + \frac{i}{\hbar}\mathbf{p}(t) \cdot (\mathbf{x} - \mathbf{x}(t)))\right) \tag{2.88}$$

where N is the dimension of the system. The integral is over all trajectories chosen randomly from the available initial phase-space as in ordinary trajectory methods. Each contribution is weighted by a phase factor $\exp(iS/\hbar)$ and multiplied with a pre-exponential factor

$$C(\mathbf{x}_0, \mathbf{p}_0, t) = \sqrt{\det\left|\frac{1}{2}\left(\frac{\partial \mathbf{p}(t)}{\partial \mathbf{p}_0} + \frac{\partial \mathbf{x}(t)}{\partial \mathbf{x}_0} - 2i\text{Im}A\frac{\partial \mathbf{x}(t)}{\partial \mathbf{p}_0} - \frac{1}{2i\text{Im}A}\frac{\partial \mathbf{p}(t)}{\partial \mathbf{x}_0}\right)\right|} \tag{2.89}$$

A similar expression can be obtained by a so-called Filinov [22] transformation of the Van Vleck propagator [21]. It has the effect of introducing a Gaussian smoothing or filtering of the highly oscillatory integrand. Hence, the solution is greatly stabilized. The smoothing depends on the value of $\text{Im}A$ but in the ideal world the result should, of course, be independent of this parameter [23].

In the scheme above, the time-evolution of the wave function is given by propagating the average position and momenta of a bundle of trajectories. Each of these initial values contribute with a weight given by the overlap integral of the initial wave function and a GWP. We note in passing that a Gauss-Hermite basis set (see below) has the same exponential factor as the GWP and, hence, the above expression may also be used with this basis set—propagating each member independently. The propagation is carried out keeping the width of each packet fixed ($\text{Re}A(t) = 0$ and $\text{Im}A(t) = $ constant). The fact that the propagator involves integration over initial variables is advantageous, since one avoids the search for trajectories satisfying double-ended boundary values. Still, problems with oscillations of the integrand, caustics, and especially with the contribution from chaotic trajectories have been reported [107, 42, 43]. In order to smooth such oscillations, various smoothing or filtering techniques have been proposed

[45]. Furthermore, various schemes [28] have been suggested in order to diminish the number of trajectories needed for an accurate evaluation of the integral.

A measure of the effectiveness of the propagator is obtained by the overlap between the exact wave function and the one defined by eq. (2.86). In general, the number of trajectories has to be increased rather drastically for long time propagation in order to preserve the norm of the wave function. However, the propagator has been successfully used to study a multidimensional system with a conical intersection coupled to a heat bath [47]. Here, approximately 10^7 trajectories were capable of yielding the accurate absorption spectra.

The propagators mentioned here are exact if the hamiltonian contains only linear and quadratic terms. All of the transformations are likewise exact in this limit (stationary phase approximation). We shall see below that this class of problems can also be handled using a local SQ (second quantization) formulation which can be introduced if a specific basis set is introduced, namely the so-called Gauss-Hermite basis set.

2.3   An orthorgonal basis set

Instead of expanding the wave function in a Gaussian basis set, it is advantageous for several reasons to use an orthorgonal basis set instead. Such a basis set is constituted by the Gauss-Hermite basis [201, 54, 53, 24, 52]. It consists of a Gauss part, which is just an ordinary GWP, and a Hermite part (see below). This basis set is uniqe in two ways, firstly because it has the ordinary GWP as the "ground state," and since the GWP generates or is connected to a classical view of quantum mechanics, we could argue that the basis set has classical mechanics as the ground state. Secondly, it can be used to introduce a discrete variable representation (DVR) [51]. In DVR schemes, one represents the wave function only at certain grid points defined as zeroes of an orthorgonal polynomial, which in the present case is a Hermite polymomial. Aside from this, we can also use the Gauss-Hermite basis set to construct a second quantization (SQ) approach to molecular dynamics. These options will be discussed later in this chapter.

Consider, for instance, a one-dimensional hamiltonian

$$H = -\frac{\hbar^2}{2m}\frac{\partial^2}{\partial x^2} + V(x) \tag{2.90}$$

where m is the particle mass and $V(x)$ the potential. We wish to solve the time-dependent Schrödinger equation (TDSE)

$$i\hbar\frac{\partial\Psi(x,t)}{\partial t} = \hat{H}(x)\Psi(x,t) \tag{2.91}$$

The wave function is now expanded as

$$\Psi(x,t) = \sum_n c_n(t)\Phi_n(x,t) \tag{2.92}$$

where the Gauss-Hermite basis functions are defined by

$$\Phi_n(x,t) = \pi^{1/4} \exp\left(\frac{i}{\hbar}(\gamma(t) + p(t)(x - x(t)) + \mathrm{Re}A(t)(x - x(t))^2)\right) \phi_n(\xi(t))$$

(2.93)

where

$$\phi_n(\xi(t)) = \frac{1}{\sqrt{n!2^n}\sqrt{\pi}} \exp(-\xi(t)^2/2) H_n(\xi(t))$$

(2.94)

and

$$\xi(t) = \sqrt{2\mathrm{Im}A(t)/\hbar}(x - x(t))$$

(2.95)

H_n is a Hermite polynomial. The functions $\phi_n(\xi(t))$ are eigenfunctions to a Schrödinger equation with a harmonic potential $\frac{1}{2}k(x - x(t))^2$, where the force constant is $k(t) = 4\mathrm{Im}A(t)^2/m$. This corresponds to a harmonic oscillator with a time-dependent frequency $\omega(t) = 2\mathrm{Im}A(t)/m$.

The basis functions Φ_n form an orthonormal set, and for $n = 0$ we have an ordinary Gaussian wave packet (GWP) in the usual Heller notation [49] centered around $x(t)$ with momentum and width parameters $p(t)$ and $A(t)$. Normalization requires that the imaginary part of the phase factor $\gamma(t)$ is related to $\mathrm{Im}A(t)$ as

$$\mathrm{Im}\gamma(t) = -\frac{\hbar}{4}\ln(2\mathrm{Im}A(t)/\pi\hbar)$$

(2.96)

We shall, furthermore, introduce

$$\dot{x}(t) = \frac{p(t)}{m}$$

(2.97)

$$\dot{p}(t) = -V_0'$$

(2.98)

$$-\dot{A}(t) = \frac{2}{m}A(t)^2 + \frac{1}{2}V_0''$$

(2.99)

where V_0' and V_0'' are undetermined so far, but would in the classical limit be the first and second derivatives of the potential. Heller [49, 50] and Karplus [52] introduced V_0' and V_0'' just as the classical forces. But in a more general derivation, the forces are obtained variationally [55], [201] using the Dirac-Frenkel variational principle (see also below). It is important to realize that the forces are, in principle, arbitrary. The GH basis set is complete no matter how it is propagated. Thus, we can let numerical convenience decide—or, even better, let the dynamics of the system decide on how to propagate the basis set.

The first equation (2.97) is not directly obtained from the Dirac-Frenkel [12] or McLachlan [13] variational principle—but simplifies the equations for the other variables. However, we notice that the equation came out of the minimum error method mentioned previously. The equation is, furthermore, necessary for a "classical mechanical" picture of the approach, that is, one in which the complete basis set is driven by "classical mechanical" equations of motion. Since the system is, in any case, overdetermined with respect to the number of parameters, this does not pose any fundamental problem.

The last equation (2.99) gives the imaginary part of the width the following equation

$$\frac{d}{dt}\text{Im}A(t) = -\frac{4}{m}\text{Re}A(t)\text{Im}A(t) \tag{2.100}$$

Thus, the value for V_0'' only affects the equation of motion for $\text{Re}A(t)$. If $\text{Re}A(t)$ is set initially to zero and remains zero, then $\text{Im}A$ is also constant, which corresponds to a frozen Gauss-Hermite basis (see appendix A). We, furthermore, introduce

$$\dot{\gamma}(t) = \frac{i\hbar A(t)}{m} - V_0 + \frac{p^2}{2m} \tag{2.101}$$

This equation for $\gamma(t)$ can be modified in the real part without problems [66]. A change of the real part only alters the phase factor in the expansion coefficients $c_n(t)$. However, the overall (including that coming from γ) phase factor is unchanged. We now insert the expansion (2.92) in the TDSE and obtain

$$i\hbar\dot{c}_n(t) =$$

$$\sum_k \left(-V_0\delta_{nk} - V_0'M_{nk}^{(1)} - \frac{1}{2}V_0''M_{nk}^{(2)} + V_{nk}^{(0)} + (2k+1)\frac{\hbar\text{Im}A(t)}{m}\delta_{nk} \right) c_k(t)$$

$$\tag{2.102}$$

where δ_{nk} is a Kronecker delta and

$$M_{nk}^{(i)} = \int d\xi \Phi_n^*(x - x(t))^i \Phi_k \tag{2.103}$$

$$V_{nk}^{(i)} = \int d\xi \Phi_n^* V(x)(x - x(t))^i \Phi_k \tag{2.104}$$

The integrals have been rewritten to integrals over ξ by assuming that

$$\sqrt{\pi}\exp(-2\text{Im}\gamma/\hbar)dx = d\xi \tag{2.105}$$

Note that a second-order expansion of the potential around $x(t)$ would give

$$V_{nk}^{(0)} = V(x(t))\delta_{nk} + V'(x(t))M_{nk}^{(1)} + \frac{1}{2}V''(x(t))M_{nk}^{(2)} \tag{2.106}$$

This expansion is of course only exact if the potential happens to be of second order. Thus, in this case, we would expect that $V_0 = V(x(t))$, $V_0' = V'(x(t))$ and $V_0'' = V''(x(t))$. That this is indeed the case is demonstrated below. We see from eq. (2.102) that in this case, the various $c_n(t)$ coefficients are not coupled, and if we initialize the wave function as a GWP $(c_n(t_0) = \delta_{n0})$ it will stay Gaussian.

However, we have not yet given any equations for V_0, V_0' and V_0''. These are now obtained by using the Dirac-Frenkel variational principle [12], that is, we minimize the integral

$$I = \int dx \left(i\hbar \frac{\partial}{\partial t}\Psi - H\Psi \right)^* \left(i\hbar \frac{\partial}{\partial t}\Psi - H\Psi \right) \tag{2.107}$$

By differentiation with respect to $\dot{\gamma}^*$, $\dot{p}(t)$ and $\dot{A}^*(t)$, we get equations for V_0, V_0' and $\frac{1}{2}V_0''$. However, it turns out that the equation for V_0 is identical to zero. Thus, the value of V_0 cannot be determined, but this factor only enters the diagonal in eqs. (2.102) and, therefore, changes the common phase of all the $c_n(t)$ coefficients. However, V_0 enters both in the equation for γ and $c_n(t)$ and, hence, the total phase factor is independent of V_0. Thus V_0 is arbitrary and we prefer simply to set it equal to $V(x(t))$, that is, the value of the potential along the path.

The values of V_0' and $\frac{1}{2}V_0''$ can, however, be determined. After a little manipulation, one obtains

$$V_0' = \frac{W^{(1)}S_{22} - W^{(2)}S_{12}}{S_{11}S_{22} - S_{12}S_{21}} \tag{2.108}$$

$$\frac{1}{2}V_0'' = \frac{W^{(2)}S_{11} - W^{(1)}S_{21}}{S_{11}S_{22} - S_{12}S_{21}} \tag{2.109}$$

where

$$W^{(1)} = \sum_{kl} c_k^* c_l \left(\sum_p M_{kp}^{(1)} V_{pl}^{(0)} - V_{kl}^{(1)} \right) \tag{2.110}$$

$$W^{(2)} = \sum_{kl} c_k^* c_l \left(\sum_p M_{kp}^{(2)} V_{pl}^{(0)} - V_{kl}^{(2)} \right) \tag{2.111}$$

$$S_{11} = \sum_{kl} c_k^* c_l \left(\sum_p M_{kp}^{(1)} M_{pl}^{(1)} - M_{kl}^{(2)} \right) \tag{2.112}$$

$$S_{12} = \sum_{kl} c_k^* c_l \left(\sum_p M_{kp}^{(1)} M_{pl}^{(2)} - M_{kl}^{(3)} \right) \tag{2.113}$$

$$S_{21} = \sum_{kl} c_k^* c_l \left(\sum_p M_{kp}^{(2)} M_{pl}^{(1)} - M_{kl}^{(3)} \right) \tag{2.114}$$

and

$$S_{22} = \sum_{kl} c_k^* c_l \left(\sum_p M_{kp}^{(2)} M_{pl}^{(2)} - M_{kl}^{(4)} \right) \tag{2.115}$$

These equations then determine the effective quantum forces to be used in the equations of motion for the center of the wave packet. Note, also, that in the limit of an infinite number of basis functions, we have $W^{(i)} = 0$ as well as $S_{ij} = 0$. Thus, in this limit the effective forces are in principle undefined. But this makes sense, since in this case the basis set covers all space and we do not need any equations of motion to move it in space. Only if the basis set is incomplete do we gain anything by moving it around.

The matrix elements $M_{kl}^{(i)}$ in eqs. (2.110–2.115) can be evaluated analytically. This is also the case for the potential matrix elements if the potential is expanded as around the trajectory $x(t)$

$$V(x) = \sum_{m=0}^{\infty} \frac{V_m}{m!} (x - x(t))^m \qquad (2.116)$$

where $V_m = \frac{d^m V}{dx^m}|_{x=x(t)}$. For the quantum forces we then have, using eqs. (2.108, 2.109)

$$V_0' = V_1 + \sum_{m=3}^{\infty} \frac{V_m}{m!} \frac{S_{22} S_{1m} - S_{12} S_{2m}}{S_{11} S_{22} - S_{12} S_{21}} \qquad (2.117)$$

$$\frac{1}{2} V_0'' = \frac{1}{2} V_2 + \sum_{m=3}^{\infty} \frac{V_m}{m!} \frac{S_{11} S_{2m} - S_{21} S_{1m}}{S_{11} S_{22} - S_{12} S_{21}} \qquad (2.118)$$

where

$$S_{nm} = \sum_{kl} c_k^* c_l \left(\sum_p M_{kp}^{(n)} M_{pl}^{(m)} - M_{kl}^{(n+m)} \right) \qquad (2.119)$$

Thus, we see that the leading term in these quantum forces (V_0' and V_0'') is actually equal to the true derivative of the potential evaluated at the trajectory $x(t)$, that is, the leading terms are the classical forces in the sense of Newton.

The coefficients $c_n(t)$ change according to the dynamics of the system or, strictly speaking, according to the anharmonicity of the potential. Thus, the result of this section leads us to the not-so-surprising result that the quantum system is driven by classical equations of motion if the potential is harmonic. If the potential is anharmonic, the "classical" equations of motion involve some more generalized effective forces given by eqs. (2.108, 2.109) or (2.117, 2.118).

The expansion of the wave function in a GH basis set generates two sets of equations, the first is the classical mechanical–looking equation,

$$m \frac{d^2 x(t)}{dt^2} = -V_0' \qquad (2.120)$$

that is, the usual Newton mechanical equation, mass times acceleration equals the force. However, the force is more general than in ordinary classical mechanics. It is not necessarily the derivative of a potential, that is, a conservative force. The force may be defined in a number of ways by truncating the G-H basis set used in the evaluation of the summation in eq. (2.117). If no terms are used, that is, $S_{nm} = 0$ then we obtain the classical force. If the potential truncates after the second-order term, we also have the classical force. In other cases, we

operate with a "quantum force" which involves information on the solution of
the quantum equations (2.102).

The important point is that the force which is used is arbitrary because
the basis set expansion is complete and all the forces do is to move the basis set
in space, that is, only the convergence pattern and speed are altered by changing
the force (see also table 2.1).

The general expression for S_{ij} is given by

$$S_{ij} = -\sum_{kl} c_k^* c_l \langle k|(x - x(t))^i Q(x - x(t))^j |l\rangle \qquad (2.121)$$

where a bracket notation has been used for the G-H basis function. Q is the
projection operator

$$Q = I - \sum_{i=0}^{n} |i\rangle\langle i| \qquad (2.122)$$

where n is the largest G-H basis function used in the expansion of the wave
function. Thus, we see that the quantum force is defined by the complementary
space of the one spanned by the basis functions. If that space is complete, we
have $Q = 0$ and, hence, $S_{ij} = 0$, that is, the forces V_0' and V_0'' are undefined.
This is consistent with the fact that if our basis set covers all space we cannot
obtain an improved convergence by moving it around. In this limit, the classical
mechanical–looking equations of motion vanish. In the other extreme we have
$Q = I$ (no basis functions). Here, the dynamics are governed entirely by classical
mechanics. Thus, we can claim that the "classical" equations of motion are
determined by the complementary space Q to the quantum space. Or, in other
words, it depends on the degree of incompleteness of the quantum description
of the system.

Table 2.1: Tunneling probability at 10τ for a
particle in a double-well potential original
located in one of the wells. N is the number of
Gauss-Hermite basis functions and $n = 0$ or
$n = 1$ defines the quantum force used,
according to eq. (2.122).

| | | Quantum force | |
N	Classical force	$n = 0$	$n = 1$
20	0.0197	0.0053	0.0061
30	0.0067	0.0124	0.0092
40	0.0185	0.0185	0.0180
50	0.0185	0.0185	0.0185

The exact number is 0.01858 at 10τ, where $1\tau = 10^{-14}$ sec.

In the limit where only one basis function is used to defined the forces, we have

$$Q = I - |0\rangle\langle 0|$$ (2.123)

which gives the following forces

$$V_0' = \frac{\langle 0|\Delta x V(x)|0\rangle}{\langle 0|\Delta x^2|0\rangle} = \frac{\langle 0|V(x)|1\rangle}{\langle 0|\Delta x|1\rangle}$$ (2.124)

$$\frac{1}{2}V_0'' = \frac{\langle 0|\Delta x^2 V(x)|0\rangle - \langle 0|\Delta x^2|0\rangle\langle 0|V(x)|0\rangle}{\langle 0|\Delta x^4|0\rangle - \langle 0|\Delta x^2|0\rangle^2}$$

$$= \frac{\langle 0|V(x)|2\rangle}{\langle 0|\Delta x^2|2\rangle}$$ (2.125)

We notice that the quantum force depends on the width parameter $\mathrm{Im}\,A(t)$.

In fig. 2.4, the quantum and classical forces are shown for a tunneling problem in a double-well potential. We notice that the quantum force allows the trajectory (see fig. 2.5) to tunnel through the barrier. The tunneling probabilities are determined from the wave function which with enough G-H basis functions is the same no matter how the basis set is propagated in space. (See also table 2.1.)

2.4 Bohmian mechanics

Since we have already introduced the concept of "quantum forces," it is natural at this point to compare the approach to the one suggested by Bohm [57] in 1952. Bohm suggested that the TDSE be reformulated to resemble equations equivalent to those for a hydrodynamic fluid. It was done by introducing the following expression for the wave function

$$\Psi(x,t) = A(x,t)\exp(iS(x,t)/\hbar)$$ (2.126)

We notice that this separation in a modulus and a phase is well known from standard treatments of quantum mechanics, treatments whose aim is studying the classical limit and/or WKB-type approximations [58]. If the expression is inserted in the TDSE

$$i\hbar\frac{\partial\Psi(x,t)}{\partial t} = \left[-\frac{\hbar^2}{2m}\nabla^2 + V(x)\right]\Psi(x,t)$$ (2.127)

the following equations are obtained

$$\frac{\partial P(x,t)}{\partial t} + \nabla\left(P\frac{\nabla S}{m}\right) = 0$$ (2.128)

$$\frac{\partial S(x,t)}{\partial t} + V(x) + \frac{(\nabla S)^2}{2m} - \frac{\hbar^2}{4m}\left[\frac{\nabla^2 P}{P} - \frac{1}{2}\frac{(\nabla P)^2}{P^2}\right] = 0$$ (2.129)

where $P(x,t) = A(x,t)^2$ is the probability density. S is, as we can see, the solution of a Hamilton-Jacobi equation. Thus, the latter equation is formally

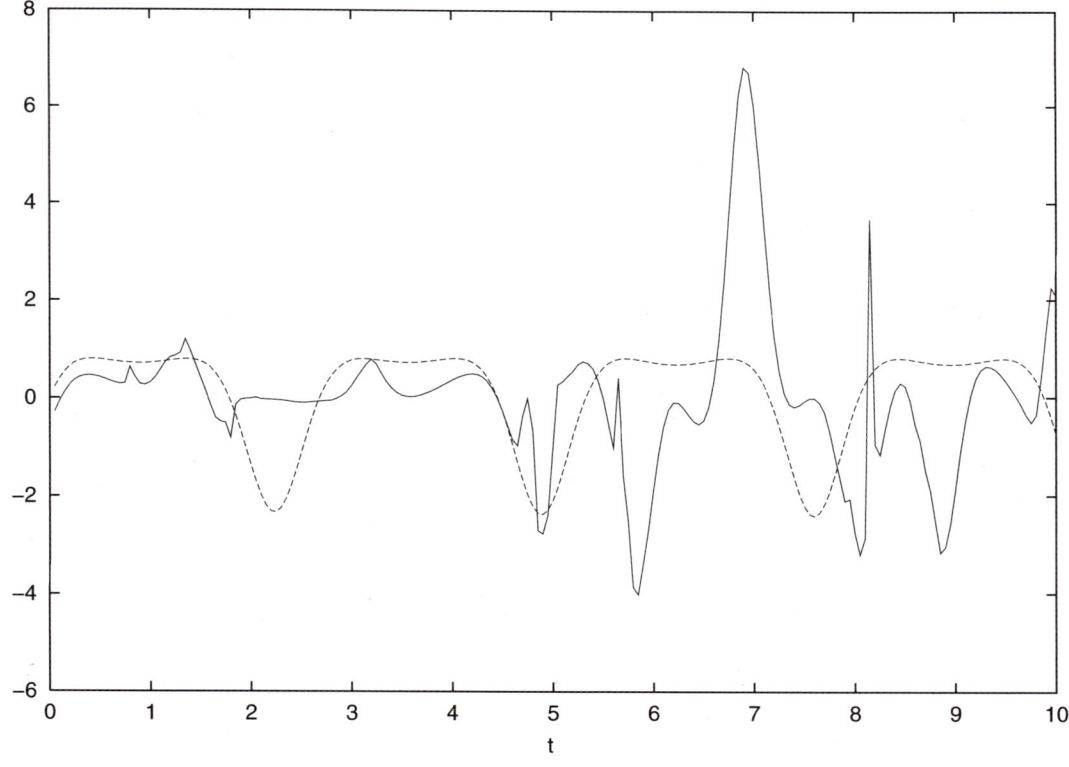

Fig. 2.4: The quantum (solid line) and classical force as a function of time in units of τ for a particle with mass 1 amu in a double-well potential with barrier height 26.23 kJ/mol at $x = 0$ and minima at $x_0 = \pm 0.5033$ Å. The initial energy of the particle is 18 kJ/mol. The quantum force is highly irregular and allows tunneling through the barrier (see fig. 2.3).

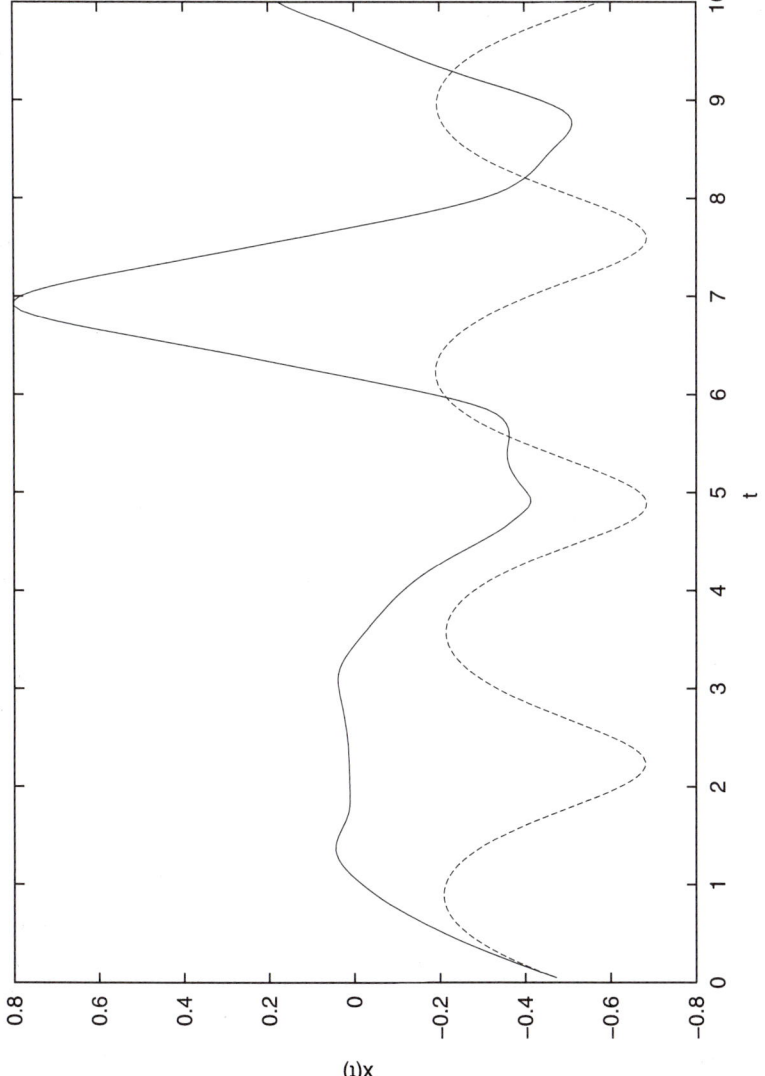

Fig. 2.5: A particle moves back and forth between the turning points in the left well of a double-well potential with barrier located at $x = 0$ if a classical force is used (dashed line). If the particle is moved by the quantum force it may penetrate through the barrier to the right well.

identical to the one known from classical mechanics but with the potential given as the classical potential $V(x)$ added to the quantum potential

$$-\frac{\hbar^2}{4m}\left[\frac{\nabla^2 P}{P}-\frac{1}{2}\frac{(\nabla P)^2}{P^2}\right]=-\frac{\hbar^2}{2m}\frac{\nabla^2 A}{A} \qquad (2.130)$$

Thus, eq. (2.129) can formally, at least, be considered as a generalization of classical mechanics, with $\nabla S/m$ being the velocity of the particle and eq. (2.128) describing conservation of particle flux. Thus, we can solve the dynamics by running trajectories governed by the equation of motion

$$m\frac{d^2 x}{dt^2}=-\nabla\left[V(x)-(\hbar^2/2m)\frac{\nabla^2 A}{A}\right] \qquad (2.131)$$

We show in fig. 2.6 the quantum potential defined for a time-independent problem through

$$V_q(x)=2(E-V(x))-\frac{\hbar^2}{m}(\nabla\Psi^*(x)\nabla\Psi(x))/(\Psi^*(x)\Psi(x)) \qquad (2.132)$$

where $V(x)=V_0/\cosh(ax)^2$. We see that the quantum potential vanishes for energies well above the barrier $V_0=100\,\text{kJ/mol}$ and gives a negative contribution at lower energies—allowing for "tunneling." A closer look at the dynamics of the quantum particles which are governed by the Bohmian potential [30] shows that the particle which reacts actually surpasses the barrier with a positive kinetic energy. However, this is a process which, due to the changes in the barrier height, is allowed in Bohmian but not in Newtonian mechanics.

The Bohmian approach to molecular quantum dynamics has recently been considered by Wyatt [30] and coworkers, as well as Askar and coworkers [31]. Problems with the long time stability of the solution have been discussed in [59] and, at least at present, the extension to many dimensions poses some problems. Solution techniques in terms of distributed approximating functions (DAFs) have been presented in ref. [60].

The quantum trajectory approach as defined by the Bohmian mechanics is conceptually appealing in the sense that the limit to classical mechanics is well defined and well behaved—it can be introduced simply by scaling the quantum potential with a scaling parameter between zero and unity. The first choice being the classical limit, the second the full quantum limit. The "quantum dressed" classical mechanics formulation, which is discussed later, has the same features and has, furthermore, the advantage that it is exact in principle even without using the quantum force to propagate the trajectories. This is not so for Bohmian mechanics. But the quantum dressed classical mechanics cannot be solved just by running trajectories—we need to solve a matrix equation with a matrix of the size of the number of grid points included in the expansion (see section 2.8). It should also not be forgotten that Bohm's aim, with his introduction of the quantum potential, was to obtain an equation which would allow a different interpretation—the so-called ontological interpretation of quantum mechanics [57]. Thus the aim was not to rewrite the SE in such a manner

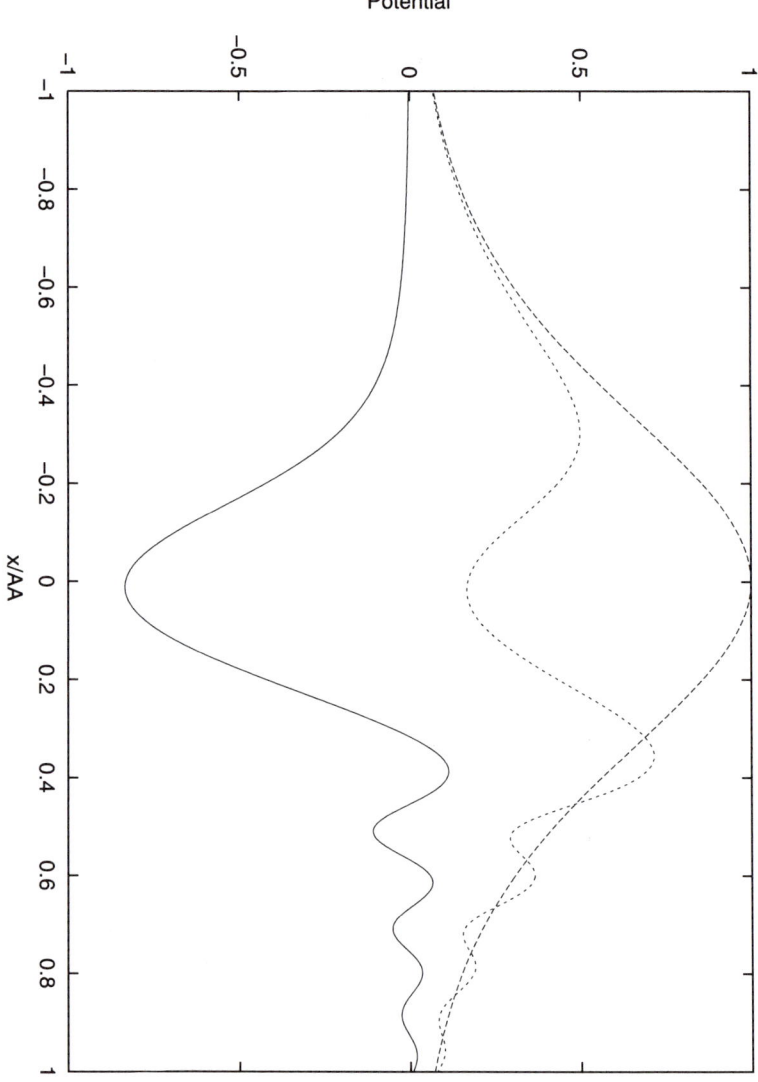

Fig. 2.6a.

Fig. 2.6(a–c): The quantum potential for three values of the energy 80, 100, 120 kJ/mol for a particle with mass 1.0 amu in an Eckart potential with barrierheight 100 kJ/mol and $a = 2 \text{Å}^{-1}$. The quantum potential is given by the full line, the Eckart potential by the dashed line and the total potential by the dash-dotted line.

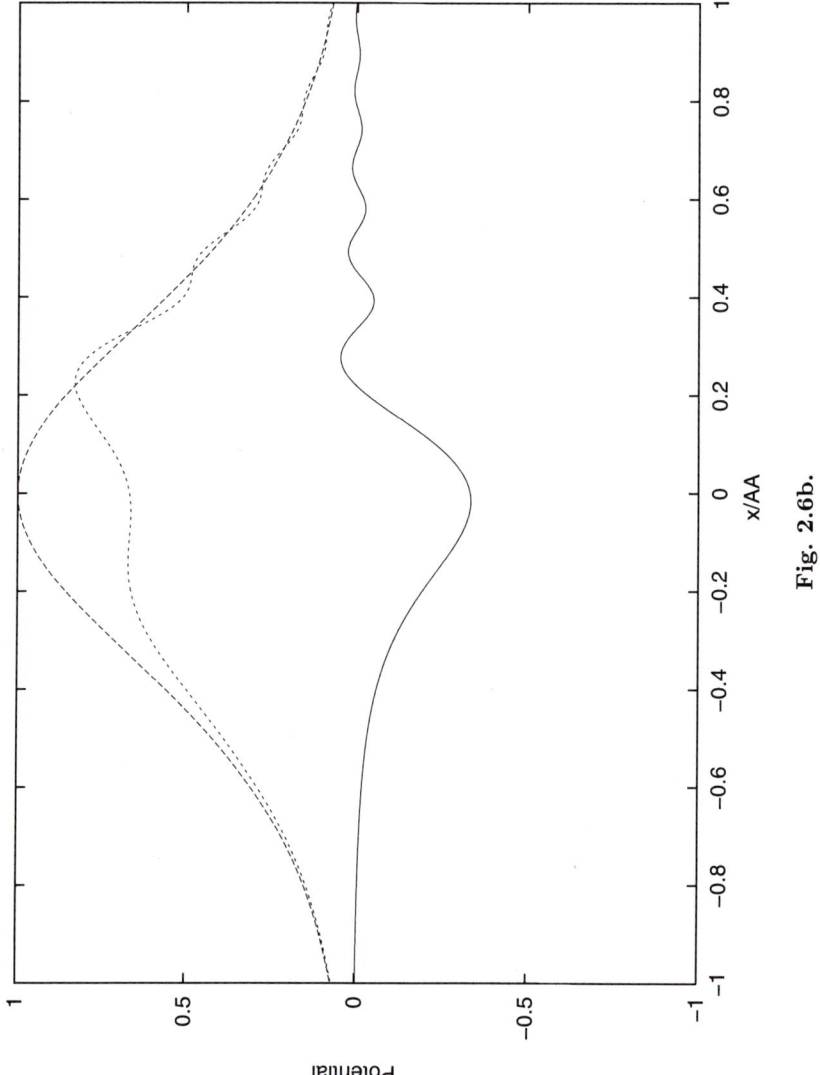

Fig. 2.6b.

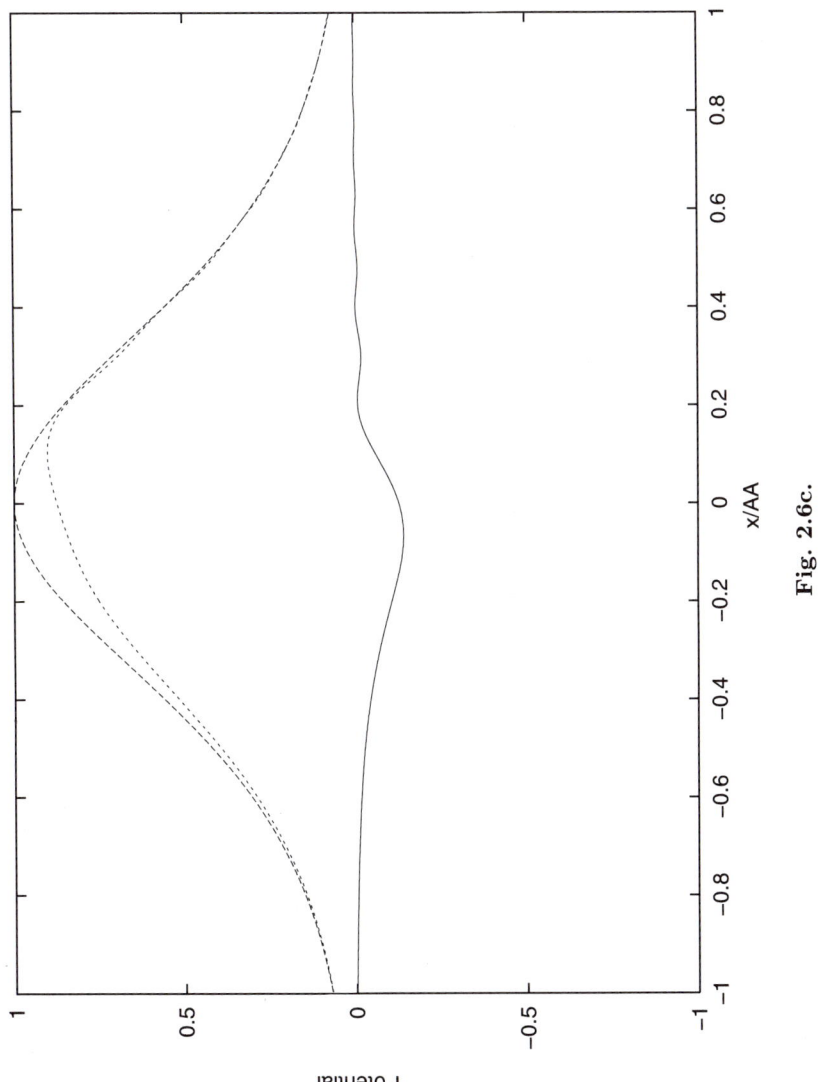

Fig. 2.6c.

that it was easier to solve—the opposite could be said about the "quantum dressed classical mechanics" approach which later will be derived from the TDGH formulation.

2.5 Mixing of Gauss-Hermite and ordinary basis sets

It is possible to use traditional time-independent basis sets for some degrees of freedom while using the G-H basis for others. For simplicity, we consider just two degrees of freedom: r the quantum and R the "classical" degree of freedom. But the methodology can easily be extended to more complex situations. Thus, we expand the wave function for the system as

$$\Psi(r, R, t) = \sum_{nk} a_{nk}(t)\phi_n(r)\Phi_k(R, t) \tag{2.133}$$

The total hamiltonian operator contains the following terms

$$\hat{H}(r, R) = \hat{H}_0 + \hat{T}_R + V(r, R) \tag{2.134}$$

The functions $\phi_n(r)$ are chosen to be eigenfunctions of H_0, that is,

$$\hat{H}_0\phi_n(r) = E_n\phi_n(r) \tag{2.135}$$

$\hat{T}_R$ is the kinetic energy operator for the R-motion and $V(r, R)$ the potential which couples the two degrees of freedom. Initially, the system is represented by the wave function

$$\Psi(r, R, t_0) = \phi_I(r)\Phi_{GWP}(R, t_0) \tag{2.136}$$

where $\Phi_{GWP}(R, t_0)$ is a Gaussian wave packet

$$\Phi_{GWP}(R, t_0) = \exp\left(\frac{i}{\hbar}(\gamma(t_0) + P_0(R - R(t_0)) + A(t_0)(R - R(t_0))^2\right) \tag{2.137}$$

centered around $R_0 = R(t_0)$ in coordinate and P_0 in momentum space. The initial condition for the parameters can be taken to be

$$R_0 \text{ large} \tag{2.138}$$

$$P_0 \text{ arbitrary} \tag{2.139}$$

$$\text{Re}A(t_0) = 0 \tag{2.140}$$

$$\text{Im}A(t_0) \text{ arbitrary} \tag{2.141}$$

$$\text{Re}\gamma(t_0) = 0 \tag{2.142}$$

$$\text{Im}\gamma(t_0) = -\frac{\hbar}{4}\ln(2\text{Im}A(t_0)/\pi\hbar) \tag{2.143}$$

The last equation follows, as mentioned previously, from normalization. The initial values of the width parameter $\text{Im}A(t_0)$ and P_0 are arbitrary in the sense that the final result is independent of where in momentum space we position the wave packet, and is also independent of its width. In practical calculations there will be a dependence due to numerical inaccuracies. But the wave packet can represent a finite energy range around the center momentum P_0 accurately. In order to extract the energy- and state-resolved transition probabilities, the initial wave function is projected on incoming plane waves times an eigenfunction for the r-system, that is, on

$$\phi_n(r)\exp(-ik_nR) \tag{2.144}$$

Likewise, the scattered wave function is projected on outgoing waves and the transition probability is obtained as the ratio between outgoing and incoming fluxes, that is,

$$P_{I\to F}(E) = \lim_{t\to\infty}\frac{k_F}{k_I}\frac{|\sum_k \int dR\exp(-ik_FR)\Phi_k(R,t)a_{Ik}(t)|^2}{|\int dR\exp(+ik_IR)\Phi_{GWP}(R,t_0)|^2} \tag{2.145}$$

where we have used that $a_{I0} = 1$. The integral over R can be evaluated analytically (see section 2.5.2). The total energy is given as

$$E = \frac{\hbar^2k_I^2}{2\mu} + E_I = \frac{\hbar^2k_F^2}{2\mu} + E_F \tag{2.146}$$

where μ is the reduced mass for the R-motion. So far, we have not specified the functions $\Phi_k(R,t)$. They are now introduced as the functions, that is,

$$\Phi_k(R,t) = \pi^{1/4}\exp\left(\frac{i}{\hbar}(\gamma(t) + P(t)(R-R(t)) + \mathrm{Re}A(t)(R-R(t))^2)\right)\xi_k(x) \tag{2.147}$$

where

$$x = \sqrt{2\mathrm{Im}A(t)/\hbar}(R-R(t)) \tag{2.148}$$

and $\xi_k(x)$ defined by

$$\xi_k(x) = \frac{1}{\sqrt{k!2^k}\sqrt{\pi}}\exp(-x^2/2)H_k(x) \tag{2.149}$$

are the harmonic oscillator wave functions. The basis functions have as mentioned two important properties. Firstly, they form an orthonormal basis and, secondly, the ground state $(n = 0)$ is the Gaussian wave packet given by eq. (2.137). We notice in passing that the normalization requires that eq. (2.143) holds at all times t and not only at t_0.

2.5.1 The classical path approximation

If we just include one basis function in the Gauss-Hermite basis set and replace a_{n0} with a_n, we have

$$\Psi(r,R,t) = \Phi_0(R,t)\sum_n a_n(t)\phi_n(r) \tag{2.150}$$

that is, a product-type wave function, which when inserted in the TDSE gives the so-called classical path equations [63]

$$i\hbar \dot{a}_n(t) = E_n a_n(t) + \sum_m V_{nm}(R(t)) a_m(t) \tag{2.151}$$

$$\dot{P}(t) = -\sum_{nm} a_n^*(t) a_m(t) \frac{\partial V_{nm}(R)}{\partial R} \bigg|_{R=R(t)} \tag{2.152}$$

$$\dot{R} = \frac{P(t)}{\mu} \tag{2.153}$$

$$\dot{A}(t) = -\frac{2}{\mu} A(t)^2 - \frac{1}{2} \sum_{nm} a_n^*(t) a_m(t) \frac{\partial^2 V_{nm}}{\partial R^2} \bigg|_{R=R(t)} \tag{2.154}$$

$$\dot{\gamma}(t) = \frac{P(t)^2}{2\mu} + \frac{i\hbar A(t)}{\mu} - V(R(t)) \tag{2.155}$$

where the matrix element V_{nm} is defined by

$$V_{nm}(R) = \langle \phi_n | V(R, r) | \phi_m \rangle \tag{2.156}$$

and $\langle \, \rangle$ indicate integration over r.

In order to obtain the classical path equations, we have made two approximations, a separability approximation between R and r and a "shape" approximation in R, that is, $\Phi_0(R, t)$ is assumed to be a GWP.

Historically, the classical path equations have played an important role for approximate calculations of mainly inelastic probabilities and cross sections. The classical path equations can be solved in various approximations and with various corrections (see further discussion in chapter 3).

They are:

1. The primitive classical path equations, where only the equations (2.151)–(2.153) are solved.
2. The variational classical path equations [63], where also the equations (2.154)–(2.155) are solved and the initial width $\mathrm{Im}A(t_0)$ and momentum P_0 parameters are determined variationally.
3. The corrected classical path equations [65]. These corrections are in the so-called SCF direction, that is, they correct the shape of the function in R but not the separability approximation.

Using the G-H basis functions, we have also the possibility of introducing

4. The Hermite corrected classical path equations. This correction is in the "MCSCF direction" and is, therefore, the most general type of correction possible. Already the first harmonic correction $k = 1$ in (2.133) gives a measure of the correlation between the two degrees of freedom. If a_{n1} is close to zero during the entire propagation, then the system is separable. This will, of course, happen if the coupling potential $V(r, R)$ is zero or small, but before this aspect is considered further we derive the equations of motion.

2.5.2   The exact equations of motion in the TDGH basis

When deriving the equations of motion in the time-dependent Gauss-Hermite (TDGH) expansion, we make use of the following equations

$$\frac{d}{dt}R(t) = P(t)/\mu \tag{2.157}$$

$$\frac{d}{dt}\mathrm{Im}A(t) = -\frac{4}{\mu}\mathrm{Re}A(t)\mathrm{Im}A(t) \tag{2.158}$$

$$\frac{d}{dt}\mathrm{Im}\gamma(t) = \frac{\hbar}{\mu}\mathrm{Re}A(t) \tag{2.159}$$

The first equation is obvious for the classical path picture to be relevant (the center of the wave packet follows a classical equation of motion) and the two other equations secure that the basis set is normalized at all times t. Since the phase factor $\mathrm{Re}\gamma(t)$, according to previous discussions, is irrelevant (see eq. (2.145)), we only need two equations of motion for the wave packet parameters—namely for $P(t)$ and $\mathrm{Re}A(t)$. We discuss this in more detail below. Inserting the expansion (2.133) in the TDSE and using equations (2.157)–(2.159), we obtain the following coupled equations for the expansion coefficients a_{ml}

$$i\hbar\dot{a}_{ml} = E_m a_{ml} + \sum_{nk} a_{nk} V_{mn}^{lk}(t)$$

$$+ a_{ml}\left(2l\frac{\hbar\mathrm{Im}A(t)}{\mu} + \frac{P(t)^2}{2\mu} - \frac{\hbar(l+\frac{1}{2})}{4\mathrm{Im}A(t)}V_{eff}''\right)$$

$$- \frac{1}{2}V_{eff}'\sqrt{\hbar/\mathrm{Im}A(t)}(\sqrt{l}a_{m,l-1} + \sqrt{l+1}a_{m,l+1})$$

$$- \frac{1}{8}\frac{\hbar}{\mathrm{Im}A(t)}V_{eff}''(\sqrt{l(l-1)}a_{m,l-2} + \sqrt{(l+1)(l+2)}a_{m,l+2}) \tag{2.160}$$

where

$$V_{mn}^{lk}(t) = \int dR\Phi_l^*(R,t)V_{mn}(R)\Phi_k(R,t) \tag{2.161}$$

and where V_{eff}' and V_{eff}'' are not specified as yet. They enter into the equations of motion for $P(t)$ and $\mathrm{Re}A(t)$, that is,

$$\dot{P}(t) = -V_{eff}' \tag{2.162}$$

$$\frac{d}{dt}\mathrm{Re}A(t) = -\frac{2}{\mu}(\mathrm{Re}A(t)^2 - \mathrm{Im}A(t)^2) - \frac{1}{2}V_{eff}'' \tag{2.163}$$

By comparing with eqs. (2.152) and (2.154), it is possible to introduce the "Ehrenfest forces," that is, to assume

$$V_{eff}' = \sum_{ml;nk} a_{ml}^*(t)a_{nk}(t)\int dR\Phi_l^*(R,t)\frac{d}{dR}V_{nm}(R)|_{R=R(t)}\Phi_k(R,t) \tag{2.164}$$

and
$$V''_{eff} = \sum_{ml;nk} a^*_{ml} a_{nk}(t) \int dR \Phi^*_l(R,t) \frac{d^2}{dR^2} V_{nm}(R)|_{R=R(t)} \Phi_k(R,t) \qquad (2.165)$$

However, it is also possible to proceed in a more rigorous manner and derive the expressions by using the Dirac-Frenkel variational principle [12]. The equations of motion are obtained by differentiation with respect to $\dot{P}$ and $\frac{d}{dt} \mathrm{Re} A(t)$. After some manipulations, one finally gets:

$$V'_{eff} = \frac{\begin{vmatrix} Q_1 & S^{(3)} - X_{12} \\ Q_2 & S^{(4)} - X_{22} \end{vmatrix}}{\begin{vmatrix} S^{(2)} - X_{11} & S^{(3)} - X_{12} \\ S^{(3)} - X_{21} & S^{(4)} - X_{22} \end{vmatrix}} \qquad (2.166)$$

and

$$\frac{1}{2} V''_{eff} = \frac{\begin{vmatrix} S^{(2)} - X_{11} & Q_1 \\ S^{(3)} - X_{21} & Q_2 \end{vmatrix}}{\begin{vmatrix} S^{(2)} - X_{11} & S^{(3)} - X_{12} \\ S^{(3)} - X_{21} & S^{(4)} - X_{22} \end{vmatrix}} \qquad (2.167)$$

where

$$Q_1 = \sum_{nk;ml} a^*_{nk} a_{ml} \left(V^{(1)}_{ml;nk} - \sum_p S^{(1)}_{kp} V^{(0)}_{np;ml} \right) \qquad (2.168)$$

$$Q_2 = \sum_{nk;ml} a^*_{nk} a_{ml} \left(V^{(2)}_{ml;nk} - \sum_p S^{(2)}_{kp} V^{(0)}_{np;ml} \right) \qquad (2.169)$$

$$X_{ij} = \sum_{nk;np} \sum_l a^*_{nk} a_{np} S^{(i)}_{kl} S^{(j)}_{lp} \qquad (2.170)$$

$$V^{(j)}_{ml;nk} = \int dR \Phi^*_l(R,t) V_{mn}(R)(R - R(t))^j \Phi_k(R,t) \qquad (2.171)$$

$$S^{(j)}_{lk} = \int dR \Phi^*_l(R,t)(R - R(t))^j \Phi_k(R,t) \qquad (2.172)$$

$$S^{(j)} = \sum_n \sum_{lk} a^*_{nk} a_{nl} S^{(j)}_{kl} \qquad (2.173)$$

We now introduce the projection operator $Q = I - P$, where I is the identity operator and

$$P = \sum_{m=0}^{N} |\Phi_m\rangle\langle\Phi_m| \qquad (2.174)$$

where N is the highest Hermite polynomial used in the expansion. Thus, we can express the quantities involved in calculating the potential derivatives as

$$Q_i = \sum_{mnkl} a^*_{nk} a_{ml} \langle \Phi_k | \Delta R^i Q V_{mn}(R) | \Phi_l \rangle \qquad (2.175)$$

$$S^{(i+j)} - X_{ij} = \sum_{nkl} a^*_{nk} a_{nl} \langle \Phi_k | \Delta R^i Q \Delta R^j | \Phi_l \rangle \qquad (2.176)$$

We notice that in the limit $P \to I$ we have $Q \to 0$, that is, both numerator and denominator approach zero. In this limit the quantum basis is complete, and we have no need for the classical equations of motion to optimize its performance. For $P = 0$ (no basis functions) we obtain the classical limit.

With the basis functions (2.147), we can evaluate the integral in eq. (2.145) over R analytically to obtain

$$P_{I \to F}(E) = \frac{k_F}{k_I} \sqrt{\frac{g_F}{g_I}} \exp(-g_F(P(t) - \hbar k_F)^2 + g_I(-P_0 + \hbar k_I)^2)$$

$$\times \left| \sum_{k=0} a_{Fk}(t)(-1)^k \frac{H_k(\Delta) \exp(-ik\theta)}{\sqrt{k!2^k}} \right|^2 \tag{2.177}$$

where H_k is a Hermite polynomial and

$$\theta = \mathrm{arctg}\left(\frac{\mathrm{Im}A(t)}{\mathrm{Re}A(t)}\right) \tag{2.178}$$

$$\Delta = \sqrt{g_F}(P(t) - \hbar k_F) \tag{2.179}$$

$$g_F = \frac{\mathrm{Im}A(t)}{2\hbar |A(t)|^2} \tag{2.180}$$

$$g_I = \frac{\mathrm{Im}A(t_0)}{2\hbar |A(t_0)|^2} \tag{2.181}$$

The expressions are evaluated in the domain where t is large. For $k = 0$ we obtain the expression valid for the classical path approach. This expression is discussed further in chapter 3.

Note that in the limit $t \to \infty$ we have $\theta \to 0$, which can be seen from the asymptotic limit, where the coupling is zero, and, hence, we have

$$\mathrm{Re}A(t) = \frac{\mu t}{2(t^2 + \tau_0^2)} \tag{2.182}$$

$$\mathrm{Im}A(t) = \frac{\mu \tau_0}{2(t^2 + \tau_0^2)} \tag{2.183}$$

Therefore, we also have

$$a_{ml}(\infty) = a_{ml}(t) \exp\left(-il\left(\frac{\pi}{2} - \mathrm{arctg}(t/\tau_0)\right)\right)$$

$$= a_{ml}(t) \exp(-il\,\mathrm{arctg}(\mathrm{Im}A(t)/\mathrm{Re}A(t))) \tag{2.184}$$

where t is taken to be so large that the interaction potential has vanished. The asymptotic behavior of $A(t)$ also shows that g_F approaches a constant for large t values.

2.6 Bound states and Gauss-Hermite functions

In order to use the Gauss-Hermite functions as a general basis set, it is necessary to consider the representation of bound state eigenfunctions in this basis. More specifically, we consider the representation of a Morse-oscillator and the

Spherical Harmonics in the G-H representation. In the previous section, the G-H basis has been used only for the translational degree of freedom. But instead of expanding the wave function in a time-independent basis in the internal degrees of freedom (vibrational and rotational degrees of freedom), it is important for the development of a general purpose time-dependent quantum molecular dynamics theory also to represent internal state wave functions in the G-H basis.

2.6.1   The Morse-oscillator in a Gauss-Hermite basis

As the Gauss-Hermite (G-H) basis set constitutes a complete and orthonormal basis, we may expand any eigenfunction of a hamiltonian in this basis set. Consider, for example, the Morse wave functions, which are eigenfunctions to the hamiltonian

$$\left[-\frac{\hbar^2}{2m} \frac{\partial^2}{\partial r^2} + v(r) \right] \Psi_n(r) = E_n \Psi_n(r) \tag{2.185}$$

where

$$v(r) = D_e(1 - \exp(-\beta(r - r_e)))^2 \tag{2.186}$$

with Morse parameters D_e, β and r_e.

Introducing now the GH basis functions, we have

$$\Psi_n(r) = \exp\left(\frac{i}{\hbar}(\gamma(t_0) + p_r(t_0)(r - r(t_0)) + \mathrm{Re} A_r(t_0)(r - r(t_0))^2 \right)$$
$$\times \sum_m c_m(t_0)\phi_m(r, t_0) \tag{2.187}$$

where

$$\phi_m(r, t_0) = \frac{1}{\sqrt{2^m m!}} \exp(-\xi(r, t_0)^2/2) H_m(\xi(r, t_0)) \tag{2.188}$$

and $\xi(r, t_0) = \sqrt{2\mathrm{Im} A_r(t_0)/\hbar}(r - r(t_0))$. Table 2.2 shows the initial expansion coefficients for a Morse-oscillator in its ground and first excited vibrational state. We now use the expansion (2.187) for t larger than t_0 and insert the expansion in the TDSE, that is, in

$$i\hbar \frac{\partial \Psi(r, t)}{\partial t} = H_0(r)\Psi(r, t) \tag{2.189}$$

where $H_0(r)$ is defined through eq. (2.185). According to the derivations of the previous sections, the potential which couples the expansion coefficients $c_m(t)$ is

$$W(r, t) = V(r) - V(t) - V'(t)(r - r(t)) - \tfrac{1}{2}V''(t)(r - r(t))^2 \tag{2.190}$$

Table 2.2: Expansion coefficients c_m^2 for Morse-oscillator states $(n = 0, 1)$ of hydrogen. The initial parameters are: $\mathrm{Im}A(t_0) = 2.5991$ a.m.u./τ, $p(t_0) = 0$, $\mathrm{Re}A(t_0) = 0$ where $\tau = 10^{-14}$ sec. The Morse parameters were chosen as $\omega_e = 5187.53\,\mathrm{cm}^{-1}$, $x_e = 0.02683$, and $r_e = 0.7416\,\text{Å}$.

m	c_m^2	
	$n = 0$	$n = 1$
0	0.98692	0.01210
1	0.01191	0.88913
2	0.00015	0.08762
3	0.00096	0.00469
4	0.00006	0.00513
5		0.00111
6		0.00012
7		0.00007
8		0.00002

where $V(t) = V(r(t))$. Introducing

$$\dot{r}(t) = p_r(t)/m \tag{2.191}$$

$$\dot{p}_r(t) = -V'(t) \tag{2.192}$$

$$-\dot{A}_r(t) = \frac{2}{m}A_r(t)^2 + \frac{1}{2}V''(t) \tag{2.193}$$

and

$$\dot{\gamma}(t) = \frac{i\hbar A_r(t)}{m} + \frac{p_r(t)^2}{2m} - V(t) \tag{2.194}$$

we obtain the following set of coupled equations in the expansion coefficients

$$i\hbar\dot{c}_n(t) = \frac{\hbar\,\mathrm{Im}A_r(t)}{m}(2n + 1)c_n(t) + \sum_k c_k(t)\langle\phi_n|W|\phi_k\rangle \tag{2.195}$$

Thus, the coefficients $c_n(t)$ are coupled through the anharmonic terms of the potential. Here, we have assumed that the equations of motion for the center of the Gauss-Hermite basis set follow classical trajectories. The expansion coefficients only affect the phase of the wave function, whereas the numerical value is unchanged. However, since the phases also change in the 2D equation (see below) we should project onto the time-propagated basis functions in order to get a time-independent projection on the bound state wave function. The time-dependence under H_0 is easily found since the matrix elements over the anharmonic terms of the potential are time-independent, the reason being that if we initialize the momentum to be zero $p_r(t_0) = 0$ and $r(t_0) = r_{eq}$ we have that the force is zero and, hence, $r(t) = r_{eq}$ and $p_r(t) = 0$. Furthermore, $\text{Re}A(t_0) = 0$ and $\text{Im}A(t_0) = (\frac{m}{4}\frac{d^2v}{dr^2}|_{r=r_{eq}})^{1/2}$. Thus, we also have $A(t) = A(t_0)$. Equation (2.195) can then be solved by diagonalizing the constant coupling matrix by the transformation matrix T and, hence, we have

$$\mathbf{c}(t) = \mathbf{T}^+\exp(-i\mathbf{D}t/\hbar)\mathbf{T}\mathbf{c}(0) \tag{2.196}$$

where $\mathbf{D}$ is the eigenvalue matrix.

2.6.2   Treatment of 2D systems in the G-H basis

Considering now a 2-dimensional system as, for instance, an atom scattered from a Morse-oscillator, we expand the wave function in the G-H basis set as

$$\Psi(R,r,t) = \sum_{n_1 n_2} c_{n_1 n_2}(t)\Phi_{n_1}(R,t)\Phi_{n_2}(r,t) \tag{2.197}$$

where

$$\Phi_{n_1}(R,t) = \left(\frac{2\text{Im}A_R(t)}{\pi\hbar}\right)^{1/4} \exp\left(\frac{i}{\hbar}(P_R(t)(R-R(t)) + \text{Re}A_R(t)(R-R(t))^2)\right)$$

$$\times H_{n_1}(\xi_R(t))\exp\left(-\frac{1}{2}\xi_R(t)^2\right) \tag{2.198}$$

and $\xi_R(t) = \sqrt{2\text{Im}A_R(t)/\hbar}(R-R(t))$. Inserting the expansion (2.197) in the TDSE and using the fact that

$$\dot{R}(t) = P_R(t)/\mu \tag{2.199}$$

$$\dot{P}_R(t) = -V_R' \tag{2.200}$$

$$-\dot{A}_R = \frac{2A_R^2}{\mu} + \frac{1}{2}V_R'' \tag{2.201}$$

together with eqs. (2.191)–(2.194), we get a matrix equation for the expansion coefficients

$$i\hbar\dot{c}(t) = \mathbf{C}c(t) \tag{2.202}$$

where the elements of the "quantum" C-matrix are given as

$$C_{n_1 n_2;n_1' n_2'} = \langle n_1 n_2|V|n_1' n_2'\rangle + \delta_{n_1 n_1'}\delta_{n_2 n_2'}[(2n_1+1)\hbar\text{Im}A_R/\mu + (2n_2+1)\hbar\text{Im}A_r/m$$

$$- V(R(t),r(t))] - \left(V_R'M^{(1)}_{n_1 n_1'} + \frac{1}{2}V_R''M^{(2)}_{n_1 n_1'}\right)\delta_{n_2 n_2'}$$

$$- \left(V_r'M^{(1)}_{n_2 n_2'} + \frac{1}{2}V_r''M^{(2)}_{n_2 n_2'}\right)\delta_{n_1 n_1'} \tag{2.203}$$

where $M^{(k)}_{n_1 n_1'} = \langle n_1|\Delta R^k|n_1'\rangle$ and $M^{(k)}_{n_2 n_2'} = \langle n_2|\Delta r^k|n_2'\rangle$. The forces entering the above equations can be taken as classical or quantum.

However, it is often convenient to introduce the so-called fixed-width approach. The fixed-width approach is obtained if we choose (the arbitrary forces) $V_R'' = 4\text{Im}A_R^2/\mu$ and $V_r'' = 4\text{Im}A_r^2/m$. Hence, we have (with $\text{Re}A_R(t_0) = 0$) that $\text{Im}A_R(t) = \text{constant}$, which can easily be seen from eq. (2.201).

The fixed-width approach can be introduced in either both or just one of the degrees of freedom. However, it is important that whatever force is used in the equations of motion, these same forces should also be used in the C-matrix elements (eq. (2.203)).

If we wish to extract information on transition amplitudes from the wave function, we expand it on the basis of plane wave states and a time-dependent Morse-oscillator wave function $\Psi_n(r,t)$ defined above, that is,

$$\Psi(R,r,t) = \sum_{nk} c_{nk}(t)\exp(ik_n R)\Psi_n(r,t)\exp(-iE_n t/\hbar) \tag{2.204}$$

and, hence, the expansion coefficients are given by

$$c_{nk}(t) = \int dR \int dr \, \exp(-ik_n R + i\omega_n t)\Psi_n(r,t)^* \Psi(R,r,t) \tag{2.205}$$

where $\omega_n = E_n/\hbar$. The integral over R can be carried out analytically, and we finally get

$$c_{nk}(t) = \frac{1}{2\pi\sqrt{\hbar}}\exp(i\omega_n t)\sum_{n_1} F_{n_1}(t)\exp(-in_1\theta)\phi_{n_1}(\Delta)g_F^{1/4} \tag{2.206}$$

where

$$F_{n_1}(t) = \sum_{n_2} c_{n_1 n_2}(t)\int dr\Psi_n(r,t)^*\Phi_{n_2}(r,t) \tag{2.207}$$

$$\Delta = \sqrt{g_F}(P_R(t) - \hbar k_n) \tag{2.208}$$

$$\theta = \text{arctg}(\text{Im}A_R(t)/\text{Re}A_R(t)) \tag{2.209}$$

$$\phi_{n_1}(\Delta) = \exp(-\tfrac{1}{2}\Delta^2)H_{n_1}(-\Delta)/\sqrt{2^{n_1}n_1!} \tag{2.210}$$

and $g_F = \text{Im}A_R(t)/(2\hbar|A_R(t)|^2)$.

The main problem with basis set expansions is the evaluation of matrix elements over the interaction potential. These integrals must, in general, be evaluated numerically, that is, many function evaluations may be needed. If, however, the potential is time-independent, we can evaluate the matrix elements analytically by introducing, for instance, the Fourier transform of the potential. The integrals over the Fourier terms are analytical, and for a 2D problem we obtain the matrix elements as

$$\langle n_1 n_2|V(r,R)|m_1 m_2\rangle = \frac{1}{\sqrt{N_1 N_2}}\sum_{k_1 k_2} c_{k_1 k_2}\exp(ia_r(r(t)-r_0)+ia_R(R(t)-R_0))$$

$$\times \exp(-\tfrac{1}{4}(b_r^2 + b_R^2))L_{m_1}^{n_1-m_1}(b_R^2/2)L_{m_2}^{n_2-m_2}(b_r^2/2)$$

$$\tag{2.211}$$

where $|m_1 m_2\rangle$ denote a G-H basis $\Phi_{m_1} \Phi_{m_2}$, the c_ks are the Fourier coefficients, N_1 and N_2 the number of gridpoints (potential values in each dimension), $a_r = 2\pi k_1/(N_1 \Delta r)$, and $b_r = a_r \sqrt{\hbar/(2A_r)}$. The potential is evaluated in the range $r = r_0 + k\Delta r$ and $R = R_0 + l\Delta R$ for $k = 0, \ldots, N_1 - 1$ and $l = 0, \ldots, N_2 - 1$.

This FFT-method is advantageous if the potential is cumbersome to evaluate. This happens, for instance, if the potential is an effective potential obtained by coupling a large system to a smaller one through the mean field approximation (see chapters 3 and 5).

Since the G-H basis set is local around the trajectories $r(t), R(t)$ if the values of $\text{Im}A_r$ and $\text{Im}A_R$ are large, we need only evaluate the potential in a region around these trajectories.

In other more special cases we may also evaluate the matrix elements analytically. If we can obtain a power series expansion around the trajectory

$$V(R) = V(R(t)) + \sum_{k=1} \frac{d^k V}{dR^k}\bigg|_{R=R(t)} (R - R(t)^k/k! \tag{2.212}$$

we can use the fact that

$$R - R(t) = \frac{1}{2}\sqrt{\frac{\hbar}{\text{Im}A_R(t)}}(a + a^+) \tag{2.213}$$

where a and a^+ are step operators. For the G-H function $|n\rangle$ we have

$$a^+|n\rangle = \sqrt{n+1}|n+1\rangle \tag{2.214}$$
$$a|n\rangle = \sqrt{n}|n-1\rangle \tag{2.215}$$

where the bracket notation for the wave functions Φ_n, has been introduced. Then by using eqs. (2.214, 2.215) we can easily evaluate the integrals of the type

$$\langle n|\Delta R^k|m\rangle = \frac{1}{2^k}(\hbar/\text{Im}A_R(t))^{k/2}\langle n|(a + a^+)^k|m\rangle \tag{2.216}$$

Also, for an exponential potential $\exp(-\alpha R)$, we can make an analytical evaluation, that is,

$$I = \langle n|\exp(-\alpha R)|m\rangle = \exp(-\alpha R(t) + y^2)(\sqrt{2}y)^{n-m}\sqrt{n!/m!}L_m^{n-m}(-2y^2)$$

where $y = -\frac{1}{2}\alpha\sqrt{\hbar/(2\text{Im}A_R)}$, L_n^m is an associate Laguerre polynomial and $n \geq m$.

2.6.3   The quantum-classical correlation

In order to minimize the number of basis functions in the translational coordinate R, we introduce a measure of the so-called quantum-classical correlation defined as

$$q = \sum_{n_1=1}\sum_{n_2=0} |c_{n_1 n_2}|^2 \tag{2.217}$$

Table 2.3: The quantum-classical correlation (q) for an atom hitting a Morse-oscillator (see also table 2.4). The correlation is shown as a function of the initial values of $\mathrm{Im}A_R$ and $\mathrm{Re}A_R$.

$\mathrm{Im}A_R$	$\mathrm{Re}A_R$	q
0.20	−0.60	0.02196
0.20	−0.50	0.05410
0.20	−0.70	0.02805
0.20	−0.65	0.02016
0.15	−0.65	0.02564
0.25	−0.65	0.02604
0.20	−0.64	0.01975
0.20	−0.63	0.01971

If this quantity is small, the classical picture for the motion in the R coordinate is valid. Thus, this degree of freedom is well described by just one single basis function $n_1 = 0$, which is a Gaussian wave packet. Hence, the number of coupled equations is diminished, and we are left with just the classical equations of motion for this degree of freedom. Since the choice of the initial value of $\mathrm{Re}A_R(t_0)$ and $\mathrm{Im}A_R(t_0)$ is arbitrary, we may monitor the quantum-classical correlation by choosing these values (see tables 2.3 and 2.4). This can be done by including just two basis functions in the R coordinate ($n_1 = 0$ and 1). Once the best set of parameters has been found, we can make a more extensive calculation including more G-H functions (see table 2.4).

Table 2.4: Transition probabilities for excitation of the ground state of a Morse-oscillator as a function of kinetic energy and number of G-H basis functions in R. The calculations were performed with initial values $\mathrm{Re}A_R = -0.63$ and $\mathrm{Im}A_R = 0.20$, giving a small quantum-classical correlation (table 2.3). Hence, the number of translational basis functions can be reduced to about 3 (0,1,2). The reduced masses are for the relative translational motion $\mu = 1.875$ amu and $m = 2/3$. The model potential used is $C \exp(-\alpha(R - \lambda r_{eq}))(1 + y + 0.5y^2 + y^3/6)$, where $y = \lambda\alpha(r - r_{eq})$ and $\alpha = 4.5167 \ \text{Å}^{-1}$, $\lambda = 2/3$ and $r_{eq} = 0.7416 \ \text{Å}$.

E_{kin} in 100 kJ/mole	$n_1 = 0$	0,1,2	0 to 5	0 to 7
2.5	4.6(−4)	1.8(−4)	1.6(−4)	2.0(−4)
3.0	1.5(−3)	6.1(−4)	7.6(−4)	7.5(−4)
3.5	3.5(−3)	2.0(−3)	2.0(−3)	1.9(−3)
4.0	6.7(−3)	4.7(−3)	4.2(−3)	4.3(−3)
4.5	1.1(−2)	8.4(−3)	8.0(−3)	8.2(−3)
5.0	1.8(−2)	1.3(−2)	1.4(−2)	1.4(−2)

Since the classical path approximation is based on using just the ground state $n_1 = 0$ for the R-degree of freedom, we can introduce the so-called Hermite correction to the classical path approximation by adding more G-H functions. This method has also been used in reactive scattering formulated in hyper-spherical coordinates (see ref. [61] and chapter 3). The Hermite correction corrects in a systematic way for the correlation, which due to the separability approximation in the classical path theory, has been omitted (see, for example, eq. (2.150)). In chapter 3 we show that the quantum-classical correlation can also be diminished by introducing a variational principle in the arbitrary initial momentum P_0.

2.6.4 Spherical harmonics in a Gauss-Hermite basis set

For a diatomic molecule, we have the rotational/vibrational motion described approximately by the product wave function

$$\psi_{njm} = \frac{1}{r} g_n(r) Y_{jm}(\theta, \phi) \tag{2.218}$$

where $g_n(r)$ is a vibrational wave function and Y_{jm} a spherical harmonics with quantum numbers j and m. In order to introduce the G-H representation instead, we use the connection between the spherical and the cartesian coordinates

$$x = r \sin \theta \cos \phi \tag{2.219}$$
$$y = r \sin \theta \sin \phi \tag{2.220}$$
$$z = r \cos \theta \tag{2.221}$$

and the product basis set

$$\phi_{n_1}(x, t_0) \phi_{n_2}(y, t_0) \phi_{n_3}(z, t_0) \tag{2.222}$$

where ϕ_n are G-H basis functions. We can now expand the vibrational/rotational wave function in the G-H basis

$$\psi_{njm}(x, y, z, t_0) = \sum_{n_1 n_2 n_3} c_{n_1 n_2 n_3}(t_0) \phi_{n_1}(x, t_0) \phi_{n_2}(y, t_0) \phi_{n_3}(z, t_0) \tag{2.223}$$

Thus, the initial value of the expansion coefficient is obtained by projecting the vibrational/rotational wave function on the G-H basis set, that is, we need to evaluate the overlap integrals:

$$O_{n_1 n_2 n_3}^{njm} = \int dx \int dy \int dz \phi_{n_1}(x, t_0)^* \phi_{n_2}(y, t_0)^* \phi_{n_3}(z, t_0)^* \frac{1}{r} g_n(r), Y_{jm}(\theta, \phi) \tag{2.224}$$

Table 2.5 shows that, depending on the value of the width parameter $\mathrm{Im}A(t_0)$, we need of the order 10–15 G-H basis functions in each dimension for a better than 90% representation. Later, we shall consider a DVR representation which facilitates the projection of the bound state wave function on the basis set (the DVR basis set), but before this is done we wish to discuss a so-called second quantization approach to molecular dynamics.

Table 2.5: The overlap $S = \sum_{n_i=0}^{n_{max}} |O_{n_1 n_2 n_3}^{njm}|^2$ as a function of $\mathrm{Im}A_x = \mathrm{Im}A_y = \mathrm{Im}A_z$ for $(njm) = (000), (010)$ and (001).

$\mathrm{Im}A_x(t_0)$	njm	$n_{max} = 5$	$n_{max} = 6$	$n_{max} = 7$	$n_{max} = 8$	$n_{max} = 9$	$n_{max} = 10$
0.10	000	0.778	0.831	0.845	0.870	0.878	0.899
	010	0.735	0.751	0.823	0.850	0.876	0.907
	001	0.316	0.398	0.445	0.542	0.576	0.639
0.20	000	0.822	0.888	0.915	0.944	0.961	0.907
	010	0.792	0.895	0.904	0.930	0.936	0.954
	001	0.439	0.599	0.660	0.726	0.773	0.825

For (001) we obtain with $\mathrm{Im}A_x(t_0) = 0.27$ and $n_{max} = 15$ an overlap of 0.972.

2.7 Second quantization and Gauss-Hermite functions

Aside from the fact that the G-H basis set generates classical mechanics in a limit, it has another rather important property, namely that the large set of equations involved when solving the equations for the expansion coefficients (2.202) may be dealt with approximately using a local second-order expansion of the potential around the trajectories. Consider, for instance, a 6-dimensional (6-D) system with the variables (x, y, z, X, Y, Z) where $\mathbf{R} = (X, Y, Z)$ and $\mathbf{r} = (x, y, z)$ are Jacobi vectors from an atom A to the center of mass of a diatomic molecule BC and from atom B to C, respectively. The initial wave function for such a system is

$$\frac{1}{R} f_{p_0}(R, t_0) Y_{lm_l}(\Theta, \Phi) \frac{1}{r} g_n(r) Y_{jm}(\theta, \phi) \tag{2.225}$$

where Θ and Φ specify the orientation of the $\mathbf{R}$-vector in a space-fixed coordinate system, $f_{p_0}(R, t_0)$ is a Gaussian wave packet centered around p_0 in momentum space. It is possible to switch to a body-fixed frame and introduce the states Y_{jl}^{JM} where J is the total angular momentum and M its projection on a space-fixed axis. The rotational and orbital angular momenta are j and l, respectively. Thus, alternatively, we have the initial wave function defined as

$$\frac{1}{Rr} Y_{jl}^{JM} f_{p_0}(R, t_0) g_n(r) \tag{2.226}$$

where [62, 83]

$$Y_{jl}^{JM} = (-1)^{j+l-M} \sqrt{(2J+1)(2l+1)/4\pi}$$
$$\times \sum_{\mu} \begin{pmatrix} j & l & J \\ \mu & 0 & -\mu \end{pmatrix} Y_{j\mu}(\eta, \xi) D_{-M,\mu,0}^{J}(\Theta, \Phi) \tag{2.227}$$

Here, η, ξ specify the orientation of the diatomic molecule in a body-fixed frame with z axis along R and $D^J_{-M,\mu}$ denotes a matrix element over the rotational operator [83].

As we have seen, the wave function can be expanded in G-H basis functions in the 6-D cartesian space. This would lead to a rather large C-matrix in the coupled equations (2.202). With 10 basis functions in each dimension, the size would be of the order 10^6 and, hence, it would be impossible or difficult to work with in practical calculations. However, if we expand the potential locally around the trajectories to second order in the 6-D space, we have

$$V(x, y, z, X, Y, Z) = V_0(x(t), y(t), z(t), X(t), Y(t), Z(t))$$

$$+ \sum_{i=1,6} \frac{\partial V}{\partial x_i}\bigg|_{x_i=x_i(t)} (x_i - x_i(t))$$

$$+ \frac{1}{2} \sum_{ij} \frac{\partial^2 V}{\partial x_i \partial x_j}\bigg|_{x_i=x_i(t)} (x_i - x_i(t))(x_j - x_j(t)) \quad (2.228)$$

It is now possible to reduce the solution of the C-matrix problem to a more manageable one. This has to do with the fact that the matrix problem can be solved in an operator representation rather than in a wave function or state representation. The operators are the usual boson operators known from second quantization formulations of quantum mechanics. Thus, we define

$$x_i - x_i(t) = \frac{1}{2}\sqrt{\frac{\hbar}{\text{Im}A_{x_i}}}(a_i + a_i^+) \quad (2.229)$$

where the boson operators have the properties

$$a_i^+ \phi_{n_i}(x_i, t) = \sqrt{n_i + 1}\phi_{n_i+1}(x_i, t) \quad (2.230)$$
$$a_i \phi_{n_i}(x_i, t) = \sqrt{n_i}\phi_{n_i-1}(x_i, t) \quad (2.231)$$

where ϕ_n denotes a GH basis function.

With the above expansion of the potential matrix, we obtain the operator representation of the C-matrix (2.203) as follows:

$$\hbar \sum_i \omega_i \left(a_i^+ a_i + \frac{1}{2}\right) + \sum_i b_i(V_{x_i}'(t) - V_{x_i}')(a_i + a_i^+)$$

$$+ \sum_i \frac{\hbar}{8\text{Im}A_{x_i}(t)}(a_i + a_i^+)^2 (V_{x_i x_i}(t) - V_{x_i}'')$$

$$+ \frac{\hbar}{8} \sum_{i \neq j} \frac{1}{\sqrt{\text{Im}A_{x_i}\text{Im}A_{x_j}}} V_{x_i x_j}(t)(a_i + a_i^+)(a_j + a_j^+) \quad (2.232)$$

where we have introduced the notation $V_{x_i}'(t) = \left(\frac{\partial V}{\partial x_i}\right)\big|_{x_i=x_i(t)}$, $V_{x_i x_j}(t) = \left(\frac{\partial^2 V}{\partial x_i \partial x_j}\right)\big|_{x_i=x_i(t),x_j=x_j(t)}$. V_{x_i}' and V_{x_i}'' denote the quantum forces. Thus, if the basis set is propagated using classical forces, the second and third term containing V' and V'' vanish. In this case, the quantum C-matrix (2.203) reduces to just the diagonal terms containing the frequencies

$$\omega_i = 2\mathrm{Im}A_{x_i}/m_i \tag{2.233}$$

and the mixed derivative off diagonal terms (the last term in eq. (2.232)). m_i is the mass connected to motion in the coordinate x_i and the matrix elements b_i in (2.232) are defined by

$$b_i = \frac{1}{2}\sqrt{\hbar/\mathrm{Im}A_{x_i}} \tag{2.234}$$

We notice that since the potential is expanded to second order in this approximation, the classical and quantum forces are identical (see eqs. (2.117, 2.118)). However, if the fixed-width approach is used, the second derivative entering the diagonal matrix is different from the classical force $V_{x_ix_i}$. We recall that we need to have $V'' = 4\mathrm{Im}A^2/m$ in order to keep $\mathrm{Im}A$ constant.

The operator representation of the C-matrix is now

$$\sum_i \hbar\omega_i(t)\left(a_i a_i^+ + \frac{1}{2}\right) + \frac{1}{2}\sum_{ij} b_i b_j (V_{x_ix_j}(t) - V_{x_i}''\delta_{ij})(a_i + a_i^+)(a_j + a_j^+) \tag{2.235}$$

Introducing a matrix representation of the operators, the equations for the C-matrix become

$$i\hbar\frac{d}{dt}\mathbf{c}(t,t_0) = \mathbf{C}(t,t_0)\mathbf{c}(t,t_0) \tag{2.236}$$

where $\mathbf{c}(t_0,t_0) = \mathbf{I}$ (the unit matrix) and

$$C_{ii} = \hbar\omega_i(t) + b_i^2(V_{x_ix_i}(t) - V_{x_i}'') \tag{2.237}$$
$$C_{ij} = b_i b_j V_{x_ix_j} \tag{2.238}$$

If we introduce the interaction representation, eq. (2.236) becomes

$$i\hbar\frac{d}{dt}\mathbf{R}(t,t_0) = \mathbf{B}(t,t_0)\mathbf{R}(t,t_0) \tag{2.239}$$

where

$$B_{ij}(t,t_0) = b_i b_j V_{x_ix_j}(t)\exp(i(\theta_i(t) - \theta_j(t))) \tag{2.240}$$
$$B_{ii} = 0 \tag{2.241}$$

and

$$\theta_i(t) = \int_{t_0}^t (\omega_i(t) + b_i^2(V_{x_ix_i} - V_{x_i}'')/\hbar)dt \tag{2.242}$$

Note that we have neglected the double quantum operators of the type $a_i a_j$ and $a_i^+ a_j^+$. We also notice that with these approximations, the C-matrix problem has been reduced to one of dimension $M \times M$ where M is the number of operators. For a 6-dimensional problem we then have $M = 6$, and so on. In order to get back to the wave function representation, all we need to do is to find the evolution operator for the system under the time-dependent hamiltonian, that is,

$$\Psi(t) = U(t, t_0)\Psi(t_0) \tag{2.243}$$

where

$$i\hbar \frac{d}{dt} U(t, t_0) = H(t, t_0)U(t, t_0) \tag{2.244}$$

We shall later show how this equation can be solved algebraically and, hence, also, how the evolution operator can be expressed in terms of the operators $a_i^+ a_j$. Having done this, we can obtain the amplitude for transition from a quantum state $nvjkJM$ to state $n'j'm'k'JM$ formally as:

$$\langle n'v'j'k'JM|S|nvjkJM \rangle$$
$$= \sum_{\{n_i, m_i\}} O_{n_1 n_2 n_3}^{n'j'm'} O_{n_4 n_5 n_6}^{k'JM} \langle \{n_i\}|U|\{m_i\}\rangle O_{m_1 m_2 m_3}^{njm} O_{m_4 m_5 m_6}^{kJM} \tag{2.245}$$

where $|k\rangle$ denotes a plane wave state and $O_{m_4 m_5 m_6}^{kJM}$ an overlap matrix between the G-H functions and the proper initial/final state wave function. The transition matrix $\langle n|U|m \rangle$ is between initial and final G-H states.

Thus, if we can solve the dynamics in the operator representation and evaluate the matrix elements over the evolution operator after the collision, we need only to sum over quantum numbers of the overlap matrix between the G-H basis functions and ordinary basis functions in order to solve the problem. We shall later give equations for the matrix elements $\langle \{n\}|U|\{m\}\rangle$ (see appendices B and C). The introduction of the G-H basis set has then made it possible to generate equations of motion for trajectories, and a local expansion around these trajectories makes the solution of the dynamics feasible even for large systems. If a given degree of freedom is treated as purely classical, then it is simply omitted from the **B**-matrix, which is reduced accordingly. If we let the translational wave function $f_{p_0}(R, t_0)$ be a Gaussian wave packet, this wave function has a distribution also in momentum space. Thus, we can in principle project on final plane wave states multiplied by an internal wave function in an energy range. However, since the unitarity of the above expression is guaranteed only if a complete basis is used, and since this basis may include also energetically closed states, we can usually get a meaningful projection only at certain energies—most conveniently defined as those where the sum of probabilities over open channels $(n'j'm')$ sum up to unity. This problem arises from the fact that all final states are, in principle, open and the problem appears in all methods in which the translational motion is coupled to the inelastic transitions in an average fashion. In chapter 3 we will discuss the problem again in the context of the classical path method.

It is, however, also possible to use the Gauss-Hermite expansion in such a fashion that we work within an exact theory. There are two ways of doing this.

One is, as we have seen, the state-expansion method. The other method, from a numerical point of view more convenient, is obtained by introducing a discrete variable representation (DVR). We have named this approach the "quantum dressed" classical mechanics method.

2.8 Quantum dressed classical mechanics

The most useful general purpose approach to quantum dynamical processes appears to be the one obtained when the G-H representation is introduced in all degrees of freedom and the equations put in a discrete variable representation (DVR). That this is at all possible is connected to the use of an orthorgonal polynomial expansion—namely, in Hermite polynomials. We can use the zeroes of these to define the grid points necessary for constructing a DVR basis. This is yet another reason for why the expansion in an orthorgonal basis set is advantageous.

2.8.1 The DVR scheme

In order to solve the TDSE for many-dimensional systems and in order to facilitate the evaluation of the kinetic energy terms, we can expand the wave function in basis sets for which such an evaluation is easy or can be carried out analytically. The G-H basis set, for instance, makes the evaluation of kinetic energy terms a simple differentiation of exponentials and Hermite polynomials. The price which is paid for this is that the matrix elements over the potential coupling terms are more cumbersome to evaluate. In grid methods, the wave function is represented as a discrete set of points and, hence, the operation with the potential on the wave function is a simple multiplication. Here, however, the kinetic energy terms pose the "problem"—they need to be evaluated by numerical techniques as, for instance, the FFT (fast Fourier transform) method or by introducing a DVR (discrete variable representation) scheme in which the algebraic manipulations involved in the evaluation of the kinetic coupling are facilitated (see, for example, [138]).

The G-H basis allows us to combine the basis set evaluation of the kinetic energy with a DVR method such that integrals over the potential are avoided. Consider the Nth Hermite polynomial with the zeros z_i (see table 2.6). The zeros can easily be determined in the following manner. For large values of N, we have the asymptotic representation of the Hermite polynomial given by [71]

$$H_N(x) \sim \cos\left(\sqrt{2N+1}x\right) \tag{2.246}$$

where N is even. For N odd, we have $\sin\left(\sqrt{2N+1}x\right)$ instead of $\cos\left(\sqrt{2N+1}x\right)$. Thus the zeros are given by $z_i = \pm i\pi/(2\sqrt{2N+1})$ with $i = 1, 3, \ldots, N-1$ for N even or $i = 0, 2, \ldots, N-1$ for N odd. In order to obtain the value accurately, we need an intial guess and then a Newton-Raphson iteration based on the formula

$$z_i^{new} = z_i^{old} - H_N(z_i^{old})/(2NH_{N-1}(z_i^{old})) \tag{2.247}$$

Table 2.6: Zeros for Hermite polynomials obtained by a Newton-Raphson iterative scheme. Only the positive values are shown.

i	z_i for $N = 10$	z_i for $N = 20$	z_i for $N = 30$
1	0.342899	0.245336	0.201122
3	1.036608	0.737464	0.603924
5	1.756687	1.234081	1.008345
7	2.532735	1.738543	1.415534
9	3.436156	2.254979	1.826748
11		2.788811	2.243398
13		3.347859	2.667138
15		3.944769	3.099927
17		4.603678	3.544450
19		5.387476	4.003915
21			4.483062
23			4.988925
25			5.533154
27			6.138273
29			6.863339

A converging scheme is obtained by starting the iteration in the following manner

$$z_i = \frac{i\pi}{2\sqrt{2N+1}} \qquad \text{for } i = 1, 3 \quad \text{or } i = 0, 2 \qquad (2.248)$$

$$z_i = z_{i-2} + 1.1(z_{i-2} - z_{i-4}) \quad \text{for } i \geq 5 \qquad \text{or } i \geq 4. \qquad (2.249)$$

where the last values are for N odd. The conceptual advantage of using an odd number of grid points is that they center around the classical trajectory. With $N = 1$, we have a classical mechanical description of that particular degree of freedom.

We now introduce the DVR basis set as [134]–[136]

$$\psi_i(x, t) = \sum_{n=0}^{N} \phi_n(z_i)\Phi_n(x, t) \qquad (2.250)$$

where

$$\Phi_n(x, t) = \pi^{1/4} \exp\left(\frac{i}{\hbar}(\gamma(t) + p_x(t))(x - x(t)) + \text{Re}A(t)(x - x(t))^2\right)\phi_n(x, t) \qquad (2.251)$$

and

$$\phi_n(\xi, t) = \frac{1}{\sqrt{n!2^n \sqrt{\pi}}} \exp(-\xi^2/2)H_n(\xi) \qquad (2.252)$$

with $\xi = \sqrt{2\text{Im}A(t)/\hbar}(x - x(t))$. The normalized DVR basis functions for $N = 10$ are shown in fig. 2.7.

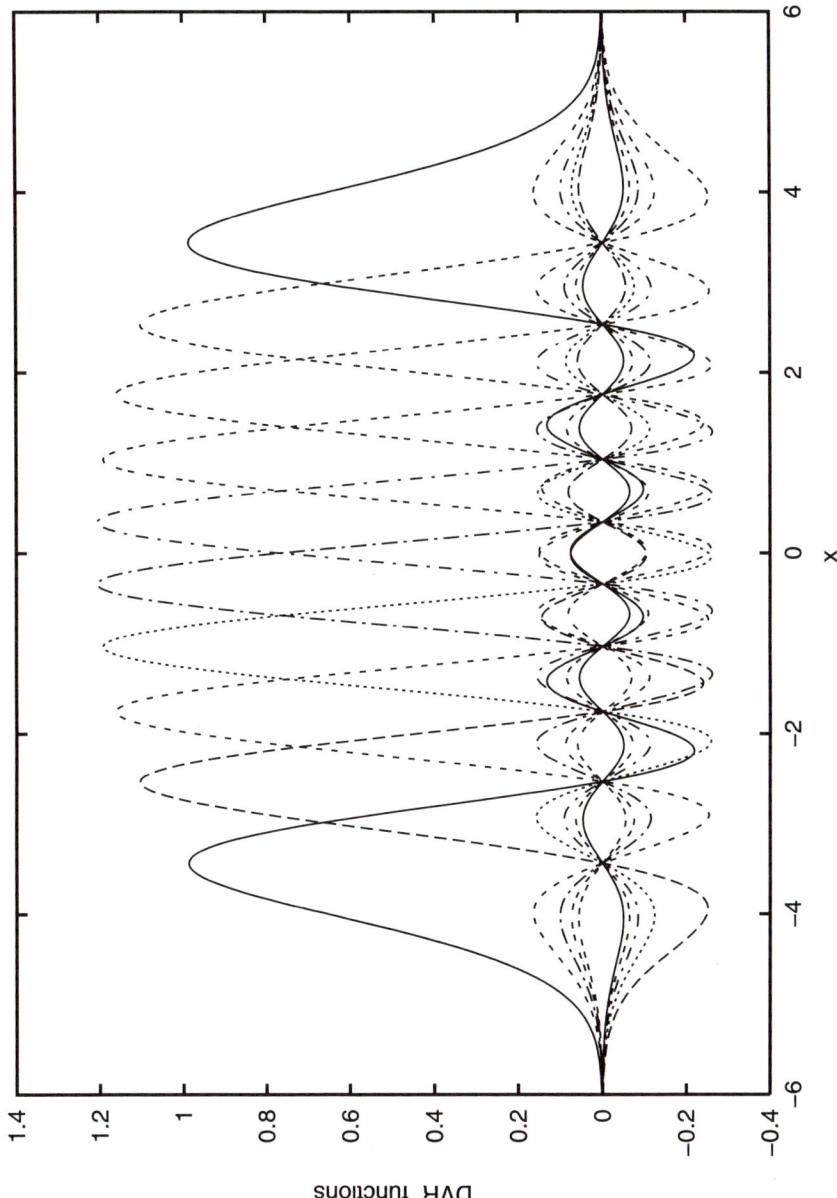

Fig. 2.7: The normalized DVR functions for $N = 10$.

The grid points z_i are defined through the zeros of the Nth Hermite polynomial and are, therefore, time-independent. However, in x, the actual coordinate the grid points are time-dependent through the equation

$$x_i(t) = x(t) + \sqrt{\hbar/2\mathrm{Im}A(t)}\, z_i \qquad (2.253)$$

Thus the grid points $x_i(t)$ are centered around the trajectory $x(t)$ (see fig. 2.8).

As mentioned previously, the forces V_0' and V_0'' in eqs. (2.259–2.261) are, in principle, arbritrary. We can, by choosing

$$V_0'' = 4\mathrm{Im}A^2/m \qquad (2.254)$$

in eq. (2.261) below, obtain $\mathrm{Im}A(t) =$ constant, provided that $\mathrm{Re}A(t_0) = 0$. With this so-called fixed-width approach, the grid points follow the classical trajectory (see fig. 2.9) at constant distance. Our DVR basis set is defined through the functions $\psi_i(x, t)$ and the wave function is expanded in this basis set as

$$\Psi(x, t) = \sum_i \psi_i(x, t) c_i(t) \qquad (2.255)$$

In fig. 2.7 we have plotted the normalized DVR functions

$$f_i(x) = \frac{1}{\sqrt{A_{ii}}} \sum_n \phi_x(z_i)\phi_n(x) \qquad (2.256)$$

where A_{ii} is defined below. We notice that they have delta-function character such that

$$f_i(x) \sim \delta(x - z_i) \qquad (2.257)$$

This property can now be utilized when setting up the equations for the expansion coefficients $d_i(t) = \sqrt{A_{ii}}c_i(t)$. By inserting the expansion (2.255) in the TDSE, we get the following set of equations

$$i\hbar \dot{\mathbf{d}}(t) = \mathbf{C}\mathbf{d} \qquad (2.258)$$

$$\dot{p}(t) = -V_0' \qquad (2.259)$$

$$\dot{x}(t) = p/m \qquad (2.260)$$

$$\dot{A}(t) = -\frac{2}{m}A^2 - \frac{1}{2}V_0'' \qquad (2.261)$$

The quantities A_{ii} are time-independent normalization constants and are defined by

$$A_{ij} = \sum_n \phi_n(z_i)\phi_n(z_j) = A_{ii}\delta_{ij} \qquad (2.262)$$

In the DVR basis, the matrix elements of the C-matrix are obtained as:

$$C_{ij} = (E_{kin} + W(x_i))\delta_{ij} + \frac{\hbar \mathrm{Im}A(t)}{m} \sum_n \phi_n(z_i)(2n + 1)\phi_n(z_j)/\sqrt{A_{ii}A_{jj}} \qquad (2.263)$$

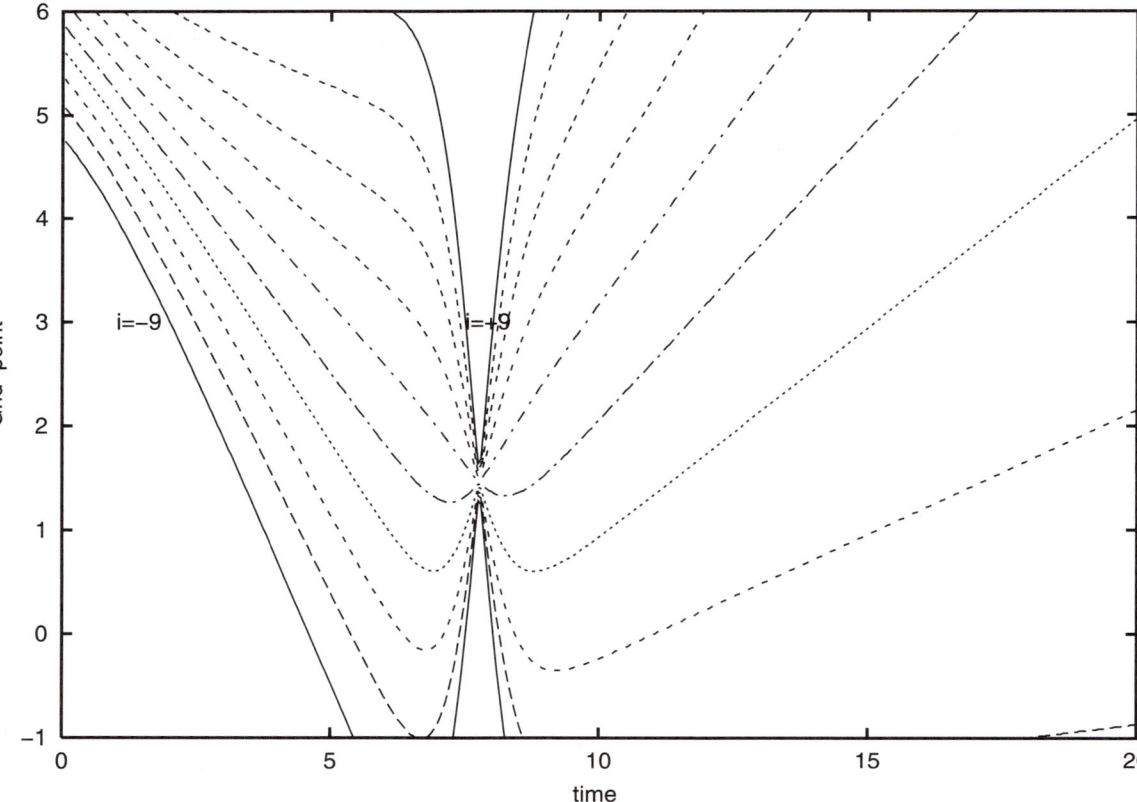

Fig. 2.8: The grid points for the translational motion in a collision between an atom and a harmonic oscillator. $N = 10$ grid points are used. The wave packet shrinks to a small region at $t = 8\tau$. If the fixed width approach is used, the grid points follow the trajectory with equal spacing.

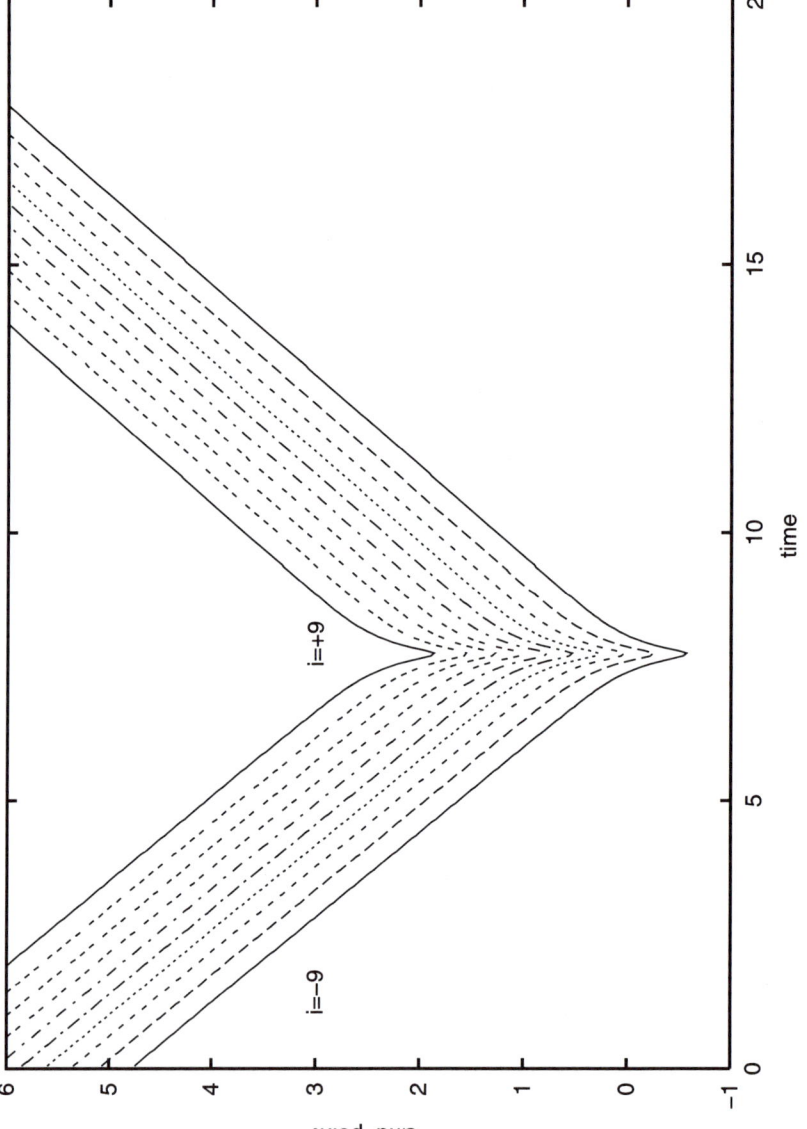

Fig. 2.9: If the fixed width approach is used, the grid points follow the trajectory with constant distance.

We notice that in the DVR representation, the potential is diagonal as usual and the coupling between the grid points comes about through the last term, which has a very simple time-dependence (through $\mathrm{Im}\,A(t)$). This term is what is left of the kinetic energy coupling—we notice that the kinetic energy operators have already worked on the basis functions generating classical equations of motion. Thus, the coupling elements $\sum_n \phi_n(z_i)\phi_n(z_j)(2n+1)/\sqrt{A_{ii}A_{jj}}$ are, therefore, time-independent and can be evaluated and stored in memory. The coupling matrix is, furthermore, diagonally dominant (see table 2.7), which facilitates the solution of the equations (2.258) in a time step. Using a Lanczos iterative scheme [137, 138], we have

$$\mathbf{d}(t+\Delta t) = \mathbf{TS}\exp\left(-\frac{i}{\hbar}\mathbf{D}\Delta t\right)\mathbf{S}^+\mathbf{T}^+\mathbf{d}(t) \qquad (2.264)$$

where $\mathbf{T}$ is an $M \times N$ matrix containing the M recursion vectors, $\mathbf{D}$ a diagonal $M \times M$ matrix, and $\mathbf{S}$ an $M \times M$ matrix which diagonalizes the tridiagonal Lanczos matrix with eigenvalues $\mathbf{D}$. The number of recursions in each time step Δt depends on the coupling and the accuracy needed. For molecular dynamics problems, 5 to 15 iterations are needed with $\Delta t = 0.1$ or 0.2 fs. However, due to the dominance of the diagonal term in the kinetic matrix, we can construct an even more efficient propagator. Thus, we define the matrix $\mathbf{E}$ as the diagonal part of the $\mathbf{C}$ matrix and split the operator such that the propagation is given as

$$\mathbf{d}(t+\Delta t) = \exp\left(-\frac{i}{2\hbar}\mathbf{E}\Delta t\right)\exp\left(-\frac{i}{\hbar}\tilde{\mathbf{C}}\Delta t\right)\exp\left(-\frac{i}{2\hbar}\mathbf{E}\Delta t\right)\mathbf{d}(t) \qquad (2.265)$$

where $\tilde{\mathbf{C}}$ contains the off-diagonal terms of $\mathbf{C}$.

Due to the splitting, the error of this propagator is of order (Δt^3). Thus, we can define the vector $\mathbf{p}(t) = \exp(-\frac{i}{\hbar}\mathbf{E}\Delta t)\mathbf{d}(t)$ and propagate it with the "smaller" $\tilde{C}$ matrix using the Lanczos propagator to get $\mathbf{p}(t+\Delta t)$, from which we obtain $\mathbf{d}(t+\Delta t) = \exp(\frac{i}{\hbar}\mathbf{E}\delta t)\mathbf{p}(t+\Delta t)$. This procedure saves about a factor

Table 2.7: Elements T_{kj} of the kinetic energy matrix with 9 grid points and a system mass of 0.504 amu. $\mathrm{Im}\,A(t_0) = 1.0$ amu τ^{-1}.

k/j	1	2	3	4	5	6	7	8	9
1	1.653	−0.295	0.085	−0.041	0.025	−0.016	0.012	−0.008	0.006
2	−0.295	1.230	−0.396	0.106	−0.049	0.028	−0.018	0.012	−0.008
3	0.085	−0.396	0.979	−0.454	0.117	−0.052	0.029	−0.018	0.012
4	−0.041	0.106	−0.454	0.842	−0.481	0.120	−0.052	0.028	−0.016
5	0.025	−0.049	0.117	−0.481	0.798	−0.481	0.117	−0.049	0.025
6	−0.016	0.028	−0.052	0.120	−0.481	0.842	−0.454	0.106	−0.041
7	0.012	−0.018	0.029	−0.052	0.117	−0.454	0.979	−0.396	0.085
8	−0.008	0.012	−0.018	0.028	−0.049	0.106	−0.396	1.230	−0.295
9	0.006	−0.008	0.012	−0.016	0.025	−0.041	0.085	−0.295	1.653

of 2 to 3 as compared to the propagation under the full C-matrix [142]. Thus, the number of Lanczos iterations can be reduced to about 5 with a step length of $\Delta t = 0.2$ fs.

In the DVR representation, the wave function may be expressed as a sum over the grid points

$$
\Psi(x,t) = \left(\frac{\hbar}{2\mathrm{Im}A(t)} \right)^{1/4}
$$

$$
\times \sum_i d_i(t) \exp\left(\frac{i}{\hbar}(p_x(t)(x - x(t)) + \mathrm{Re}A(t)(x - x(t))^2) \right)
$$

$$
\times \delta(\xi - z_i)/\sqrt{A_i} \tag{2.266}
$$

where $\delta(\xi - z_i)$ is a delta function. This expression demonstrates the discrete representation of the wave function at the grid points. The variable ξ is defined as

$$
\xi = \sqrt{2\mathrm{Im}A(t)/\hbar}(x - x(t)) \tag{2.267}
$$

and, hence, $\xi = z_i$ gives $x = x_i$. The amplitude at a given grid point is $d_i(t)$ and we have

$$
\sum_i |d_i(t)|^2 = 1 \tag{2.268}
$$

Thus, the total probability for a tunneling process, for instance, can be obtained by "grid summation." For a Gaussian wave packet started left ($x < 0$) of a barrier located at the origin $x = 0$ we have

$$
\langle P_{tunn} \rangle = \lim_{t \to \infty} \sum_{i>i^*} |d_i(t)|^2 \tag{2.269}
$$

where i^* is defined by $x_{i^*} > 0$. We may, however, also project on plane waves to get energy-resolved transmission probabilities, that is, express Ψ as

$$
\Psi(x,t) = \sum_k c_k \exp(ikx)/\sqrt{2\pi} \tag{2.270}
$$

This gives the following energy-resolved transmission amplitudes

$$
c_k = \frac{1}{\sqrt{2\pi}} \left(\frac{\hbar}{2\mathrm{Im}A(t)} \right)^{1/4} \sum_{i>i^*} \frac{d_i(t)}{\sqrt{A_i}}
$$

$$
\times \exp\left(\frac{i}{\hbar}(p(t)(x_i - x(t)) + \mathrm{Re}A(t)(x_i - x(t))^2 - kx_i) \right) \tag{2.271}
$$

where x_i is defined by eq. (2.253) and where we have used the delta-function property of the DVR functions. The tunneling probability as a function of energy is finally obtained as

$$
P_{tunn}(E) = \frac{|c_k|^2}{P(k, k_0)} \tag{2.272}
$$

where $P(k, k_0)$ is the probability distribution of the wave numbers k in the initial wave packet centered around k_0 (see eq. 2.59). The connection between the energy-resolved and the average tunneling probability is

$$\langle P_{tunn}\rangle = \int P_{tunn}(E)P(k, k_0)dk \tag{2.273}$$

where $E = \hbar^2 k^2/2m$. Thus, it is possible to obtain $P_{tunn}(E)$ by determining the average probability at a number of k_0 values and then invert the above expression. This is advantageous if the projection on asymptotic plane waves involves many more grid points than is necessary for determining the average probabilities [122].

2.8.2   Arbitrarily sized systems

The extension of the DVR method to many dimensions d is now straightforward. Introducing the total number of grid points as

$$N = \Pi_{i=1}^d N_i \tag{2.274}$$

where N_i is the number of grid points in mode i. The potential is represented at these points as

$$V(x_1^{(1)}, \dots, x_{N_1}^{(1)}, x_1^{(2)}, \dots, x_{N_2}^{(2)} \dots) \tag{2.275}$$

The effective potential W is also diagonal in the grid representation and is obtained by subtracting first and second derivative terms evaluated at the trajectory $\mathbf{x}(t) = x^{(1)}(t), \dots, x^{(d)}(t)$. Thus, we have

$$W\left(x_1^{(1)}, \dots, x_{N_d}^{(d)}\right) = V\left(x_1^{(1)}, \dots, x_{N_d}^{(d)}\right) - \sum_{k=1}^d \sum_{i=1}^{N_k} V_k'\left(x_i^{(k)} - x^{(k)}(t)\right)$$
$$- \frac{1}{2}\sum_{k=1}^d \sum_{i=1}^{N_k} V_k''\left(x_i^{(k)} - x^{(k)}(t)\right)^2 \tag{2.276}$$

where, with a classical propagation of the trajectories, we have

$$\dot{x}_k = p_k(t) \tag{2.277}$$
$$\dot{p}_k = -V_k' \tag{2.278}$$

and $V_k' = (\partial V/\partial x^{(k)})$. We notice that it is often convenient to keep the width of the wave packet fixed, or in the DVR representation to keep the spacing between the grid points constant. This can be achieved by using

$$V_k'' = \frac{4}{m_k}\mathrm{Im}A_k(t)^2 \tag{2.279}$$

and choosing $\mathrm{Re}A_k(t) = 0$ initially. If this is done, these values of the second derivative enter the expression (2.276) for the effective potential W.

The "kinetic" energy coupling terms are defined by

$$T_{kj}^{(p)} = \Pi_{l \neq p} \delta_{i^{(l)} k^{(l)}} \frac{\hbar \mathrm{Im} A_p(t)}{m_p} \sum_{n=0}^{N_p-1} \tilde{\phi}_n(z_k)(2n+1)\tilde{\phi}_n(z_j) \tag{2.280}$$

for the modes $p = 1, \ldots, d$. N_p is the number of grid points for mode (p) and k, j are grid points in mode p. The grid points in mode l are denoted $i^{(l)}$ and

$$\tilde{\phi}_n(z_k) = \phi_n(z_k)/\sqrt{A_{kk}^{(p)}} \tag{2.281}$$

where $A_{kk}^{(p)}$ is the normalization constant defined in eq. (2.262). The kinetic energy term couples grid points within a particular mode. If a given degree of freedom is treated "classically" we have just a single grid point in that mode, that is, $N_i = 1$. Hence, this mode appears in the dynamical equations only through the classical equations of motion for the trajectory and the effect of the classical value in the potential W. We notice that the method is extremely simple to program (see flow diagram in fig. 2.10). We need the classical equations of

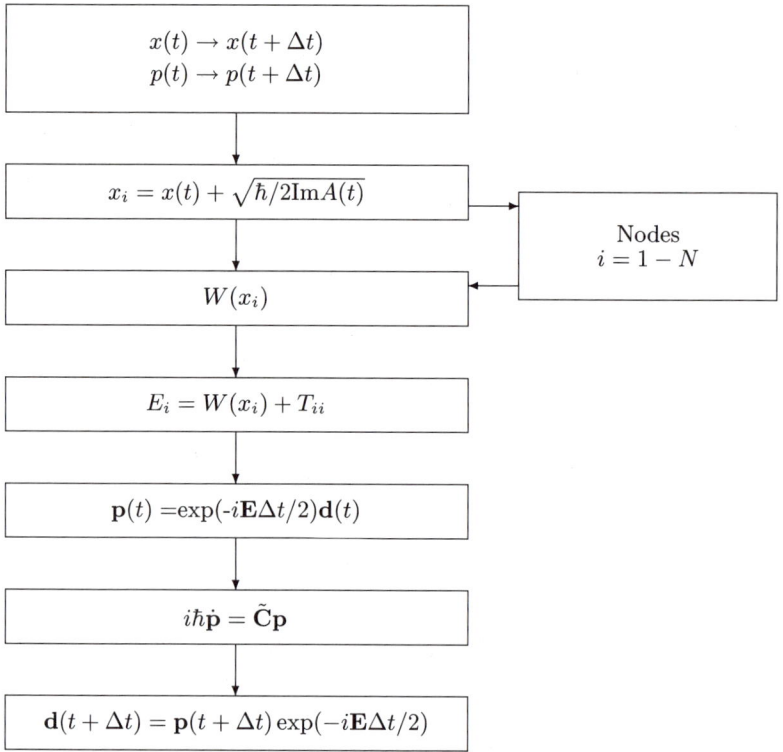

Fig. 2.10: Flow diagram for a time-dependent DVR approach to molecular dynamics. In a direct dynamics approach, the information on the potential $V(x_i)$ is calculated at node i.

motion for each degree of freedom (DOF). A set of grid points for each DOF and the dimension of the coupling matrix is a product of the number of grid points for those DOFs treated quantally. Since the only quantity which needs evaluation in each time step is the potential at each grid point, the amount of storage needed is minimal. When using Lanczos propagating scheme for solving the equations for $d_i(t)$, we usually store the M recursion vectors, that is, the storage requirement is then $M \times N$. At the expense of extra CPU or I/O to the disc, we may either calculate the recursion vectors twice or read them from the disc when needed, in each time step. Thereby, the storage requirement in fast memory is reduced to just 3 recursion vectors, that is, a total of $3N$ variables.

For large systems, the advantage of the method comes about if a large part of the system can be treated with just one or two grid points. The formulation offers a straightforward way of testing whether a classical mechanical description of a given DOF is adequate.

Since the kinetic coupling matrix is sparse, it is advantageous to utilize this in the programming of the Lanczos iteration scheme. Thus a table of non-zero elements can be set up, storing the matrix element indices and their value (except perhaps for the factor $\text{Im} A(t)$). Also, the fact that the matrix is symmetric (see table 2.7) can be used to reduce the numerical labor in each time step. Table 2.8 shows that typically less than a fraction of a per cent of the off-diagonal matrix elements are different from zero. In this respect, the method is similar to the distributed approximating functions (DAFs) of Kouri and coworkers [56].

Since we need to calculate the potential in each time step, the method scales, in many applications, roughly linearly with the number of grid points. How close we are to linear scaling depends on the efficiency with which the sparse matrix can be multiplied onto a vector. As mentioned, the number of non-zero matrix elements is small. It follows the formula

$$N = \Pi_{i=1}^{d} N_i \times \sum_{i=1}^{d}(N_i - d) \tag{2.282}$$

where N_i is the number of DVR points in dimension i and d is the number of dimensions treated within the DVR representation. In conventional grid methods [138], the effect of the kinetic energy operators is evaluated using FFT technique. Here, the CPU-time scales as a constant $\times N \log N$, where the constant depends on the machine architecture and whether N is an integral power of 2 or not.

Table 2.8: Number of non-zero off-diagonal elements (M) in a 4D-quantum problem treated by the DVR-scheme. The fraction follows roughly the expression d/N^{d-1} for $N_i = N$ ($i = 1, \ldots, d$).

Mode	1	2	3	4	Dimension	M	Percentage
N_p	11	11	11	11	14641	585640	0.27
	5	13	13	13	10985	439400	0.36
	5	15	15	15	16875	776250	0.27

Typical values of the constant are about 5. The DVR method scales as $N \sum_i N_i$, which is larger than the FFT scaling by a factor $N_i/\log N_i$, which for typical values of $N_i \sim 20\text{–}40$ is of the order 5–10. Thus, the two methods behave similarly as far as the kinetic energy term is concerned.

But the real advantage of the time-dependent DVR method is that the DVR points follow the dynamics of the system. Hence, we do not need to include imaginary absorbing potentials in order to avoid having the wave function approach the edges of the grid. Another advantage is that using one grid point corresponds to classical mechanics, that is, to a meaningful description of the dynamics. In ordinary grid methods use of one grid point would be nonsense. This construction of the theory also means that the number of grid points in each dimension can often be kept smaller than in ordinary methods. As far as the storage in RAM memory is concerned, we may, by writing the recursion vectors on disc, reduce it to just the three vectors in each recursion step. This means that 6D quantum calculations with about 15–20 grid points in each mode can be handled on a PC computer. The programming of the effect of the kinetic energy matrix on the **d**-vector is very simple (see an algorithm in table 2.9). The TDGH-DVR method has so far been used for treating up to 6 dimensions with more than one grid point [152]. The grid can be tailored to the problem at hand. If we, for instance, are interested in state-resolved cross sections, a certain number of points are needed to represent the vibrational rotational state of a molecule (see table 2.10). If, on the other hand, we are interested in dissociation or reaction probabilities, we just sum the squared amplitudes for bond lengths larger than a given minimum value (2.269), and, hence, the number of points for the molecule degrees of freedom can be diminished. Consider, for instance, a hydrogen molecule scattered from a solid. The molecule has 6 degrees of freedom (X, Y, Z) for the center-of-mass motion and (x, y, z) for the molecule. In order to obtain the probability for dissociation at the surface we just sum over the amplitudes for grid points having the bond distance r larger than a critical value r^*, that is,

$$P_{diss} = \sum_{p=1}^{N_d} |d_p|^2 h(r_p - r^*) \tag{2.283}$$

where $h(x)$ is a Heaviside function, unity for positive and zero for negative arguments. The value of r_p is obtained as

$$r_p = \sqrt{x_i^2 + y_j^2 + z_l^2} \tag{2.284}$$

where i, j and l run over the grid points in the three coordinates. Table 2.11 shows that about 15 grid points for the molecule modes give sufficiently accurate dissociation probabilities P_{diss}. The calculations were done using an effective potential (see section 5.3) to couple the molecule motion to excitation processes in the solid. Since the grid points follow the trajectory, it will be advantageous to solve the classical dynamics first. Depending upon the outcome of the trajectory, whether it is, for instance, reactive or non-reactive, we can decide in which modes to place the grid points. These modes do not necessarily have to coincide

Table 2.9: Fortran program for calculation of the effect of the kinetic energy matrix on a vector in the TDGH-DVR method for quantum molecular dynamics. The number of dimensions is n and the number of DVR-points in each dimension is ngrid(i), i=1,n. The matrix h(i,j,im) contains the elements $\hbar \mathrm{Im} A_{im}/\mathrm{mass}(im)$ $\times \sum_{m=0}^{ngrid(im)-1}(2m+1)\tilde{\phi}(z_i)\tilde{\phi}(z_j)$ for modes "im." The array ind(i,j) (j=1, ... , n) contains the grid point numbering and $nn(k)=\Pi_{i=k}^{n}ngrid(i)$. u(j) contains the wave function at the jth grid point and y(i) is the result after multiplying with the off-diagonal part of the kinetic energy matrix. The diagonal part is included together with the potential in the split-Lanczos algoritm (see fig. 2.10).

Initialization of ind(i,j)
```
        i=0
        do 1 i1=1,ngrid(1)
        do 1 i2=1,ngrid(2)
        ...
        do 1 in=1,ngrid(n)
        i=i+1
        ind(i,1)=i1
        ind(i,2)=i2
        ...
        ind(i,n)=in
1       continue
```
The kinetic energy coupling
```
        i=0
mode 1  do 1 i1=1,ngrid(1)
        iis(1)=i1
mode 2  do 1 i2=1,ngrid(2)
        iis(2)=i2

        ...
mode n  do 1 in=1,ngrid(n)
        iis(n)=in
        i=i+1
        su=0
        do 2 k=1,n
        do 3 l=1,n
3       jjs(l)=iis(l)
        ii=ind(i,k)
        do 2 ki=1,ngrid(k)
        jjs(k)=ki
        j=jjs(n)+nn(n)(jjs(n-1)-1)+···+nn(2)(jjs(1)-1)
        if (i.eq.j) go to 2
        jj=ind(j,k)
        su=su+h(ii,jj,k)*u(j)
2       continue
1       y(i)=su
```

Table 2.10: Representation of the initial vibrational/ rotational states $n, j, m = 0, 0, 0$ and 1, 10, 0 of hydrogen. Various choices of the width parameter $\operatorname{Im} A_x(t_0) = \operatorname{Im} A_y(t_0) = \operatorname{Im} A_z(t_0)$ and grid points in cartesian coordinates $n_x = n_y = n_z$ have been used. The width is in units of amu/τ, where $\tau = 10^{-14}$ sec. The table shows $\sum_i |d_i|^2$ obtained using the 3-dimensional generalization of eq. (2.266) with $\operatorname{Re} A_x = 0$ and $\Psi(x, y, z) = \phi_n(r) Y_{jm}(\theta, \phi)/r$.

n, j, m	0,0,0	0,0,0	0,0,0	0,0,0	1,10,0
$n_x/\operatorname{Im} A_x(t_0)$	0.16	0.20	0.25	0.30	0.25
10	0.946	1.075	0.951	0.856	0.754
12	1.039	0.988	1.017	0.964	0.880
14	1.004	1.003	0.992	1.021	1.012
16	1.012	0.987	1.006	0.996	0.963
18	0.986	1.007	0.999	0.999	0.919
20	0.995	0.999	1.001	1.000	0.936
21	0.998	1.000	1.002	0.998	1.022

with the coordinates in which the classical equations of motion are solved. The method may be used for a direct dynamics approach to molecular processes by letting a cluster of computers (PCs) calculate the potential surface $V(x_i)$ at the grid points x_i and the force at $x(t)$ (see fig. 2.10), so as to be able to propagate the trajectory and obtain the vector $\mathbf{d}(t + \Delta t)$. In the TDGH-DVR method, quantum-classical correlation can be accounted for by including more than one grid point in the degrees of freedom which are strongly coupled. In many cases, it will be advantageous to mix the method with ordinary state expansion methods.

Finally, we mention that the TDGH-DVR method (or quantum dressed classical mechanics method) gives the answer to the question:

Table 2.11: Average dissociation probability for hydrogen over a unit cell of a copper crystal at E_{kin}=100 kJ/mol as a function of DVR basis set in the DOFs (X, Y, Z, x, y, z).

Basis	Number of DVR points	P_{diss}	$\operatorname{Im} A_X$	$\operatorname{Im} A_Y$	$\operatorname{Im} A_Z$	$\operatorname{Im} A_x = \operatorname{Im} A_y = \operatorname{Im} A_z$
7,7,5,13,13,13	538265	0.128	0.20	0.20	2.0	0.16
7,7,5,14,14,14	672280	0.154	0.20	0.20	2.0	0.16
9,9,5,13,13,13	889785	0.201	0.20	0.20	2.0	0.16
9,9,5,14,14,14	1111320	0.195	0.20	0.20	2.0	0.16
9,9,5,15,15,15	1366875	0.241	0.20	0.20	0.10	0.30
9,9,7,15,15,15	1913625	0.244	0.20	0.20	0.10	0.30

Given that you have described the molecular system using classical dynamics— what then is the systematic way of correcting the description so as to approach the quantum limit?

The quantum dressed classical mechanics method is exact in the limit of many grid points. In the other limit (one grid point), it has a theory, which is extremely useful from a dynamical point of view, namely classical mechanics.

2.9 The MCTDH approach

The multi-configuration time-dependent Hartree (MCTDH) approach is also in principle an exact theory which, with enough basis functions, will give the correct answer to a dynamical problem. The method has been formulated primarily by Cederbaum et al. [69]. It is based on a variationally optimized time evolution of expansion coefficients as well as basis functions. In this respect it is similar to the Gauss-Hermite methodology but the scope is different. Whereas the G-H method formulates quantum dynamics such that it obtains a classical mechanical limit, the MCTDH method optimizes the basis functions so as to describe a single particle (mode) or a group of degrees of freedom as well as possible. Thus, the wave function is expanded as

$$\Psi(x_1, \ldots, x_d, t) = \sum_{i_1=1}^{n_1} \cdots \sum_{i_d=1}^{n_d} A_{i_1,\ldots,i_d}(t) \Pi_{k=1}^d \phi_{i_k}^{(k)}(x_k, t) \tag{2.285}$$

where n_i is the number of single-particle functions for mode i. Application of the Dirac-Frenkel variational principle gives a set of coupled equations in the expansion coefficients and the single particle functions

$$i\hbar \dot{A}_{i_1,\ldots,i_d}(t) = \sum_{j_1,\ldots,j_d} A_{j_1,\ldots,j_d} \langle \Phi_{i_1,\ldots,i_d} | H_1 | \Phi_{j_1,\ldots,j_d} \rangle \tag{2.286}$$

$$i\hbar \dot{\phi}_n^{(k)} = h_k \phi_n^{(k)} + (1 - P^{(k)}) \sum_{ml} \rho_{nl}^{-1} H_{ml}^{(k)} \phi_l^{(k)} \tag{2.287}$$

where we have introduced the notation

$$\Phi_{i_1,\ldots,i_d} = \Pi_{k=1}^d \phi_{i_k}^{(k)}(x_k, t) \tag{2.288}$$

and assumed that the hamiltonian can be split in two parts, a separable part and a coupling part

$$H = \sum_{k=1}^d h_k(x_k) + H_1(x_1, \ldots, x_d, t) \tag{2.289}$$

The projection operator is defined by

$$P^{(k)} = \sum_m |\phi_m^{(k)}\rangle\langle\phi_m^{(k)}\rangle \tag{2.290}$$

The matrix elements $H_{ml}^{(k)}$ are actually operators, since they depend on x_k, that is,

$$H_{ml}^{(k)}(x_k) = \langle \psi_m^{(k)} | H_1 | \psi_l^{(k)} \rangle \tag{2.291}$$

where

$$\psi_m^{(k)}(x_1, \dots, x_{k-1}x_{k+1}, \dots, x_d) = \sum_{i_1=1}^{n_1} \cdots \sum_{i_{k-1}}^{n_{k-1}} \sum_{i_{k+1}}^{n_{k+1}} \cdots \sum_{i_d=1}^{n_d} A_{i_1,\dots,i_{k-1}ki_{k+1},\dots,i_d}(t)$$
$$\times \phi_{i_1}^{(1)} \cdots \phi_{i_{k-1}}^{(k-1)} \phi_{i_{k+1}}^{(k+1)} \cdots \phi_{i_d}^{(d)} \qquad (2.292)$$

Finally, the reduced density matrix entering eq. (2.287) is defined through the equation

$$\rho_{nl}(t) = \langle \psi_n^{(k)} | \psi_l^{(k)} \rangle \qquad (2.293)$$

Although the MCTDH method in its formulation given above does not aim at a quantum-classical description of the dynamics, it has some similarity to the theory obtained using the expansion of the wave function in the G-H basis set. The theory can then be combined with more approximate schemes for the bulk of a system involving coupling to, for instance, a "heat bath" [46]. In the MCTDH approach, one has the possibility of constructing optimized basis sets according to the nature of the various degrees of freedom. Often, a small number of optimized basis functions can be used and, hence, many-dimensional systems can more easily be treated [70]. On the contrary, the G-H method works with the same product wave functions independent of the system under investigation.

2.10 Summary

The general way of solving the TDSE for a molecular dynamics problem is to expand the wave function in a basis set. The task is, of course, to keep the basis set as small as possible. Obviously, it is advantageous to put the basis functions or grid points where the "action" is, that is, where the wave function is localized. If, for instance, we consider a tunneling problem, the wave packet is initially localized on one side of the barrier, and it moves in a localized fashion towards the potential barrier where it spreads over a larger range. However, when leaving the barrier it is again localized now in two wave packets, one reactive and one non-reactive. In such a situation we need only many basis functions when the wave packet bifurcates and then a prescription for how to localize the basis sets, that is, how and where to put the basis functions (see also appendix A). The simplest scheme, both conceptually and programming-wise, is obtained by letting classical mechanics determine where to put the basis functions. This idea is the basic one behind expansions in moving Gaussians or Gauss-Hermite basis functions. It may not be the optimal procedure from a convergence point of view. Here schemes such as the MCTDH where, as in electronic structure calculations, dedicated basis functions may be constructed for a given problem—will often have faster convergence. However, as a general-purpose and easy-to-program method we prefer, for instance, the time-dependent TDGH-DVR (quantum dressed classical mechanics) approach. The fact that this method has (from a dynamical point of view) an extremely useful theory—classical dynamics—as the limit, makes it appealing. This and the programming convenience makes up for the fact that more basis functions might be needed in some DOFs as compared to methods where much more effort is put in the construction of dedicated basis sets.

For approximate but still accurate calculations, we will need specialized methods which aim at solving the dynamics for specific situations. These computational methods are of significant value for the more engineering approach to molecular dynamics. In the next chapters some of these methods are described in enough detail to allow for a numerical implementation of the methods.

Chapter 3

Approximate theories

Many calculations of rates or cross sections will have to be carried out using reliable approximate theories, simply because the number of systems and the number of detailed rate constants needed for interpreting experimental data or performing chemical kinetic simulations is too large for an attempt to produce these numbers completely from first principles. Aside from this, the information on potential energy surfaces is also a problem, which makes it necessary, in order to estimate whether the surfaces are reliable, to perform easy-to-do calculations before deciding whether more elaborate calculations can be justified using the surface at hand. However, this situation sets up some requirements for approximate theories, before they can be used for large-scale calculations:

Requirements for approximate dynamical calculations:

- They should "reproduce" as closely as possible exact benchmark calculations on simpler systems or in lower dimensions.
- They should be extendable to larger systems and dimensions.
- They should be able to treat inelastic, as well as reactive, processes.
- They should be extendable to multi-surface problems.
- They should be easier to do than the exact quantum calculations.
- They should be able to take advantage of an increase in computer speed.

A number of benchmark calculations, often performed with model potential surfaces, are now available. The first of such, and still popular for comparison, are the calculations on colinear inelastic collisions performed in 1966 by Secrest and Johnson [73]. Since then, calculations on colinear reactive processes for A+BC [74] and inelastic AB+CD systems appeared [76]. By the late eighties, calculations on atom-diatom reactions in full dimensionality (4 dimensions) have been possible, and benchmark calculations on systems as $D+H_2$ [100] and $F+H_2$ [77] are available. The "exact" calculations are performed by expanding the wave function in basis sets consisting of eigenfunctions to part of the hamiltonian. These "channels" are coupled by the remaining part of the hamiltonian, and the calculations are, therefore, often referred to as coupled channel (CC) calculations.

The approximate methods are developed to treat the many degrees of freedom in larger systems. "Larger systems" means, with the present facilities, more than 4 to 5 atoms, multi-surface problems or systems involving heavy masses. In these cases, the number of coupled channels becomes astronomical and the "experimental" resolution is often without enough detail to justify exact calculations, even if they could be performed. Another aspect which should not be forgotten is that the potential energy surface is only approximately known for such systems and, hence, sets a limit for the accuracy of the information that one can obtain.

On the other hand, the computer speed and performance does in fact increase and becomes cheaper and cheaper. Thus, an approximate method should be of such a nature that it is easy to take advantage of this development.

The methods which will be described in the present chapter all obey the requirements mentioned above. When constructing approximate dynamical theories, one should be especially careful not to introduce constraints which prevent energy or probability from flowing freely according to the dynamical forces of the system. It has been popular to introduce frozen bonds, angles, and so on, in order to reduce the "dynamical" degrees of freedom to a number which can be handled by quantum mechanical methods. Although certain bonds may indeed be "spectator" bonds, that is, maintain a constant value in the course of a chemical reaction, it is dangerous to introduce ad hoc constraints which prevent the system from finding the lowest energetically allowed pathway in phase-space. Thus, constrained dynamics will usually underestimate cross sections and rate constants. A dramatic example is shown in fig. 3.1. The VCC-IOS (vibrational coupled channel—infinite order sudden) method treats

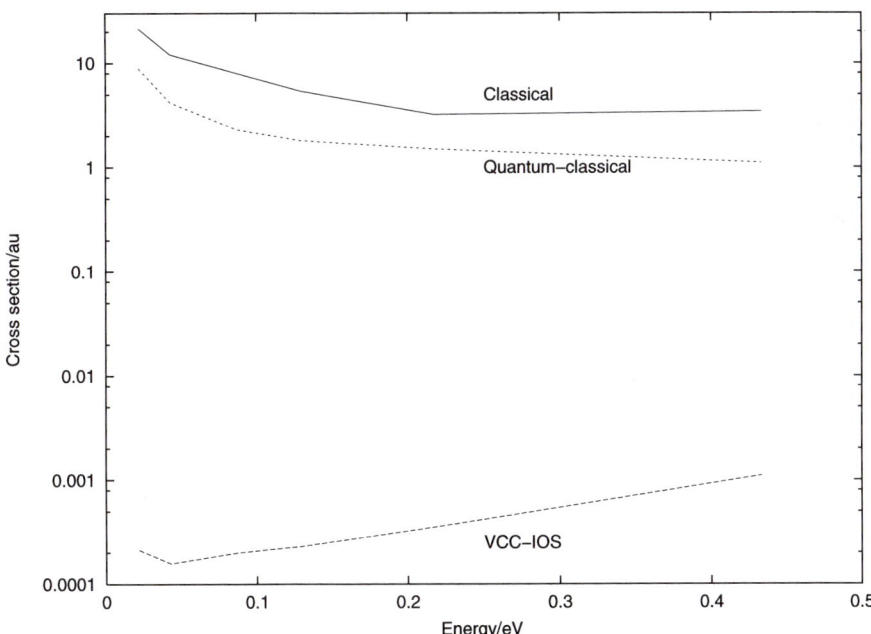

Fig. 3.1:   Cross sections for vibrational relaxation of O_2^+ in collisions with Kr atoms as a function of kinetic energy. Data from [72].

the vibrational degree of freedom exactly, but uses an infinite order sudden approximation for the rotational motion. This approximation corresponds (in classical mechanical terms) to freezing the molecule's orientation during the collision. This approximation is expected to be valid for fast collisions and potentials which are not very anisotropic. Both of these approximations are not fulfilled for the Kr-O_2^+ system. If the molecule is allowed to rotate, it can align itself according to the intermolecular forces of the system and thereby find a position more favorable for vibrational excitation (for further discussion, see ref. [72]). The quantum-classical results shown in fig. 3.1 were obtained using the $V_q R_c T_c$ method described in this chapter. The classical results are obtained using classical dynamics for all degrees of freedom. In these calculations, the conservation of zero-point energy is not automatically fulfilled and may create problems. The example shows that it is not advisable to use theories with dynamical constraints, unless the nature of these is well understood.

3.1   Time-dependent SCF

By the time-dependent self-consistent field (TDSCF) treatment of nuclear dynamics, we shall denote a theory in which one assumes that the wave function can be expressed as a product—a Hartree product wave function. Thus, for just two degrees of freedom we have a "trial" function of the type

$$\Psi(x_1, x_2, t) = \phi_1(x_1, t)\phi_2(x_2, t) \tag{3.1}$$

Obviously, this is an approximation which can only be justified if the two systems are nearly separable (being exact if the systems are separable, that is, if $V(x_1, x_2) = 0$). The approximation is, furthermore, applicable for short time dynamics [37] where the correlation between the x_1 and the x_2 degrees of freedom is expected to be sufficiently weak.

If we insert this trial function in the TDSE and multiply from the left by $\Psi(x_1, x_2, t)^*$ and integrate over the coordinates x_2 and x_1, respectively, we obtain

$$i\hbar \frac{\partial \phi_1(x_1, t)}{\partial t} = (H_1(x_1) + \langle \phi_2|V(x_1, x_2)|\phi_2\rangle)\phi_1(x_1, t)$$
$$- i\hbar \left\langle \phi_2 \left| \frac{\partial \phi_2}{\partial t} \right. \right\rangle \phi_1(x_1, t) \tag{3.2}$$

$$i\hbar \frac{\partial \phi_2(x_2, t)}{\partial t} = (H_2(x_2) + \langle \phi_1|V(x_1, x_2)|\phi_1\rangle)\phi_2(x_2, t)$$
$$- i\hbar \left\langle \phi_1 \left| \frac{\partial \phi_1}{\partial t} \right. \right\rangle \phi_2(x_2, t) \tag{3.3}$$

where the brackets indicate integration over x_1 in the first and over x_2 in the second equation. The hamiltonian for the system is of the form $H = H_1(x_1) + H_2(x_2) + V(x_1, x_2)$. Thus, we have reduced the problem from a two-dimensional one to two one-dimensional problems. Each system feels a potential averaged over the other coordinate, that is, the motion of one coordinate takes place in the mean field of the other. This scheme can easily be extended to any number of dimensions and is the basis for the most common approximation in multi-dimensional systems. It is well known that if the correlation between the two systems is large, the SCF approximation breaks down. This is especially the case if a bifurcation of the wave function between two reactive channels or two electronic states occurs. In such cases, the interaction between the systems is large and to describe it in a mean fashion is not possible. The problem with the amount of correlation between various degrees of freedom has been discussed in connection with the derivation of the classical path method, where we introduced a measure for the amount of correlation (see chapter 2, section 6). More often, the problem has been investigated by comparing the results obtained with the TDSCF and the MCTDSCF (multi-configuration time-dependent self-consistent field method). Such comparisons are abundant in the literature (see, for instance, refs. [37, 38]). A review of the TDSCF method is given in ref. [39] and, since the TDSCF method offers a good first solution to the problem, it has been used to generate a set of time-dependent basis functions [40] and a time-dependent DVR scheme [41]. Although TDSCF has obtained a reputation as the "poor man's" quantum mechanics, the results obtained with it are actually much better than its reputation would indicate.

3.2 The classical path theory

The classical path (CP) theory denotes an approach in which some degrees of freedom are treated classically, others quantally. The coupling between the two sets of degrees of freedom is treated in a simplified and often ad hoc manner. Early examples of this approach treat the nuclear motion classically, that is, the nuclei follow trajectories, whereas the electrons are quantized. Sometimes, these theories are also called "semi-classical" or quantum-classical theories. In the simplest versions of the theory, the trajectories are taken to be independent of the specific state of the quantum system, that is, independent of whether a transition in the quantum system is actually induced or what transition is introduced. Obviously, this will be a good approximation if the energy of the nuclear motion is large compared with the energy spacing between the levels of the quantum system. This type of approach was used, for instance, in the celebrated 1948 paper on stopping power by Niels Bohr [10]. Also in molecular scattering the CP approach has been popular. For a review of the early theories, see the 1969 paper by Rapp and Kassal [78].

At these high energies, the so-called impact parameter picture is convenient (see fig. 3.2). At large values of the impact parameter, we can assume that the trajectory is only affected slightly by the collision, that is, a simple relation

$$R(t)^2 = b^2 + (vt)^2 \tag{3.4}$$

holds. Here, $R(t)$ is the center of mass distance between the two molecules, b the impact parameter, and v the velocity of the relative center of mass motion.

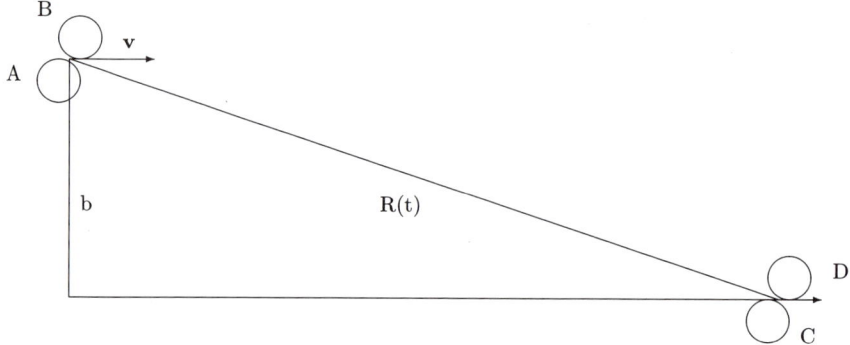

Fig. 3.2: Impact parameter picture.

Assuming that the energy transfer is dominated by the long-range multi-pole interactions, we can obtain a simple first-order expression for the cross sections and rate constants in the impact parameter picture [82]. Thus, we have [157]

$$V_{LR} = \sum_{l_1 l_2 l} A_{l_1 l_2 l}(\Omega, \omega_A, \omega_B) \frac{Q_{l_1} Q_{l_2}}{R^{l_1 + l_2 + 1}} \sqrt{\frac{(2l_1 + 2l_2)!}{(2l_1 + 1)!(2l_2 + 1)!}} \quad (3.5)$$

where Ω, ω_A and ω_B, respectively, specify the orientation of R and the two molecule axes in a space-fixed coordinate system. Q_{l_i} are the multi-pole moments, which depend on the intramolecular bond distances, where l_i is 1 for a dipole and 2 for a quadrupole, etc. The expansion coefficients are defined as

$$A_{l_1 l_2 l}(\Omega, \omega_A, \omega_B) = \sum_{m_1, m_2, m} \begin{pmatrix} l_1 & l_2 & l \\ m_1 & m_2 & m \end{pmatrix} Y_{l_1, m_1}(\omega_A) Y_{l_2, m_2}(\omega_B) Y_{lm}(\Omega) \quad (3.6)$$

that is, in terms of a 3-j symbol [83] and spherical harmonics Y_{lm}. In the first-order treatment of the excitation process by collisions between two diatomic molecules, this potential gives [82] the following expression for the rate-constant

$$k_{\alpha, \alpha'}(T) = \sqrt{8k_B T / \pi \mu} \frac{8\pi^2}{\hbar^2 \omega^2} (2j_1' + 1)(2j_2' + 1) \sum_{l_1 l_2 l} (2p\omega)^{2l}$$

$$\times |\langle n_1 | Q_{l_1} | n_1' \rangle|^2 |\langle n_2 | Q_{l_2} | n_2' \rangle|^2$$

$$\times \begin{pmatrix} j_1 & l_1 & j_1' \\ 0 & 0 & 0 \end{pmatrix} \begin{pmatrix} j_2 & l_2 & j_2' \\ 0 & 0 & 0 \end{pmatrix} f_m(z) / [(2l_1 + 1)!(2l_2 + 1)!]$$

where μ is the reduced mass for the relative motion of the two diatoms while, α and α' are collective quantum numbers $\alpha = (n_1 j_1 n_2 j_2)$ for the vibrational/rotational states of the two molecules. The quantity p is defined as

$$p = \sqrt{\mu / 2k_B T} \quad (3.7)$$

Q_{l_i} are the multi-pole moments which depend on the bond distance r_i, and $f_m(z)$ is an integral

$$f_m(z) = \int_0^\infty dt\, t^m \exp(-t^2 - z/t) \tag{3.8}$$

with $z = 2p\omega b_0$, where b_0 is the smallest value of the impact parameter for which the above simple straight line approach is valid. Typical values for b_0 are about 3 Å or more.

Another approximate path $R(t)$ which has been used for colinear $A + BC$ encounters with an exponential repulsive potential $C\exp(-\alpha R_{AB})$, where $R_{AB} = R - \lambda r$ and $\lambda = m_C/(m_B + m_C)$, is

$$\exp(-\alpha R(t)) = \mathrm{sech}^2(\tfrac{1}{2}\alpha v_0 t) \tag{3.9}$$

This trajectory is obtained by solving the equations of motion

$$\dot{R}(t) = \frac{P_R(t)}{\mu} \tag{3.10}$$

$$\dot{P}_R(t) = -\frac{\partial}{\partial R} C\exp(-\alpha R) \tag{3.11}$$

The vibrational transition probability from state n to m with the path (3.9) and a first-order perturbational treatment of the excitation is obtained as

$$P_{nm} = \frac{U_{nm}^2}{\hbar^4} \frac{(E_n - E_m)^2 4\pi^2 \mu^2}{\alpha^4} \sinh^{-2}(\pi(E_n - E_m)/v_0\alpha\hbar) \tag{3.12}$$

where μ, v_0 are the reduced mass and velocity of the translational motion, E_n a vibrational energy, and U_{nm} a matrix element

$$U_{nm} = \langle \phi_n | \exp(\alpha\lambda r) | \phi_m \rangle \tag{3.13}$$

where $\phi_n(r)$ is a vibrational wave function and $\lambda = m_C/(m_B + m_C)$ for a colinear $A + BC$ collision. In these two examples, the path is independent of the transition it induces in the molecules.

We now wish to examine a simple collision problem in more detail. Consider an atom hitting a diatomic molecule where we have the eigenstates $\phi_n(r)$ for the isolated molecule, that is,

$$H_0\phi_n(r) = E_n\phi_n(r) \tag{3.14}$$

Initially, the molecule is assumed to be in state I and at later times we could expand the wave function as

$$\Psi(r,t) = \sum_n a_n(t)\phi_n(r) \tag{3.15}$$

The intermolecular potential is $V(r, R)$, and if we assume that the molecule is excited by perturbation through a trajectory $R(t)$ we could determine the amplitudes $a_n(t)$ by solving the TDSE

$$i\hbar\frac{\partial\Psi(r,t)}{\partial t} = (H_0 + V(r, R(t)))\Psi(r,t) \tag{3.16}$$

or by inserting the above expansion (3.15) we obtain

$$i\hbar \dot{a}_n(t) = E_n a_n(t) + \sum_m V_{nm}(t) a_m(t) \tag{3.17}$$

where $V_{nm}(t) = \langle \phi_n | V(r, R(t)) | \phi_m \rangle$ and the bracket indicate integration over r. A rather obvious way of coupling the equation for the trajectory $R(t)$ to the motion of the molecule would be to introduce the interaction through the "Ehrenfest" average potential, that is,

$$\dot{R} = P_R/\mu \tag{3.18}$$

$$\dot{P}_R(t) = - \left\langle \Psi \left| \frac{\partial V(r, R)}{\partial R} \right|_{R=R(t)} \right| \Psi \right\rangle \tag{3.19}$$

Strictly speaking, the above equations are not related to Ehrenfest's theorems,

$$\frac{d}{dt} \langle R \rangle = \left\langle \frac{\partial H}{\partial P_R} \right\rangle \tag{3.20}$$

$$\frac{d}{dt} \langle P_R \rangle = - \left\langle \frac{\partial H}{\partial R} \right\rangle \tag{3.21}$$

where the brackets denote averages over the wave function (depending on R). The mean values do not, in general, follow classical mechanics. Furthermore, (in eq. (3.19)) the average is over the r coordinate only and not R. Nevertheless, the mean field–type potential used in eq. (3.19) is often referred to as the result of an Ehrenfest-type approach. We will, therefore, often use this designation in the following.

The above mean field approach has, however, one problem. It does not properly fulfill the principle of detailed balance (or microscopic reversibility). As a side effect to this, energetically closed states might also be populated in a collision. Hence, the first comparisons of results obtained with this method and exact calculations [73, 78, 79] were disappointing.

As a matter of fact, the comparison with exact calculations leads to the result that all the simple, mainly first-order, theories developed in the fifties had to be abandoned.

According to the reversibility principle, we have the following requirement

$$P_{nm}(E) = P_{mn}(E) \tag{3.22}$$

that is, the probability for the transition n to m should be identical to the reverse probability at a given total energy E, where for the above simple system

$$E = E_n + \frac{P_R(-\infty)^2}{2\mu} = E_m + \frac{P_R(+\infty)^2}{2\mu} \tag{3.23}$$

It turns out that the above equations (3.17, 3.18, and 3.19) do not meet this requirement. At low collision energies, we would have the following relation

$$P_{nm}(E_{kin}) \sim P_{mn}(E_{kin}) \tag{3.24}$$

where $E_{kin} = P_R(-\infty)^2/2\mu$. In order to improve the situation, it was, therefore, suggested to use as an initial relative velocity [79]

$$\frac{1}{\mu}P_R(-\infty) = \frac{1}{2}(v_n + v_m) \tag{3.25}$$

where $\frac{1}{2}\mu v_n^2 = E - E_n$. Thus, each transition $n \to m$ has its own best initial velocity given by eq. (3.25). This so-called symmetrized Ehrenfest approach turned out to be surprisingly accurate for inelastic rotational and vibrational transitions (see tables 3.1–3.3). The agreement between exact numbers and the results obtained with the simplest version of the quantum-classical theory is

Table 3.1: Comparison of quantum and classical path excitation probabilities for a colinear collision between an atom and a diatomic molecule interacting through an exponential potential. The energy is in units of $\frac{1}{2}\hbar\omega$ and the upper table is for a Morse-oscillator, the lower for a harmonic oscillator case. The CP numbers are obtained using the so-called symmetrized Ehrenfest approach. The VCP numbers are obtained using the variational principle for determining the best initial momentum. Data from ref. [64].

Energy	Transition	CP	Quantum	VCP
10.0	0-1	3.56(−5)	3.88(−5)	3.58(−5)
	1-0	3.86(−5)	3.88(−5)	3.88(−5)
	0-2	0.83(−10)	1.06(−10)	0.84(−10)
	2-0	1.17(−10)	1.06(−10)	1.15(−10)
	1-2	1.01(−5)	1.09(−5)	1.02(−5)
	1-4	0.82(−22)	2.20(−22)	0.98(−22)
	4-1	2.19(−22)	2.20(−22)	1.82(−22)
20.0	0-1	3.59(−3)	3.88(−3)	3.61(−3)
	1-0	3.82(−3)	3.88(−3)	3.84(−3)
	0-2	3.73(−6)	4.61(−6)	3.75(−6)
	2-0	4.85(−6)	4.61(−6)	4.90(−6)
	3-4	2.58(−3)	2.90(−3)	2.59(−3)
4.9455	0-1	1.08(−4)	1.12(−4)	1.11(−4)
	1-0	1.07(−4)	1.12(−4)	1.11(−4)
6.9455	0-1	2.80(−3)	2.93(−3)	2.84(−3)
	1-2	2.12(−4)	2.30(−4)	2.18(−4)
8.9455	0-1	1.45(−2)	1.53(−2)	1.49(−2)
	0-2	2.04(−5)	2.30(−5)	2.18(−5)
	1-2	5.51(−3)	5.97(−3)	5.59(−3)
16.7882	0-1	0.178	0.197	0.207
	0-2	1.31(−2)	1.62(−2)	1.52(−2)
	1-2	0.214	0.237	0.239

(-3) means $\times 10^{-3}$.

Table 3.2: Comparison of classical path and quantum probabilities for vibrational transitions in diatom-diatom collisions. Data from ref. [75]. Energy in units of the harmonic oscillator spacing in N_2 or H_2. The exact numbers are from ref. [76].

Energy	Transition	System N_2+CO CP	(HO) Quantum	System N_2+CO CP	(MO) Quantum	N_2+O_2 CP	(HO) Quantum
1.25	00-01	2.75(−11)	2.88(−11)	0.36(−11)	0.44(−11)	5.23(−7)	5.14(−7)
	01-10	1.54(−3)	1.57(−3)	1.72(−3)	1.76(−3)	4.54(−7)	5.21(−7)
2.25	00-01	1.36(−6)	1.39(−6)	1.24(−7)	1.35(−7)	3.30(−4)	3.33(−4)
	00-10	8.88(−8)	8.82(−8)	7.86(−9)	8.65(−9)	9.75(−8)	9.64(−8)
	01-10	1.74(−2)	1.83(−2)	1.73(−2)	1.83(−2)	4.75(−4)	5.79(−4)
	01-11	2.45(−12)	2.7(−12)	4.5(−13)	2.6(−13)	2.16(−5)	2.11(−5)
3.25	00-01	6.41(−5)	6.53(−5)	5.48(−6)	5.95(−6)	4.57(−3)	4.65(−3)
	00-10	7.15(−6)	7.15(−6)	5.90(−7)	6.49(−7)	8.88(−6)	8.91(−6)
	00-02	6.57(−11)	7.09(−11)	7.8(−13)	9.0(−13)	2.08(−6)	2.27(−6)
	01-10	3.46(−2)	3.79(−2)	3.33(−2)	3.64(−2)	2.97(−3)	3.83(−3)
	01-02	4.00(−6)	3.46(−6)	5.14(−7)	5.68(−7)	1.89(−3)	1.93(−3)
	01-11	2.22(−7)	2.27(−7)	1.90(−8)	2.20(−8)	2.70(−6)	3.11(−6)
5.25	00-10	2.21(−3)	2.36(−3)	1.88(−4)	2.03(−4)	5.22(−2)	5.81(−2)
	00-02	2.35(−8)	1.68(−8)	5.68(−9)	7.49(−9)	8.67(−4)	9.33(−4)
	00-11	2.02(−7)	2.30(−7)	1.00(−9)	1.48(−9)		
	01-11	1.21(−4)	1.51(−4)	1.15(−5)	1.37(−5)	8.13(−4)	9.99(−4)

HO = Harmonic oscillator, MO = Morse oscillator.

		N$_2$+O$_2$		H$_2$+H$_2$		H$_2$+H$_2$	
		CP	(MO) Quantum	CP	(HO) Quantum	CP	(MO) Quantum
1.25	00-01	$1.43(-7)$	$1.46(-7)$	$1.33(-4)$	$1.41(-4)$	$6.23(-5)$	$7.79(-5)$
	01-10	$6.46(-7)$	$7.44(-7)$	$2.19(-2)$	$2.34(-2)$	$2.50(-2)$	$2.69(-2)$
2.25	00-01	$7.92(-5)$	$8.08(-5)$	$1.39(-2)$	$1.52(-2)$	$4.06(-3)$	$5.00(-3)$
	00-02	$1.53(-10)$	$1.62(-10)$	$9.77(-7)$	$1.11(-6)$	$4.37(-7)$	$8.50(-7)$
	01-10	$5.26(-4)$	$6.31(-4)$	0.100	0.135	0.091	0.118
	01-02	$7.78(-6)$	$8.01(-6)$	$2.44(-4)$	$2.65(-4)$	$4.51(-4)$	$5.76(-4)$
	01-11	$2.24(-10)$	$2.45(-10)$	$1.30(-4)$	$1.42(-4)$	$7.05(-5)$	$1.01(-4)$
3.25	00-01	$1.07(-3)$	$1.09(-3)$	$5.98(-2)$	$6.87(-2)$	$1.70(-2)$	$2.10(-2)$
	00-02	$1.48(-7)$	$1.56(-7)$	$4.51(-4)$	$5.57(-4)$	$4.54(-5)$	$1.05(-4)$
	01-10	$3.15(-3)$	$3.92(-3)$	0.148	0.245	0.135	0.200
	01-02	$5.54(-7)$	$5.70(-4)$	$2.31(-2)$	$2.39(-2)$	$1.22(-2)$	$1.51(-2)$
	01-11	$6.69(-7)$	$7.64(-7)$	$1.32(-2)$	$1.44(-2)$	$3.66(-3)$	$5.16(-3)$
5.25	00-01	$1.34(-2)$	$1.37(-2)$	0.185	0.196	$4.91(-2)$	$7.20(-2)$
	00-02	$5.38(-7)$	$5.74(-5)$	$1.24(-2)$	$2.09(-2)$	$1.26(-3)$	$2.74(-3)$
	01-10	$1.38(-2)$	$1.84(-2)$	0.124	0.247	0.193	0.283
	01-02	$1.56(-2)$	$1.62(-2)$	0.149	0.128	$1.42(-2)$	$2.82(-2)$
	01-11	$1.81(-4)$	$2.14(-4)$	$9.34(-2)$	$9.41(-2)$	$2.02(-2)$	$4.44(-2)$

In the CP results up to 36 states have been included when expanding the wave function.

Table 3.3: Comparison of classical path and quantum partial cross sections for rotational excitation of hydrogen from $j = 0$ to 2 in collisions with He-atoms at 0.10 eV. Units a_0^2. J is the total angular momentum. The classical path results (CP) were calculated using the coupled state approximation (see later) and should therefore be compared with the quantum CS (coupled states) results. The table shows good agreement with the exact CC (coupled channel) results from ref. [67].

J	CP	CC	CS
0	0.0064	0.0059	0.0061
5	0.0570	0.0532	0.0560
10	0.0549	0.0571	0.0538
15	0.0188	0.0248	0.0180
20	0.0014	0.0026	0.0013
Total	0.706	0.730	0.690

typically better than 10–30 %. However, there are a few exceptions for some low-mass diatom-diatom transitions (see table 3.2). For such systems, we recommend that both forward and reverse transition probabilities are calculated. Usually the classical path probability in the "exothermic" direction is superior (see table 3.1). We notice that the comparison is carried out for several mass combinations and for probabilities varying over twenty orders of magnitude. The accuracy which can be obtained with the symmetrized Ehrenfest approach is for many practical "engineering purposes" sufficient.

In practical calculations, we would calculate the transition probabilities at a given value of the kinetic energy $U = P_R^2(-\infty)/2\mu$ and then assign a total energy to a given transition $n \to m$. Using the formula (3.25), we obtain

$$E = U + \frac{1}{2}\Delta E + \frac{\Delta E^2}{16U} \tag{3.26}$$

where $\Delta E = E_m - E_n$. The allowed values of U are $U \geq \frac{1}{4}\Delta E$. For $U = \frac{1}{4}\Delta E$ we have $E = 0$ for an exothermic transition $\Delta E < 0$ and $E = \Delta E$ in the endothermic direction $\Delta E > 0$.

We have seen that the classical path theory "works," but we have not explained why. This will be done in the following section.

We notice that the above equations (3.17, 3.18, 3.19) are identical to those derived in section 2.5 eqs. (2.151)–(2.153), but with the important difference that in section 2.5 the equations were obtained without any specific assumptions on treating part of the system classically. Only an approximate wave function (2.150)—a product-type wave function with a GWP as trial function for the R-system was assumed.

As mentioned previously, the GWP has two parameters: $P_0 = P_R(t_0)$, the momentum at which the GWP is centered in momentum space, and the so-called width parameter $\alpha_0 = \text{Im}A(t_0)$. Strictly speaking, the value of $\text{Re}A(t_0)$

can also be thought of as a parameter—but it does not enter the equations below and we will, therefore, assume that the initial value of $\mathrm{Re}\,A(t_0)$ is zero. The product-type wave function (2.150) is correct asymptotically when there is no interaction between the two systems, that is, for $V(r,R) \sim 0$. As soon as the two systems start to interact, it becomes an approximation. By choosing an initial translational wave function with a certain width, we are able to cover a range of energies, that is, to obtain the state-resolved probabilities for a number of energies, using eq. (2.177), for instance. Usually, the results for momenta near P_0 are most accurate, since the initial wave packet has its maximum here. In principle, however, P_0 and α_0 can be arbitrary—the probabilities $P_{I \to F}(E)$ will simply not depend on where the GWP is centered initially in momentum space and what the width parameter is. If, on the other hand, we use the approximate product-type wave function as a trial function at all times (a requirement for obtaining the CP equations), and perform a projection on plane incoming and outgoing waves similar to what was done in section 2.5, we obtain

$$P_{I \to F}(P_0, \alpha_0) = \frac{k_F}{k_I} \sqrt{\frac{g_F}{g_I}} \exp(-g_F(P_R(t) - \hbar k_F)^2 + g_I(-P_0 + \hbar k_I)^2)|a_F|^2$$

$$(3.27)$$

where g_I and g_F are defined in section 2.5 and $E = E_I + (\hbar k_I)^2/2\mu = E_F + (\hbar k_F)^2/2\mu$ defines the wave numbers k_I and k_F.

We notice that the probabilities now do depend on the two parameters α_0 and P_0. However, we can now introduce a semi-classical ansatz, which states that *the best value of the parameter P_0 is the one which makes the probability independent of α_0* (the other parameter). Thus we define the best momentum variationally by

$$\left. \frac{\partial P_{I \to F}(P_0, \alpha_0)}{\partial \alpha_0} \right|_{P_0 = P_0^*} = 0 \qquad (3.28)$$

(see fig. 3.3). For practical reasons, we define α_0 as the value of $\mathrm{Im}\,A(t)$ at the turning point for the motion in $R(t)$, that is, when $P_R(t) = 0$. It turns out that this momentum derived variationally is very close to the arithmetic mean value mentioned above. That this is the case can be expected by the notion that for $g_I \sim g_F$ we get the probability independent of g_I for $P_0 = \frac{1}{2}(\hbar k_I + \hbar k_F)$, since the exponential function in eq. (3.27) vanishes. This definition of the best momentum is possible only because the derivation of the equations has been made from first principle, which defines P_0 as a free parameter. But it also defines it as a parameter on which the probability should not depend. Table 3.1 shows that this variational determination of the momentum does, in fact, give rather accurate numbers for systems for which the product-type wave function is believed to be adequate, that is, for situations where the coupling between the r and R motion is weak. This derivation gives a practical recipe for calculating state-to-state rates or cross sections for energy transfer between diatomic or polyatomic molecules. The condition for which the scheme is valid is then that the coupling between the quantum and the classical degrees of freedom is weak. Thus, *strongly*

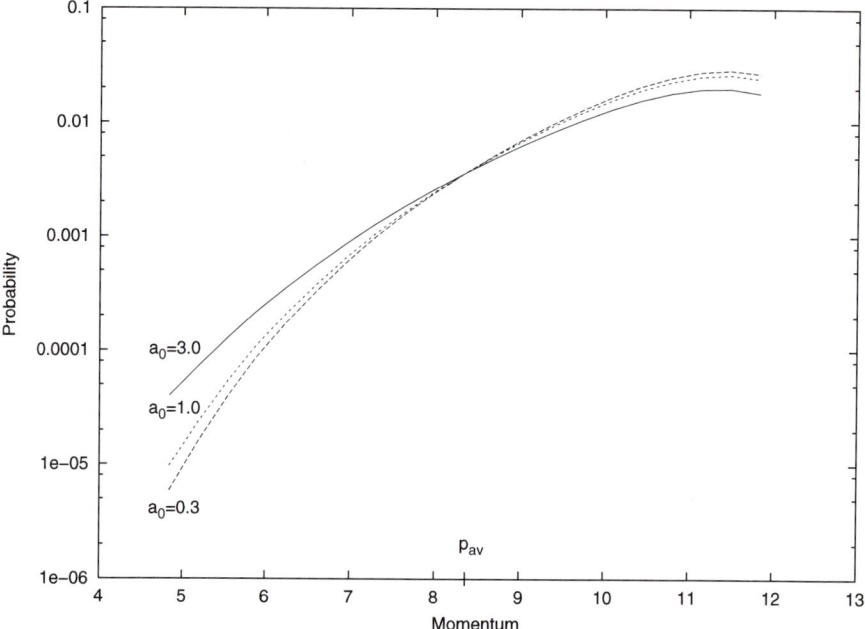

Fig. 3.3: Probability for excitation of a Morse-oscillator from $v = 0$ to $v = 1$ as a function of the initial momentum in a collision with an atom. For a certain best momentum, the probability is independent of the width parameter $a_0 = \text{Im}A(t = 0)$ defined at the turning point for the motion of $R(t)$. This best momentum parameter is close to the arithmetic mean value p_{av} indicated on the figure.

coupled degrees of freedom should be treated within the same dynamics, classically or quantally. We see that the mere division in two sets of degrees of freedom, a classical set and a quantum set, is itself an approximation, which we have called the separability approximation.

In order to improve the description we can introduce more than one trial function, that is, we can expand the wave function in, for instance, a Gauss-Hermite basis. With just one additional G-H function, is it possible to monitor the separability of the two systems. Since the ground state of the G-H basis generates classical mechanical equations of motion in the R coordinate, we have named the amount of excitation to the higher order G-H basis a measure of the quantum-classical correlation, that is, it is defined as

$$\sum_{k=1}\sum_{n}|a_{nk}|^2 \qquad (3.29)$$

where a_{nk} are the expansion coefficients in eq. (2.133). This correlation will remain small if the system is nearly separable. It is advantageous to look for coordinates in which the system is nearly separable. This is the motivation for introducing hyper-spherical variables (see section 3.2.6) for the description of chemical reactions.

Table 3.4: Quantum-classical theory according to the coupling between the degrees of freedom.

Energy	Translation	Rotation	Vibration	Method
Low energies	C	Q	Q	$T_c R_q V_q$
High energies	C	C	Q	$T_c R_c V_q$

C = Classical and Q = Quantum treatment

In the next sections we describe methods for treating energy transfer problems. Thus, we deal with the following degrees of freedom: translation, rotation, and vibration. At low energies we expect the coupling between the translation and the rotation and vibrational degrees of freedom to be weak, especially for systems with small moments of inertia, leading to large rotational energy spacings. At higher energies and/or heavier systems, the coupling between rotation and translation becomes strong. This, then, suggests the division given in table 3.4. This division is, furthermore, convenient from a numerical point of view, since at higher energies we have to introduce too many rotational/vibrational basis functions for obtaining convergence of the scattering calculations. Thus, this avenue can only be used for some light mass systems, the reason being the number of states needed for convergence increases as

$$N \sim \frac{IE^2}{\omega \hbar^3} \qquad (3.30)$$

for an atom-diatom system. E is the energy available for the molecule, I is the moment of inertia, and ω is the vibrational frequency. For two diatomic molecules, we have the following estimate of the number of states

$$N \sim \frac{I_1 I_2 E^4}{6\omega_1 \omega_2 \hbar^6} \qquad (3.31)$$

and the situation is even worse for polyatomic molecules (see also table 3.5).

Table 3.5: Number of open vibration/rotation states $(n_1 j_1 m_1 n_2 j_2 m_2)$ for diatom-diatom systems as a function of energy.

Energy kJ/mol	$H_2 + H_2$	$N_2 + N_2$
1	1	30
10	70	3×10^5
100	7×10^5	3×10^9

We notice that only the internal states of the molecules have been included. In the so-called coupled channel (CC) calculations, the number of channels is much larger, due to additional labeling of these channels by orbital angular momentum quantum numbers.

3.2.1 Relation to other theories

We have seen in the previous section that the classical path theory can be obtained by assuming a product-type wave function and a simple GWP for the one describing the translational motion.

Thus, two approximations were introduced: a separability approximation and a shape approximation for the translational wave function. By introducing a variational determination of the free parameters, we have seen that it is possible to improve the theory. The result of this variational approach is that we can justify the use of a mean relative velocity. This mean velocity concept has also been suggested by other related theories, such as the Eikonal approach [80], the Wigner [81] theory, and some WKB-type theories derived by Delos and Thorson [2] in the seventies. It was also noted that results obtained with the first-order classical path expression (3.12) were improved significantly by introducing a mean velocity instead of v_0 [78].

Below we shall consider some of these approaches in more detail.

The Eikonal theory

We consider the time-independent hamiltonian for a two dimensional system with coordinates r, R, where r is an internal vibrational coordinate and R a translational coordinate. Thus, we have the following Schrödinger equation

$$-\frac{\hbar^2}{2\mu}\frac{\partial^2}{\partial R^2}\Psi(R,r) - \frac{\hbar^2}{2m}\frac{\partial^2}{\partial r^2}\Psi(R,r) + [v(r) + V(R,r)]\Psi(R,r) = E\Psi(R,r)$$

(3.32)

where μ is the mass connected to the relative motion, m is the mass connected to the vibrational motion, $v(r)$ is the oscillator, and $V(r,R)$ is the intermolecular potential.

We can now expand the wave function in eigenfunctions $\phi_n(r)$ of the hamiltonian $H_0 = -\frac{\hbar^2}{2m}\frac{d^2}{dr^2} + v(r)$ with eigenvalues E_n, that is,

$$\Psi(R,r) = \sum_n c_n(R)\phi_n(r)$$

(3.33)

With this expansion, we then obtain

$$-\frac{\hbar^2}{2\mu}\frac{d^2}{dR^2}c_n(R) + E_n c_n(R) + \sum_m V_{nm}(R)c_m(R) = E c_n(R)$$

(3.34)

where $V_{nm}(R) = \langle \phi_n | V(R,r) | \phi_m \rangle$. We now introduce $c_m(R) = A_m(R)\exp(iS_n(R)/\hbar)$ and the momentum

$$\frac{1}{2\mu}P_n^2 = \frac{1}{2\mu}(\nabla S_n)^2 - \frac{i\hbar}{2\mu}\nabla^2 S_n$$

(3.35)

where $\nabla = d/dR$.

In the Eikonal approximation, we neglect the term $\nabla^2 A_n$, that is, we assume that the variation in $A(R)$ with R is weak. This approximation is fulfilled if $\nabla^2 A/A \ll k^2$, where $2\mu E/\hbar^2 = k^2$. With this approximation, we then obtain

$$\frac{i\hbar}{\mu}\nabla S_n \nabla A_n - \frac{P_R^2}{2\mu}A_n = E_n A_n + \sum_k V_{nk}(R)A_k(R)\exp(i(S_k(R) - S_n(R))/\hbar)$$

$$(3.36)$$

If we now make a semi-classical approximation and neglect the $\hbar$-term in eq. (3.35), we have $P_n = \nabla S_n$ and that

$$\frac{1}{2\mu}P_n^2 = E - E_n - V_{nn}(R)$$

$$(3.37)$$

we obtain

$$i\hbar v_n(R)\frac{dA_n}{dR} = \sum_k V_{nk}(R)A_k(R)\exp(i(S_k(R) - S_n(R))/\hbar)$$

$$(3.38)$$

The time-dependent equations can now be derived by assuming $v_n(R) = dR/dt$ and rewriting $S_k - S_n = \int dR(P_k - P_n)$ as $\int dt P_R(P_k - P_n)/\mu$, where $P_R = \mu dR/dt$. Furthermore, using

$$P_R = \tfrac{1}{2}(P_k + P_n)$$

$$(3.39)$$

we finally have

$$i\hbar\dot{A}_n = \sum_k V_{nk}A_k\exp\left(\frac{i}{\hbar}\left(\int dt(E_n + V_{nn} - E_k - V_{kk})\right)\right)$$

$$(3.40)$$

These equations are identical to eqs. (3.17) with $A_n(t) = a_n(t)\,\exp(\int dt(E_k + V_{kk}))$. In order to obtain these equations, we had to introduce the Eikonal approximation, a semi-classical approximation, and we had to assume $P_R = \frac{1}{2}(P_n + P_k) \sim P_k$, that is, a condition which is fulfilled in the high energy limit. This way of deriving the coupled equations for the expansion coefficients led to the postulate that the classical path equations are valid only in the high energy limit, where the above derivation holds. In the previous section, we have, however, seen that the equations also provide very good results at energies where the above conditions are not fulfilled. The reason for this apparent contradiction is that in the present way of deriving the equations we have missed the important point, namely that the initial momentum $P_0 = P_R(t_0)$ is actually a parameter, that is, undetermined by the equations. If P_0 is a parameter and if the results depend on it, one needs a prescription for choosing it. The failure of early classical path theories was that this prescription was taken from classical mechanics, that is, P_0 was set equal to $\sqrt{(E - E_I)/2\mu}$, where I is the initial vibrational state. This choice appeared to be natural because of the apparent classical mechanical treatment of the motion in R. However, in chapter 2 we derived the equations of motion without making any reference to classical mechanics, and, hence, we are justified to find another prescription. By introducing a variational principle for determining the parameter P_0, we obtained the possibility of fulfilling the quantum boundary conditions, namely that our final result for the transition probability or S-matrix element should indeed be independent of the parameters P_0 and $\text{Im}A(t_0)$. The introduction of an average energy appears not only in the Eikonal treatment of inelastic collisions but also, for instance, in the equation derived from the SE for the Wigner function for the transition n to m (for a definition of the Wigner function, see section 5.6).

The expression for the transition Wigner function contains the classical hamiltonian and quantum terms of order $\hbar^2$ and higher, that is,

$$[H_{cl}(p,q) + \text{terms } \hbar^2, \hbar^4 \ldots]\Gamma_{nm}(p,q) = \tfrac{1}{2}(E_n + E_m)\Gamma_{nm}(p,q) \qquad (3.41)$$

Thus, for the transition n to m, it appears as if the "classical" energy is one-half the energy of the initial and final state (see, for example, Dahl [68]).

3.2.2   Energy transfer

Energy transfer in molecular collisions plays an important role in the understanding of non-equilibrium processes, excitation pathways in connection with unimolecular reactions, the speed with which equilibration is achieved, the heating of materials, the dynamics of energy flow in the upper atmosphere, gas lasers, isotopic separation by vibration-vibration transitions, and so on. Energy transfer between molecules depends on intramolecular forces, energy mismatch, and other molecular properties such as masses, moments of inertia, and molecule charges. In general, the energy transfer is large for transitions with small energy mismatch, that is, the energy gap between final and initial state, but this is only so if the transition is supported by the potential. Multi-quantum transitions can usually be induced with significant probability by long-range forces only if these are allowed to first order. This requirement is not necessary for short-range induced transitions. However, this does not mean that first-order theories are especially useful for determining rates or transition probabilities. Likewise, the introduction of dynamical constraints of any sort should be avoided.

The quantum-classical theory is, in a way, ideal for treating inelastic energy transfer processes. We can let the important or interesting part of the problem be quantized and treat the remaining parts classically. Thus the quantum mechanical part can be kept small enough to allow for an infinite order solution of the equations. It should, however, be remembered that the separation in a quantum and a classical part is itself an approximation—a separability approximation. This division should, therefore, not be introduced between degrees of freedom which are strongly coupled.

3.2.3   The $V_q R_q T_c$ method

For systems with light masses at moderate energies, we can easily achieve convergence by expanding the wave function in a product vibrational/rotational basis set. For a collision between an atom and a diatomic molecule, we have

$$\Psi(r,\theta,\phi,t) = \sum_{njm} a_{njm}(t)\phi_n(r)Y_{jm}(\theta,\phi) \qquad (3.42)$$

where n denotes the vibrational quantum number and j, m the rotational and projection quantum numbers, respectively. Thus, the vibration/rotational functions obey the Schrödinger equation

$$\hat{H}_q^0 \frac{\phi_n(r)}{r} Y_{jm}(\theta, \phi) = E_{nj} \frac{\phi_n(r)}{r} Y_{jm}(\theta, \phi) \tag{3.43}$$

where

$$\hat{H}_q^0 = -\frac{\hbar^2}{2m_r} \frac{1}{r^2} \frac{\partial}{\partial r} \left(r^2 \frac{\partial}{\partial r} \right) - \frac{\hbar^2}{2I} \hat{L}^2(\theta, \phi) \tag{3.44}$$

$$\hat{L}^2(\theta, \phi) = \frac{1}{\sin\theta} \frac{\partial}{\partial\theta} \left(\sin\theta \frac{\partial}{\partial\theta} \right) + \frac{1}{\sin^2\theta} \frac{\partial^2}{\partial\phi^2} \tag{3.45}$$

r is the bond distance and $\phi_n(r)$ a vibrational wave function, for instance, a Morse function. I is the moment of inertia and m_r the reduced mass of the diatomic molecule. Coupling between rotation and vibration occurs through the r dependence of I. This coupling has, for simplicity, been omitted here. The orientation of the diatomic molecule in a space-fixed coordinate system is given by the angles θ and ϕ. The wave functions for the rotational motion of the molecule are eigenfunctions to $\hat{L}^2(\theta, \phi)$, the spherical harmonics $Y_{jm}(\theta, \phi)$.

In our quantum-classical framework, the hamiltonian would be

$$H = H_{cl} + H_q^0 + V(R, r, \gamma) - V_{eff}(R, t) \tag{3.46}$$

where

$$H_{cl} = \frac{P_R(t)^2}{2\mu} + \frac{\hbar^2 l(l+1)}{2\mu R^2} + V_{eff}(R, t) \tag{3.47}$$

The effective potential is obtained as the Ehrenfest average, that is,

$$V_{eff}(R, t) = \langle \Psi | V(R, r, \gamma) | \Psi \rangle \tag{3.48}$$

where the bracket indicates integration over r, θ, and ϕ. The intermolecular potential is most conveniently expressed as a function of R, the distance between the atom and the center of mass of the molecule and γ (the angle between r and R). In order to evaluate the matrix elements over the vibrational coordinate r and the angle γ, it is, furthermore, convenient to use a rotational basis set which refers to a body-fixed instead of the space-fixed coordinate system used above. Thus, we introduce a rotation of the basis set

$$Y_{jm}(\gamma, \phi') = \sum_{m'} D_{mm'}^j(\Phi, \Theta, 0) Y_{jm'}(\theta, \phi) \tag{3.49}$$

where γ, ϕ' specify the orientation of the diatomic molecule in a coordinate system with z axis along R (see fig. 3.4), θ, ϕ give the orientation in a space-fixed system and Θ, Φ the orientation of R in a space-fixed system. The matrix elements $D_{mm'}^j$ are over the rotational operator [83] for the rotation with the

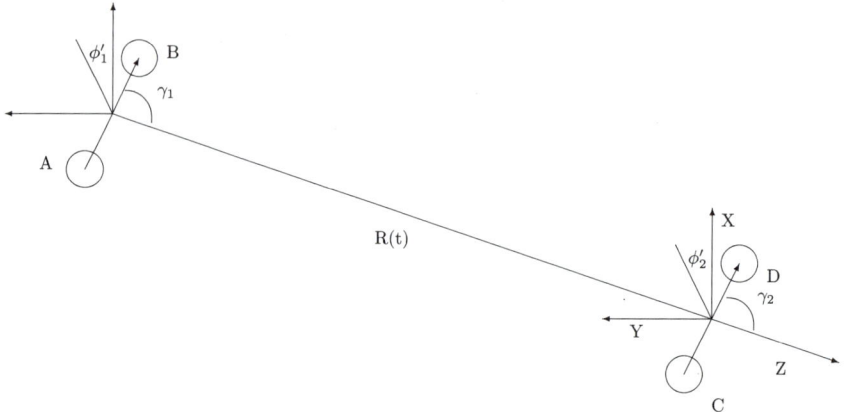

Fig. 3.4: Rotating coordinate system for AB + CD collisions. The orientation of the diatomic molecules is specified by the angles γ_i, ϕ'_i in a coordinate system with z-axis along $R(t)$.

angles Θ and Φ. The matrix elements are now easy to evaluate if we introduce the expansion

$$V(R, r, \gamma) = \sum_{ik} V_{ik}(R)(r - r_{eq})^i P_k(\cos\gamma) \qquad (3.50)$$

where r_{eq} is the equilibrium bond length of the diatomic molecule. The matrix element over the rotational wave functions can be evaluated to be [83]

$$\langle Y_{jm}|P_k|Y_{j'm'}\rangle = \sqrt{\frac{2j'+1}{2j+1}} \langle kj'00|j0\rangle\langle j'km'0|j0\rangle \qquad (3.51)$$

where $\langle ab\alpha\beta|c\gamma\rangle$ are the so-called Wigner-coupling elements. They are related to the 3-j symbols by [83]

$$\langle ab\alpha\beta|c - \gamma\rangle = (-1)^{a-b-\gamma}\sqrt{2c+1} \begin{pmatrix} a & b & c \\ \alpha & \beta & \gamma \end{pmatrix} \qquad (3.52)$$

The 3-j symbols are zero unless $\alpha + \beta + \gamma = 0$. If $\alpha = \beta = \gamma = 0$ then we need $a + b + c$ to be even [83] in order to obtain a non-zero value of the 3-j symbol. Thus, for homonuclear diatomics, where k is even, we have a selection rule on the rotational transitions which may be induced by the collision. They are either even-even or odd-odd transitions. For the expansion coefficients, we obtain the following set of coupled equations

$$i\hbar\frac{d}{dt}a_{njm}(t) = E_{nj}a_{njm}(t) + \sqrt{\frac{2j'+1}{2j+1}} \sum_{n'j'm'} a_{n'j'm'}(t)\sum_{ik} V_{ik}$$

$$\times\, (R(t)\langle\phi_n|(r - r_{eq})^i|\phi_{n'}\rangle\langle kj'00|j0\rangle\langle j'km'0|jm\rangle)$$

$$-\, i\hbar \sum_{n'j'm'} a_{n'j'm'}(t)(\dot{\Theta}U^{(1)}_{mm'} + \dot{\Phi}U^{(2)}_{mm'})\delta_{jj'}\delta_{nn'} \qquad (3.53)$$

We notice that the potential coupling is diagonal in the magnetic quantum numbers m, due to the selection rules imposed by the 3-j symbols. Transitions among these states are induced by the rotation matrix $\mathbf{U}$. These couplings are time-dependent, due to the rotation of the space-fixed coordinate system by the angles $\Theta(t)$ and $\Phi(t)$ such that the R axis is along z in the rotated coordinate system. Thus, we have [89]

$$U_{mm'}^{(1)} = \tfrac{1}{2}(\sqrt{(j'-m'+1)(j'+m')}\,\delta_{m,m'-1} - \sqrt{(j'+m'+1)(j'-m')}\,\delta_{m,m'+1})$$
(3.54)

and

$$U_{mm'}^{(2)} = -im'\delta_{mm'}$$
(3.55)

In the so-called coupled states (CS) approximation, we neglect the coupling among the m-states, that is, the coupling induced by the matrix $\mathbf{U}$.

The effective potential, which governs the classical equations of motion, is now obtained as

$$V_{e\!f\!f}(R,t) = \langle \Psi|V|\Psi \rangle$$

$$= \sum_{njmn'j'm'} a_{njm} a_{n'j'm'}^{*} \sum_{ik} V_{ik}(R)\langle \phi_n|(r-r_{eq})^i|\phi_{n'}\rangle$$

$$\times \sqrt{\frac{2j+1}{2j'+1}}\langle kj00|j'0\rangle\langle jkm0|j'm'\rangle$$
(3.56)

We notice that the effective potential only depends on R, and, hence, the relative motion is governed by the hamiltonian

$$H_{cl} = \frac{1}{2\mu}\left(P_R^2 + \frac{L^2}{R^2}\right) + V_{e\!f\!f}(R,t)$$
(3.57)

where $L = \hbar(\ell+1/2)$ has been introduced. Since $P_\Theta = L$, we have the following equations of motion

$$\dot{R} = P_R/\mu$$
(3.58)

$$\dot{P}_R = \frac{L^2}{\mu R^3} - \frac{\partial}{\partial R}V_{e\!f\!f}$$
(3.59)

$$\dot{\Theta} = \frac{bv_0}{R(t)^2}$$
(3.60)

$$\dot{\Phi} = 0$$
(3.61)

where L, the orbital angular momentum, is assumed to be constant. This is only correct in the "semi-classical" limit where $J \gg j$, that is, when $J \sim \ell$. The relation to the impact parameter is given by [190]

$$b = \hbar(\ell+1/2)/(\mu v_0) = L/(\mu v_0)$$
(3.62)

where v_0 is the initial relative velocity. The initial kinetic energy is $U = \frac{1}{2}\mu v_0^2$ and, according to the symmetrization principle, we have $v_0 = \frac{1}{2}(v_{nj} + v_{n'j'})$, which can be reformulated to

$$E = \tfrac{1}{2}(E_{nj} + E_{n'j'}) + U + (E_{n'j'} - E_{nj})^2/16U \tag{3.63}$$

for the transition $nj \to n'j'$. The inelastic cross section is finally obtained as

$$\sigma_{nj \to n'j'}(E, J) = \frac{\pi}{(2j+1)k_{nj}^2} \sum_{J=0}^{\infty} \sum_{m'=-\bar{j}}^{+\bar{j}} (2J+1)|a_{n'j'm'}(\infty)|^2 \tag{3.64}$$

where $k_{nj}^2 = 2\mu(E - E_{nj})/\hbar^2$ and $\bar{j} = \min(j, j')$.

Table 3.6 shows a comparison between cross sections obtained with the $V_q R_q T_c$ method and accurate quantum calculations.

The large j-limit

In the above scheme we introduced the decoupling approximation—in which the matrix $\mathbf{U}$ is neglected. In the limit of large j values, we may proceed in a different manner and still include the effect of the $\mathbf{U}$-matrix. Using the fact that $\dot{\Phi} \sim 0$, let us introduce the quantities b_n by

$$\mathbf{b} = \exp(\Theta \mathbf{U})\mathbf{a} \tag{3.65}$$

and, hence, the equations for $b_n(t)$ are

$$i\hbar \dot{b}_k(t) = E_k b_k(t) + \sum_{k'} W_{kk'} b_{k'}(t) \tag{3.66}$$

where k and k' are collective indices $k = (n, j, m)$ and W_{km} is an element of the matrix $\mathbf{W}$ given by

$$\mathbf{W} = \exp(\Theta \mathbf{U})\mathbf{V}\exp(-\Theta \mathbf{U}) \tag{3.67}$$

Table 3.6: Cross sections for rotational excitation of hydrogen colliding with He-atoms at 0.1 and 0.9 eV total energy. The quantum-classical numbers are obtained using the $V_q R_q T_c$ method. Data from ref. [87].

Energy	Transition	Cross section in a_0^2	
		Quantum-classical	Quantum
0.1 eV	0-2	0.706	0.690
0.9 eV	0-2	12.35	12.00
0.9 eV	0-4	0.83	0.80
0.9 eV	2-4	3.75	3.60

Thus, we have

$$W_{njm;n'j'm'} = \sum_{ik} V_{ki}(R)M_{nn'}^{(i)}\langle kj'00|j0\rangle \sum_{q=-k}^{k} d_{q0}^{k}(\Theta)\langle kjqm|j'm'\rangle \qquad (3.68)$$

where $d_{q0}^{k}(\Theta)$ is a rotational coupling element defined, for example, in ref. [83]. Thus, the m states are coupled in this scheme. However, in the limit where $j \gg k$ and $j' \gg k$, we can introduce the approximation [83]

$$\langle kjqm|j'm'\rangle \sim (-1)^{k+j'-j}d_{j'-j,q}^{k}(\beta) \qquad (3.69)$$

where $\cos(\beta) = m'/j'$. The closure relation for the rotation matrix elements now gives [83]

$$\sum_{q=-k}^{k} d_{q0}^{k}(\Theta)d_{j'-j,q}^{k}(\beta) = d_{j'-j,0}^{k}(\Theta+\beta) \qquad (3.70)$$

and, hence, we obtain the matrix elements as

$$W_{nj;n'j'}(\beta) = \sum_{ik} V_{ik}(R)M_{nn'}^{(i)}\langle kj'00|j0\rangle d_{j'-j,0}^{k}(\beta+\Theta) \qquad (3.71)$$

Thus, the coupling elements depend only parametrically on the m-quantum numbers, through the angle β. The m states are decoupled as in the CS-approximation. The calculations are carried out using a random choice of β in the range $[0:\pi]$.

If the kinetic energy is not too high compared with the rotational level spacing, we expect the separability between the translational degree of freedom and the rotational/vibrational degrees of freedom to be fulfilled. Hence, the $V_qR_qT_c$ is the method of choice. It is also feasible to use the $V_qR_qT_c$ scheme for calculating energy transfer in diatom-diatom collisions if the CS-approximation is introduced [85].

As the energy and/or the moment of inertia increase, we will have stronger coupling between the translational and rotational motion, that is, we should here switch to a method where the separability is introduced between the translational/rotational motion and the vibrational degrees of freedom. Such a method is the $V_qR_cT_c$ method.

3.2.4   The $V_qR_cT_c$ method

If the rotational degree of freedom is also treated classically, the complexity is decreased even further. We will first consider the collision between an atom and a diatomic molecule, then treat diatom-diatom collisions, and, finally, in chapter 5, treat polyatomic molecules. For atom-diatom collisions, we then only have one quantum degree of freedom left and, hence, the wave function is expanded as

$$\Psi(r,t) = \sum_{n} a_n(t)\phi_n(r,t) \qquad (3.72)$$

where $\phi_n(r, t)$ are rotationally perturbed vibrational eigenfunctions. Considering the hamiltonian, we obtain

$$H = \frac{p_r^2}{2m} + v(r) + \frac{j(t)^2}{2mr^2} + \frac{1}{2\mu}(P_R^2 + \ell^2/R^2) + V(R, r, \gamma) \tag{3.73}$$

where $j(t)$ is the classical rotational angular momentum. We now quantize the r-coordinate and define $\hat{H}_0$ as

$$\hat{H}_0 = -\frac{\hbar^2}{2m}\frac{\partial^2}{\partial r^2} + v(r) + \frac{j(t)^2}{2m}(r^{-2} - r_{eq}^{-2}) \tag{3.74}$$

that is, we have included all the static r dependent terms, and also the centrifugal stretch coupling in $\hat{H}_0$. The hamiltonian for the classical motion then becomes

$$H_{cl} = \frac{j^2}{2mr_{eq}^2} + \frac{1}{2\mu}(P_R^2 + \ell^2/R^2) + V_{eff}(R, \gamma, t) \tag{3.75}$$

where

$$V_{eff}(R, \gamma, t) = \langle \Psi | V(R, r, \gamma) | \Psi \rangle \tag{3.76}$$

The brackets indicate integration over the quantum coordinate r. The interaction between the quantum and classical system is taken care of by the hamiltonian

$$H_{int} = V(R, r, \gamma) - V_{eff}(R, \gamma, t) \tag{3.77}$$

The wave functions $\phi_n(r, t)$ are time-dependent through the angular momentum $j(t)$ appearing in $\hat{H}_0$. We can get an estimate of this wave function by using a first-order perturbational solution. Thus, we have

$$\phi_n(r, t) = \phi_n^0(r) + \sum_{m \neq n} \phi_m^0(r) \frac{H_{mn}^{(1)}}{E_n^0 - E_m^0} \tag{3.78}$$

where the matrix elements are given as

$$H_{mn}^{(1)} = \langle \phi_m^0 | r^{-2} - r_{eq}^{-2} | \phi_n^0 \rangle \frac{j(t)^2}{2m} \tag{3.79}$$

where the brackets indicate integration over r.

From the TDSE, we obtain a set of coupled equations for the amplitudes $a_n(t)$

$$i\hbar\dot{a}_m(t) = E_m a_m + \sum_n a_n(t)\left[\langle \phi_m | V(R, r, \gamma)|\phi_n\rangle - 2i\hbar\frac{d\ln j(t)}{dt}\frac{H_{mn}^{(1)}}{E_n^0 - E_m^0}\right] \tag{3.80}$$

These equations have to be solved together with the classical equations for the relative translational and rotational motion. The angles conjugate to the actions $j(t)$ and $\ell(t)$ are q_j and q_ℓ, respectively. The effective potential depends upon the variables R (conjugate to P_R) and the angle γ. If the classical equations of motion are solved in action/angle variables, the connection between γ and the action-angle variables is needed. It is [90]

$$\cos\gamma = -\cos q_\ell \cos q_j + \frac{\ell^2 + j^2 + J^2}{2\ell j}\sin q_j \sin q_\ell \qquad (3.81)$$

where J is the total angular momentum.

It is, however, often more convenient to use cartesian coordinates, that is, to use

$$H_{cl} = \frac{1}{2m}(p_x^2 + p_y^2 + p_z^2) + \frac{1}{2\mu}(P_X^2 + P_Y^2 + P_Z^2)$$
$$+ V_{eff}(R, \gamma, t) + \lambda(r_{eq}^2 - x^2 - y^2 - z^2) \qquad (3.82)$$

where we have introduced a Lagrange multiplier λ. This is necessary to keep the molecule rotating at a fixed bond length. Note that the Coriolis coupling or, strictly speaking, the centrifugal stretch coupling is incorporated in the quantum part of the hamiltonian.

The equations of motion in cartesian coordinates are:

$$\dot{X} = P_X/\mu \qquad (3.83)$$

$$\dot{P}_X = -\frac{\partial V_{eff}}{\partial R}\frac{X}{R} - \frac{\partial V_{eff}}{\partial \cos\gamma}\frac{\partial \cos\gamma}{\partial X} \qquad (3.84)$$

$$\dot{x} = \frac{p_x}{m}\left(1 + \sum_{nm} a_n^* a_m [\mathbf{F}, \mathbf{M}]_{nm}\right) \qquad (3.85)$$

$$\dot{p}_x = 2\lambda x - \frac{\partial V_{eff}}{\partial \cos\gamma}\frac{\partial \cos\gamma}{\partial x} \qquad (3.86)$$

plus the corresponding equations for the Y and Z components. In this derivation, we have used the fact that

$$R = \sqrt{X^2 + Y^2 + Z^2} \qquad (3.87)$$

and the relation

$$\cos\gamma = \frac{xX + yY + zZ}{Rr_{eq}} \qquad (3.88)$$

The Lagrange multiplier is defined as [86]

$$\lambda = \frac{p_x^2 + p_y^2 + p_z^2}{2mr^2}h(t)\frac{\partial V_{eff}}{\partial \cos\gamma} \qquad (3.89)$$

where the function $h(t)$ is

$$h(t) = 1 + \sum_{nm} a_n^* a_m [\mathbf{F}, \mathbf{M}]_{nm} \qquad (3.90)$$

Thus, we need to evaluate the commutator between the matrices $\mathbf{F}$ and $\mathbf{M}$ with the matrix elements given by

$$M_{nm} = \frac{1}{E_n^0 - E_m^0} \left\langle \phi_n^0 \left| -\frac{2\Delta r}{r_{eq}} + \frac{3\Delta r^2}{r_{eq}^2} - \ldots \right| \phi_m^0 \right\rangle \quad (n \neq m) \quad (3.91)$$

$$F_{nm} = \langle \phi_n^0 | H_{int} | \phi_m^0 \rangle + E_n^0 \quad (3.92)$$

The $V_q R_c T_c$ method is extremely convenient for studying energy transfer to heavier molecules since the computational effort scales only linearly with the number of vibrational states included in the expansion of the wave function. For atom-diatom collisions, we expect it to be as accurate as the quantum coupled states method (see, for instance, table 3.7).

Instead of expanding in vibrational states, another possibility would be to use a grid in the vibrational coordinate. In this manner, convergence problems with the vibrational basis set for energies near the dissociation limit are avoided. Both the state-expansion and the grid method are straightforward to extend to diatom-diatom collisions.

3.2.5 The $V_q R_c T_c$ method for diatom-diatom collisions

For two diatomic molecules, we could proceed in exactly the same manner as for atom-diatom collisions, that is, we expand the wave function as

$$\Psi(r_1, r_2, t) = \sum_{n_1 n_2} a_{n_1 n_2}(t) \phi_{n_1}(r_1, t) \phi_{n_2}(r_2, t) \quad (3.93)$$

where the vibrational wave functions depend upon time through the vibration/rotation coupling terms. The classical dynamics should be solved for the rotational motion of both diatomic molecules and the relative translational motion.

Table 3.7: Rate constant $k_{10}(T)$ in units of cm^3/sec for vibrational relaxation of nitrogen colliding with He as a function of temperature. The quantum-classical results were obtained using the $V_q R_c T_c$ method and the quantum values using the coupled states approximation.

T	Quantum-classical [91]	Quantum [92]	Experiment [93]
100	5.9(−20)	5.6(−20)	3.2 ± 1.5(−20)
132	1.8(−19)	1.5(−19)	1.6 ± 0.15(−19)
149	2.9(−19)	2.4(−19)	2.7 ± 0.2(−19)
156	3.1(−19)	2.9(−19)	3.3 ± 0.3(−19)
175	5.8(−19)	4.6(−19)	5.0 ± 0.3(−19)
210	1.1(−18)	9.9(−19)	1.2 ± 0.05(−18)
262	3.2(−18)	2.7(−18)	3.2 ± 0.1(−18)
291	5.7(−18)	4.3(−18)	5.1 ± 0.1(−18)

$(−20)$ means 10^{-20}.

As before, we may introduce an effective Ehrenfest average-type potential for the classical motion. Thus, we have

$$H_{cl} = \frac{1}{2\mu}(P_X^2 + P_Y^2 + P_Z^2) + \sum_{i=1,2} \frac{1}{2m_i}(p_{x_i}^2 + p_{y_i}^2 + p_{z_i}^2)$$

$$- \lambda_i((r_{eq}^i)^2 - x_i^2 - y_i^2 - z_i^2) + V_{eff}(R, c1, c2, sc1, sc2, ss1, ss2, t) \quad (3.94)$$

where the first term is the kinetic energy of the relative motion, the next takes care of the rotational motion of the two diatomic molecules, and the final term is an effective potential. As in the case of atom-diatom collisions, we have introduced a Lagrange multiplier in order to keep the bond length fixed. The effective potential can be written as a function of the center-of-mass distance R and the orientation of the two molecules (see fig. 3.4). We have also introduced the notation $ci = \cos(\gamma_i)$, $sci = \sin(\gamma_i)\cos(\phi_i)$, and $ssi = \sin(\gamma_i)\sin(\phi_i)$ ($i = 1, 2$).

The effective potential has been obtained as the expectation value of the total interaction potential over the wave function Ψ, that is, the dependence of the quantum coordinates has been averaged out. In order to propagate the solution in cartesian coordinates, we need the connection between the angles and the coordinates. This connection is given as

$$ci = (x_i X + y_i Y + z_i Z)/Rr_i \quad (3.95)$$

$$sci = \frac{1}{r_i R}\left(\frac{XZx_i}{\sqrt{X^2+Y^2}} + \frac{YZy_i}{\sqrt{X^2+Y^2}} - z_i\sqrt{X^2+Y^2}\right) \quad (3.96)$$

$$ssi = \frac{y_i X - x_i X}{r_i\sqrt{X^2+Y^2}} \quad (3.97)$$

Using these equations, we get the following equations of motion

$$\dot{X} = P_X/\mu \quad (3.98)$$

$$\dot{P}_X = -\left(\frac{\partial V_{eff}}{\partial R}\frac{X}{R} + \sum_{i=1,2}\frac{\partial V_{eff}}{\partial ci}\frac{\partial ci}{\partial X} + \frac{\partial V_{eff}}{\partial sci}\frac{\partial sci}{\partial X} + \frac{\partial V_{eff}}{\partial ssi}\frac{\partial ssi}{\partial X}\right) \quad (3.99)$$

$$\dot{x}_i = \frac{p_{x_i}}{m_i}h_i \quad (3.100)$$

$$\dot{p}_{x_i} = 2\lambda_i x_i - \frac{\partial V_{eff}}{\partial ci}\frac{\partial ci}{\partial x_i} - \frac{\partial V_{eff}}{\partial sci}\frac{\partial sci}{\partial x_i} + \frac{\partial V_{eff}}{\partial ssi}\frac{\partial ssi}{\partial x_i} \quad (3.101)$$

The Lagrange multiplier is, in this case, defined through the expression

$$2\lambda_i = \frac{1}{r_{eq,i}^2}\left[x_i\frac{\partial V_{eff}}{\partial x_i} + y_i\frac{\partial V_{eff}}{\partial y_i} + z_i\frac{\partial V_{eff}}{\partial z_i}\right] - \frac{h_i}{m_i}(p_{xi}^2 + p_{yi}^2 + p_{zi}^2) \quad (3.102)$$

where

$$h_i = 1 + \sum_{\alpha\beta} a_\alpha^* a_\beta[\mathbf{F}, \mathbf{M}^{(i)}] \quad (3.103)$$

Here, $\alpha = n_1, n_2$ and $\beta = m_1 m_2$ are collective indices and the commutator is between matrices $\mathbf{F}$ and $\mathbf{M}$ with elements

$$F_{\alpha\beta} = \langle \alpha|V|\beta \rangle + E_\alpha \delta_{\alpha\beta} \tag{3.104}$$

$$M_{\alpha\beta}^{(i)} = \left\langle \phi_{n_i} \left| -\frac{2\Delta r_i}{r_{eq,i}} + \frac{3(\Delta r_i)^3}{r_{eq,i}^2} + \cdots \right| \phi_{m_i} \right\rangle \delta_{n_j,m_j}(E_{n_i} - E_{m_i})^{-1} \tag{3.105}$$

where $i \neq j = (1,2)$. We have also introduced the notation $\Delta r_i = r_i - r_{eq,i}$.

For the expansion coefficients, we obtain the following set of equations

$$i\hbar \dot{a}_{n_1 n_2} = E_{n_1 n_2} a_{n_1 n_2} + \sum_{m_1 m_2} a_{m_1 m_2}(t)(\delta_{n_2 m_2} M_{n_1 m_1}^{(1)} f_{VT}^{(1)} + \delta_{n_1 m_1} M_{n_2 m_2}^{(2)} f_{VT}^{(2)}$$
$$\times \tfrac{1}{2}\delta_{n_2 m_2} M_{n_1 m_1}^{(2)} g_{VT}^{(2)} + \tfrac{1}{2}\delta_{n_1 m_1} M_{n_2 m_2}^{(2)} g_{VT}^{(2)}$$
$$+ M_{n_1 m_1}^{(1)} M_{n_2 m_2}^{(1)} h_{VV}) \tag{3.106}$$

where

$$f_{VT}^{(i)} = \left.\frac{\partial V}{\partial r_i}\right|_{eq} + 2i\hbar j_i \frac{dj_i}{dt} \frac{1}{m_i r_{eq,i}^3 (E_{n_i}^0 - E_{m_i}^0)} \tag{3.107}$$

$$g_{VT}^{(i)} = \left.\frac{\partial^2 V}{\partial r_i^2}\right|_{eq} \tag{3.108}$$

$$h_{VV} = \left.\frac{\partial^2 V}{\partial r_1 \partial r_2}\right|_{eq} \tag{3.109}$$

and where $M_{nm}^{(k)} = \langle \phi_n^0|(r - r_{eq})|\phi_m^0 \rangle$. The method described in this section is available through the program DIDIEX (see ref. [88]) and has been used for the calculation of data tables of vibration-vibration (VV) and vibration-translation/rotation (VT/R) rates for systems as $CO+CO$, N_2+N_2, N_2+CO, O_2+O_2, $HF+HF$, $DF+DF$, N_2+CO, $HF+DF$, H_2+CO and H_2+H_2 [94] (see also table 3.8).

Diatom-diatom collisions using a 2D grid method

Near the dissociation limit, or for energies where chemical reactions may occur, it is advantageous to change to a grid method instead of the state-expansion method used in the previous section. For a collision of two diatomic molecules, we express the wave function as $\Psi(r_1, r_2, t)$ and using, for instance, Jacobi variables to describe the system, we would obtain a TDSE for the wave function as

$$i\hbar \frac{\partial \Psi}{\partial t} = \left[\sum_{i=1,2} -\frac{\hbar^2}{2m_i}\frac{\partial^2}{\partial r_i^2} + \frac{1}{2m_i r_i^2}\left(p_{\theta_i}^2 + \frac{1}{\sin\theta_i^2} p_{\phi_i}^2 \right) + V(r_1, r_2, t) \right] \Psi \tag{3.110}$$

where $m_i(i = 1, 2)$ are the reduced masses of the two diatomic molecules. The orientation of the two molecules is specified by the time-dependent angles θ_i, ϕ_i. The time-dependence of the potential arises from its dependence on the classical

Table 3.8: Rate constants for vibrational relaxation of O_2 colliding with O_2 molecules as a function of temperature. Numbers obtained using the $V_q R_c T_c$ method.

Temperature	20-10	10-00	10-00 (experiment)
300	6.6(−18)	2.7(−18)	
500	1.3(−16)	5.1(−17)	2.7(−17)
750	1.0(−15)	4.1(−16)	
1000	4.6(−15)	1.9(−15)	1.6(−15)
2000	1.9(−13)	8.6(−14)	6.7(−14)
5000	9.1(−12)	4.8(−12)	4.7(−12)

Data from refs. [95, 96].

variables. As before, we assume that the effective potential for the classical degrees of freedom is determined by

$$H_{eff} = \langle \Psi | H_{sc} | \Psi \rangle \tag{3.111}$$

that is, we have

$$
\begin{aligned}
H_{eff} = {} & \frac{1}{2\mu} P_R^2 + \frac{1}{2\mu R^2} \left(P_\Theta^2 + \frac{1}{\sin^2\Theta} P_\Phi^2 \right) \\
& + \sum_{i=1,2} \frac{1}{2m_i} \left(p_{\theta_i}^2 + \frac{1}{\sin^2\theta_i} p_{\phi_i}^2 \right) \langle \Psi | 1/r_i^2 | \Psi \rangle \\
& + V_{eff}(R, \Theta, \Phi) + \langle \Psi | \hat{T}_{kin} | \Psi \rangle
\end{aligned} \tag{3.112}
$$

where the last term is the expectation value of the quantum kinetic energy.

The probability for a given vibrational transition is obtained by projecting the final wave function on the product vibrational state, that is,

$$P_{n_1 n_2 \to n_1' n_2'} = \lim_{t \to \infty} \int dr_1 \int dr_2 \Psi(r_1, r_2, t)^* \phi_{n_1'}(r_1) \phi_{n_2'}(r_2) \tag{3.113}$$

The method can also be used to estimate the total reaction or dissociation probability by calculating the flux over a dividing surface in the bond which breaks, that is, we have

$$
\begin{aligned}
P_d = {} & \frac{i\hbar}{2m_2} \int dt \int dr_1 \left[\frac{\partial \Psi(r_1, r_2, t)^*}{\partial r_2} \Psi(r_1, r_2, t)|_{r_2 = r_2^*} \right. \\
& \left. - \frac{\partial \Psi(r_1, r_2, t)}{\partial r_2} \Psi^*(r_1, r_2, t)|_{r_2 = r_2^*} \right]
\end{aligned} \tag{3.114}
$$

which is the integrated flux over the boundary r_2^* (see fig. 3.5).

Once the probability has been determined, we obtain the cross sections as

$$\sigma_{n_1 n_2 \to n_1' n_2'} = \frac{\pi}{k_{n_1 n_2}^2} \frac{1}{(2j_1 + 1)(2j_2 + 1)} \sum_{\ell=0}^{\ell_{max}} (2\ell + 1) \langle P_{n_1 n_2 \to n_1' n_2'} \rangle \tag{3.115}$$

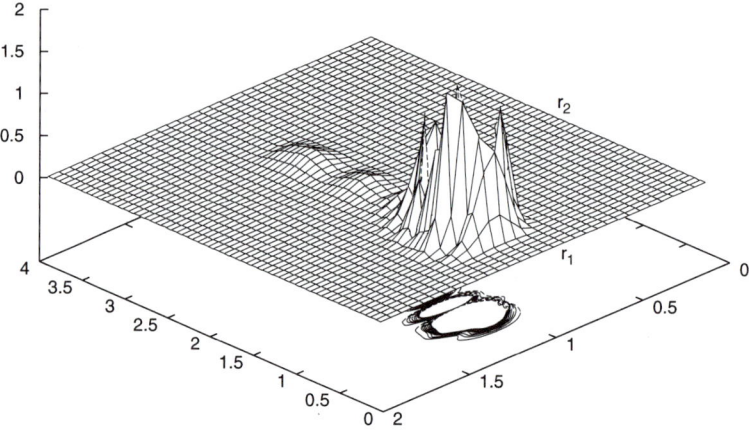

Fig. 3.5: Quantum-classical treatment of AB + CD collisions where the two bond distances r_1 and r_2 are quantized and the 2D wave packet represented and propagated on a grid. The figure shows a wave packet for the $H_2 + OH \rightarrow H_2O + H$ reaction. The wave packet shows dissociation in the hydrogen bond r_2 but is confined around the OH bond near the equilibrium value. The initial vibrational states are $n_1 = n_2 = 0$.

where we have averaged over the initial rotational states and summed over the orbital angular momentum ℓ. The average probability is obtained by averaging over the remaining classical variables picked randomly, such that, for N trajectories we have

$$\langle P_{n_1 n_2 \to n_1' n_2'} \rangle = \frac{1}{N} \sum_{i=1}^{N} P_{n_1 n_2 \to n_1' n_2'}(i) \qquad (3.116)$$

where the probability depends on the initial values of the classical variables for the ith trajectory. Although rotational state-resolved cross sections are in principle obtainable by a classical treatment of the rotational motion, the quantum-classical theory is most often used in cases where we wish to average over the "uninteresting" degrees of freedom—those treated approximately by classical mechanics. If this is the case, we would, for instance, be interested in vibrationally state-resolved rate constants, where a Boltzmann distribution over initial rotational states and kinetic energy is introduced. Thus, we have

$$k_{n_1 n_2 \to n_1' n_2'}(T) = \sqrt{8k_B T/\pi\mu} \frac{1}{Q_{rot}} \int d(\beta E_{kin}) \int dj_1 (2j_1 + 1) \int dj_2 (2j_2 + 1)$$
$$\times \beta E_{kin} \exp(-\beta E_{kin} - \beta E_{rot}) \sigma_{n_1 n_2 \to n_1' n_2'} \qquad (3.117)$$

By introducing the above expression for the cross section, noting that

$$\frac{\hbar^2 k_{n_1 n_2}^2}{2\mu} = E_{kin} \qquad (3.118)$$

and introducing the rotational partition functions for diatomic molecules as

$$Q_{rot} = \frac{2I_1 k_B T}{\hbar^2} \frac{2I_2 k_B T}{\hbar^2} \tag{3.119}$$

where I_i are moments of inertia, we obtain

$$k_{n_1 n_2 \to n_1' n_2'}(T) = \sqrt{\frac{8k_B T}{\pi\mu}} (T_0/T)^3$$

$$\times \int d(\beta E_{kin}) \exp(-\beta(E - E_{n_1 n_2})) \langle \sigma_{n_1 n_2 \to n_1' n_2'}(E_{kin}, T_0) \rangle \tag{3.120}$$

where $\beta = 1/k_B T$ and the "average cross section" is defined as

$$\langle \sigma_{n_1 n_2 \to n_1' n_2'}(E_{kin}, T_0) \rangle$$
$$= \frac{\pi\hbar^6}{8\mu(k_B T_0)^3 I_1 I_2} \int d\ell \int dj_1 \int dj_2 (2\ell + 1)(2j_1 + 1)(2j_2 + 1) \langle P_{n_1 n_2 \to n_1' n_2'} \rangle \tag{3.121}$$

T_0 is an arbitrary reference temperature which vanishes from the final expression for the rate constant but gives the correct unit for the average cross section, that is, Å^2.

The energy $E - E_{n_1 n_2}$ is related to the semi-classical energy U through the equation

$$E - E_{n_1 n_2} = U + \frac{1}{2}\Delta E + \frac{1}{16}\Delta E^2/U \tag{3.122}$$

where $\Delta E = E_{n_1' n_2'} - E_{n_1 n_2}$, U is the sum of the kinetic and rotational energy, that is, the classical energy at which the trajectories are run. When the rates are calculated, we introduce the semi-classical symmetrization principle by using the above equation (3.122). In practical calculations, we therefore, carry out the integration at a number of energies U and define the total energy using eq. (3.122).

In this chapter, we have so far treated atom-diatom and diatom-diatom systems within the classical path theory. As mentioned above, the $V_q R_c T_c$ method can be used to treat energy transfer to polyatomic molecules. However, we shall return to polyatomic systems in chapter 5. The treatment of these systems will be facilitated by introducing the second quantization concepts described in chapter 4.

3.2.6 Reactive scattering

A. Reactions in hyper-spherical coordinates

In order to treat chemical reactions within the primitive quantum-classical theory, we need to introduce coordinates which allow a separation in degrees of freedom with coordinates where the coupling between the subsets is as weak as possible. Then, the separability assumption is better fulfilled, and we may hope that some of the degrees of freedom can be treated approximately through trajectories introduced by using the Gauss-Hermite expansion mentioned in chapter 2. This is actually the motivation for introducing the so-called hyper-spherical coordinates in the quantum-classical theory. The success of the quantum-classical method in its simple form lies in the ability to identify degrees of freedom which are strongly coupled. Take, for instance, a chemical reaction A+BC→ AB+C. Although the two coordinates R from the center of mass of BC to A and the BC bond distance r are not strongly coupled initially and also not in low energy inelastic scattering, they would be strongly coupled if a reaction takes place. The translational coordinate in one reaction channel is the vibrational coordinate (roughly speaking) in another channel. Hence, we must introduce a coordinate system where the separability between the modes is more obvious. Such a coordinate system for reactive systems is the hyper-spherical.

The hyper-spherical coordinates involve, for a triatomic system, the use of two angles (the hyper-angles θ and ϕ) and a distance the hyper-radius ρ. These coordinates can be used instead of, for instance, the three interatomic distances or the Jacobi coordinates R, r, and η. Here η is the angle between R and r. The connection between the Jacobi and the hyper-spherical coordinates is given by [97]

$$r^2 = \frac{d_1^2}{2}\rho^2(1 + \sin\theta\cos\phi) \tag{3.123}$$

$$R^2 = \frac{1}{2d_1^2}\rho^2(1 - \sin\theta\cos\phi) \tag{3.124}$$

$$\cos\eta = -\frac{\sin\theta\sin\phi}{\sqrt{1 - \sin^2\theta\cos^2\phi}} \tag{3.125}$$

where d_1 is a constant, which depends on the mass combination of the A+BC system, r is the BC bond distance and R the distance from A to the center of mass of BC. We notice that the hyper-radius can be expressed as

$$\rho^2 = d_1^2 R^2 + d_1^{-2} r^2 \tag{3.126}$$

that is, ρ goes to infinity if one of the coordinates approaches that limit. The hyper-spherical coordinates treat the three rearrangement channels evenhandedly. The hyper-radius is a measure of the size of the ABC triangle, whereas the angles θ and ϕ determine the shape. The interatomic distances are given by [97]

$$r_{AB}^2 = \tfrac{1}{2}d_3^2\rho^2[1 + \sin\theta\cos(\phi + \epsilon_3)] \tag{3.127}$$

$$r_{BC}^2 = \tfrac{1}{2}d_1^2\rho^2[1 + \sin\theta\cos(\phi)] \tag{3.128}$$

$$r_{AC}^2 = \tfrac{1}{2}d_2^2\rho^2[1 + \sin\theta\cos(\phi - \epsilon_2)] \tag{3.129}$$

where $d_k^2 = (m_k/\mu)(1 - m_k/M)$, m_1 the mass of atom A, m_2 the mass of atom B, and m_3 the mass of atom C. M is the total mass $m_1 + m_2 + m_3$ and $\mu = \sqrt{m_1 m_2 m_3/M}$. The angles ϵ_2 and ϵ_3 define the two channels B+AC and C+AB, respectively. They are defined by

$$\epsilon_2 = 2\arctan(m_3/\mu) \tag{3.130}$$

$$\epsilon_3 = 2\arctan(m_2/\mu) \tag{3.131}$$

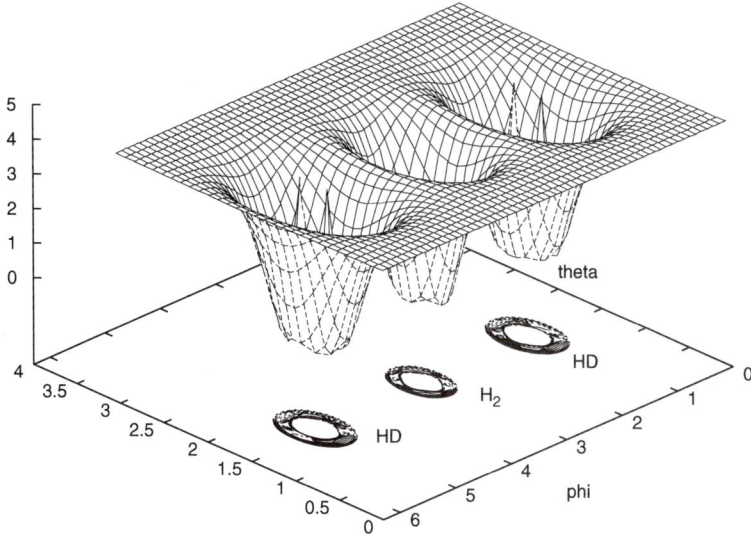

Fig. 3.6: Potential contour for the D + H$_2$ system at $\rho = 5$ Å as a function of the hyper-angles θ and ϕ. The three channels for H$_2$ and HD are isolated, and no reaction can occur. The potential is shown in units of 100 kJ/mol.

Figures 3.6 and 3.7 show the potential for the D+H$_2$ system as a function of θ and ϕ at two values of the hyper-radius. At large values of the hyper-radius, the three reaction channels form isolated potential wells (fig. 3.6). As the value of ρ diminishes, the barrier between the wells are lowered—allowing for a reaction.

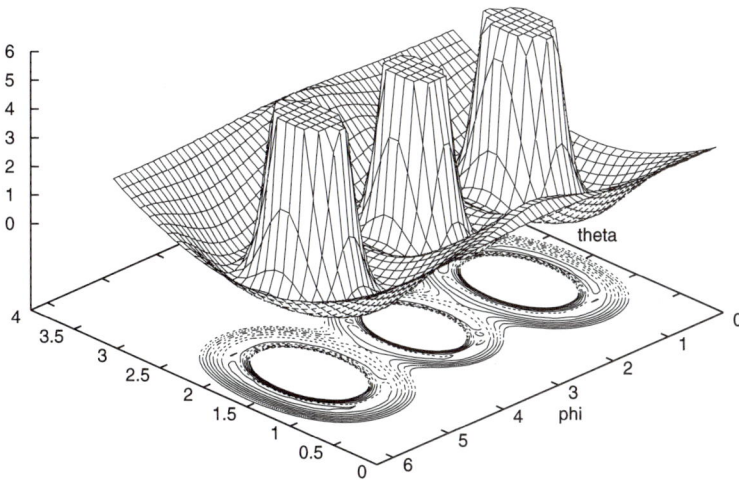

Fig. 3.7: The potential contour at $\rho = 2$ Å. The three channels are now coupled, and reaction can take place.

Consider now the above expressions for the three interatomic distances. We note that for $\rho \to \infty$, in order to keep the distance finite, we need to have $\theta \to \pi/2$ and $\phi \sim \pi$ in the A+BC channel, $\phi \sim \pi + \epsilon_2$ in the B+AC channel, and $\phi \sim \pi - \epsilon_3$ in channel C+AB. Thus, for $\rho \to \infty$ the physically interesting region in θ, ϕ space shrinks to three regions around $(\theta_0, \phi_0) = (\frac{\pi}{2}, \pi)$, $(\frac{\pi}{2}, \pi + \epsilon_2)$ and $(\frac{\pi}{2}, \pi - \epsilon_3)$.

In order to investigate this limit further, we introduce

$$\theta = \theta_0 - \xi \sin\eta \tag{3.132}$$

$$\phi = \phi_0' + \xi \cos\eta \tag{3.133}$$

where the angle η lies in the range 0 to π and ξ is a quantity which approaches zero as ρ goes to infinity such that the product $x = \rho\xi$ stays finite. From the above equations we can obtain, to order ξ^2,

$$\sin\theta = 1 - \tfrac{1}{2}\xi^2 \sin^2\eta \tag{3.134}$$

$$\cos\phi = \cos\phi_0'(1 - \tfrac{1}{2}\xi^2 \cos^2\eta) - \sin\phi_0' \xi \cos\eta \tag{3.135}$$

where the angle ϕ_0' is not yet defined. Substituting these expressions in eq. (3.128), we obtain

$$r_{BC}^2 = \tfrac{1}{2}d_1^2 \rho^2 [1 + \cos\phi_0'(1 - \tfrac{1}{2}\xi^2 + \tfrac{1}{4}\xi^4 \sin^2\eta\cos^2\eta) \\ - \sin\phi_0'(\xi\cos\eta - \tfrac{1}{2}\xi^3 \sin^2\eta\cos\eta)] \tag{3.136}$$

From this expression, we see that the limit $\rho \to \infty$, $\theta \to \pi/2$ and $\xi \to 0$ with x finite gives r_{BC} finite if $\phi_0' = \phi_0 = \pi$. In this limit, we then get

$$r_{BC}^2 = \tfrac{1}{4}d_1^2 x^2 \tag{3.137}$$

The same considerations can be carried out for the other channels [98] (see also table 3.9). Thus we see that for $\rho \to \infty$, we approach the three arrangement channels without discriminating between them.

Aside from the three variables (ρ, θ, ϕ), we need some angles (Euler angles) (α, β, γ) to describe the rotation of the ABC triangle in space. Before considering the quantum and quantum-classical hamiltonian, it is convenient to have a look

Table 3.9: Asymptotic $\rho \to \infty$ behavior in the three reaction channels of an A+BC reaction.

Channel	r_{BC}	r_{AB}	r_{AC}
A+BC	$\tfrac{1}{2}d_1 x$	∞	∞
AB+C	∞	$\tfrac{1}{2}d_3 x$	∞
AC+B	∞	∞	$\tfrac{1}{2}d_2 x$

$x = \rho\xi$.

at the classical hamilton function in hyper-spherical coordinates. It is for a three atomic system given by [98]

$$H_{cl} = \frac{P_\rho^2}{2\mu} + \frac{2}{\mu\rho^2}\left(P_\theta^2 + \frac{P_\phi^2}{\sin^2\theta}\right) + \frac{P_\gamma(P_\gamma - 4\cos\theta P_\phi)}{2\mu\rho^2\sin^2\theta}$$
$$+ \frac{P_J^2 - P_\gamma^2}{\mu\rho^2\cos^2\theta}(1 + \sin\theta\cos2\gamma) + V(\rho,\theta,\phi) \qquad (3.138)$$

We see that the Euler angles α and β do not appear in the hamilton function. Thus, the corresponding momenta, the total angular momentum $P_\alpha = P_J$, and its projection on a space-fixed axis P_β, respectively [97], are constants of motion. The first term in eq. (3.138) is the kinetic energy along the hyper-radius. The significance of the second term will be investigated below. The third and fourth terms are coupling terms between the overall rotational motion and the internal motion (Coriolis coupling terms). Let us now take the asymptotic limit $\rho \to \infty$ and $\xi \to 0$ in the above hamiltonian function. We then obtain

$$H_{cl} \to \frac{P_\rho^2}{2\mu} + V(x) + \frac{P_\gamma^2}{2\mu\rho^2} + \frac{2}{\mu\rho^2}(P_\theta^2 + P_\phi^2) + \frac{P_J^2 - P_\gamma^2}{\mu x^2}\frac{(1 + \cos2\gamma)}{\sin^2\eta} \qquad (3.139)$$

We now introduce new momenta P_x and P_η instead of P_θ and P_ϕ. This can be done using the F_3 generator [101], which is a function of the old momenta and the new coordinates x, η, that is, $F_3(P_\theta, P_\phi, x, \eta)$. That is

$$F_3 = -\theta(x,\eta)P_\theta - \phi(x,\eta)P_\phi \qquad (3.140)$$

and we get the new momenta as

$$P_x = -\frac{\partial F_3}{\partial x} \qquad (3.141)$$

$$P_\eta = -\frac{\partial F_3}{\partial \eta} \qquad (3.142)$$

Thus, using eqs. (3.132)–(3.133), we get

$$P_x = \frac{1}{\rho}(\sin\eta P_\theta + \cos\eta P_\phi) \qquad (3.143)$$

$$P_\eta = \frac{x}{\rho}(\cos\eta P_\theta - \sin\eta P_\phi) \qquad (3.144)$$

Introducing this in the asymptotic hamiltonian, we have

$$H_{cl} = \frac{P_\rho^2}{2\mu} + \frac{P_\gamma^2}{2\mu\rho^2} + \frac{2}{\mu}\left(P_x^2 + \frac{P_\eta^2}{x^2}\right) + V(x) + \frac{P_J^2 - P_\gamma^2}{\mu x^2}\frac{(1 + \cos2\gamma)}{\sin^2\eta} \qquad (3.145)$$

The first two terms are the kinetic and centrifugal energy connected to the motion along the hyper-radius. Introducing $x = 2r/d_i$ and $d_i = \sqrt{\mu/m}$ where

m is the reduced mass of the diatomic fragment, we get

$$H_{vib} = \frac{2}{\mu}P_x^2 + V(x) = \frac{P_r^2}{2m} + V(r) \qquad (3.146)$$

which is the vibrational hamiltonian for the diatomic fragment. For a description of the rotational motion of the diatomic molecule, one needs two polar angles, η and ξ, where η is the angle between the r and the R axis. Thus, η and ξ give the orientation of the diatomic molecule in a body-fixed coordinate system with the z axis along R. Introducing in the last term, $P_\xi = \sqrt{P_J^2 - P_\gamma^2 \cos\gamma}$, we get the additional term for the rotational energy of the diatomic molecule, that is, we have

$$H_{rot} = \frac{1}{2mr^2}\left(P_\eta^2 + \frac{P_\xi^2}{\sin^2\eta}\right) \qquad (3.147)$$

However, to show this rigorously, we need to consider the asymptotic form of the quantum mechanical hamiltonian [207].

In hyper-spherical coordinates, the quantum mechanical hamiltonian takes the following form [206]

$$\hat{H} = -\frac{\hbar^2}{2\mu}\frac{\partial^2}{\partial\rho^2} + \frac{2}{\mu\rho^2}\hat{L}^2(\theta,\phi) + \frac{1}{\mu\rho^2}\left[\frac{\hat{J}_x^2}{1-\sin\theta} + \frac{\hat{J}_y^2}{1+\sin\theta} + \frac{\hat{J}_z^2}{2\sin^2\theta}\right]$$
$$- \frac{2\cos\theta\hat{J}_z\hat{P}_\phi}{\mu\rho^2\sin^2\theta} + V(\rho,\theta,\phi) + \Delta V(\rho,\theta) \qquad (3.148)$$

The operators are defined by [83]

$$\hat{J}_x = -i\hbar\cos\gamma\left[\frac{1}{\sin\beta}\frac{\partial}{\partial\alpha} - \text{tg}\gamma\frac{\partial}{\partial\beta} - \cot\beta\frac{\partial}{\partial\gamma}\right] \qquad (3.149)$$

$$\hat{J}_y = -i\hbar\sin\gamma\left[\frac{1}{\sin\beta}\frac{\partial}{\partial\alpha} + \cot\gamma\frac{\partial}{\partial\beta} - \cot\beta\frac{\partial}{\partial\gamma}\right] \qquad (3.150)$$

and

$$\hat{J}_z = -i\hbar\frac{\partial}{\partial\gamma} \qquad (3.151)$$

$$\hat{P}_\phi = -i\hbar\frac{\partial}{\partial\phi} \qquad (3.152)$$

$$\hat{L}^2(\theta,\phi) = -\hbar^2\left[\frac{\partial^2}{\partial\theta^2} + \frac{1}{\sin^2\theta}\frac{\partial^2}{\partial\phi^2}\right] \qquad (3.153)$$

The potential $\Delta V(\rho,\theta)$ is called the extra potential and it arises from the kinetic energy term when transforming from cartesian (or Jacobi) to hyper-spherical coordinates. It is given by

$$\Delta V(\rho,\theta) = -\frac{\hbar^2}{2\mu\rho^2}\left[\frac{1}{4} + \frac{4}{\sin^2 2\theta}\right] \qquad (3.154)$$

The genuine potential is expressed as $V(\rho, \theta, \phi)$. The remaining operators are rotational momentum operators $\hat{J}_x$ and $\hat{J}_y$. Candidates for a classical mechanical treatment are now the Euler angles—the overall rotation is a slow motion expected to be only weakly coupled to the other degrees of freedom. The motion along the hyper-radius is a translational type of motion with mass μ and it could also be treated classically.

Since the hamiltonian is independent of the two Euler angles α and β, we are left with just four variables. Thus, an exact treatment of the three-body problem is a four-dimensional (4D) quantum mechanical problem and the wave function is $\Psi(\rho, \theta, \phi, \gamma)$. But with a classical treatment of the angle γ and perhaps also ρ, the quantum problem is reduced to a 3D or a 2D one. Wave functions $\Psi(\theta, \phi, t)$ in 2 and $\Psi(\rho, \theta, \phi, t)$ in three dimensions are routinely propagated using, for example, grid expansion techniques (see below). As we have seen from the asymptotic considerations above, a quantum treatment of the hyper-angles θ and ϕ corresponds asymptotically to a quantum v, j labeling of the wave function, that is, the vibrational and the rotational motions are quantized, but the rotational projection, as well as the motion along the hyper-radius, are classical quantities. Thus, in a 2D quantum+2D classical theory, we would have the mixed quantum-classical hamiltonian

$$\hat{H} = \frac{P_\rho^2}{2\mu} + \hat{H}_0(\theta, \phi; \rho) + \hat{H}_1(\theta, \phi; \rho, \gamma, P_J, P_\gamma) + V(\rho, \theta, \phi) + \Delta V(\rho, \theta) \quad (3.155)$$

where

$$\hat{H}_0 = -\frac{2\hbar^2}{\mu\rho^2} \left[\frac{\partial^2}{\partial\theta^2} + \frac{1}{\sin^2\theta} \frac{\partial^2}{\partial\phi^2} \right] \quad (3.156)$$

$$\hat{H}_1 = \frac{P_\gamma(P_\gamma - 4\cos\theta\hat{P}_\phi)}{2\mu\rho^2\sin^2\theta} + \frac{P_J^2 - P_\gamma^2}{\mu\rho^2\cos^2\theta}(1 + \sin\theta\cos2\gamma), \quad (3.157)$$

$$\hat{P}_\phi = -i\hbar\frac{\partial}{\partial\phi} \quad (3.158)$$

If we again take the asymptotic limit $\rho \to \infty$, we obtain

$$\hat{H} = \frac{P_\rho^2}{2\mu} - \frac{2\hbar^2}{\mu} \left[\frac{\partial^2}{\partial x^2} + \frac{1}{x}\frac{\partial}{\partial x} - \frac{1}{4x^2} \right] + V(x)$$

$$- \frac{2\hbar^2}{\mu x^2} \left[\frac{\partial^2}{\partial\eta^2} + \frac{1}{4\sin^2\eta} + \frac{1}{4} \right] + \frac{P_J^2 - P_\gamma^2}{\mu x^2 \sin^2\eta}(1 + \cos(2\gamma)) \quad (3.159)$$

The eigenfunctions to this hamiltonian are

$$\phi_{vj}(x, \eta; \nu) = \frac{1}{\sqrt{x}} \sqrt{\sin\eta} g_{vj}(x) P_j^\nu(\cos\eta) \quad (3.160)$$

where P_j^ν is an associated Legendre polynomial and $g_{vj}(x)$ a vibrational wave function. The volume element on which the functions are normalized is $xdxd\eta$. The wave function is labeled by the vibrational and rotational quantum numbers v, j. The index ν is, however, a continuous quantity defined as

$$\nu = \frac{\cos\gamma}{\hbar}\sqrt{P_J^2 - P_\gamma^2} \qquad (3.161)$$

The basis set forms an orthorgonal basis set in the limit of large values of j where

$$\phi_{vj}(x,\eta;\nu) = \sqrt{\frac{2}{\pi x}}g_{vj}(x)\cos\left(\left(j+\frac{1}{2}\right)\eta - \frac{\pi}{4} + \frac{\nu\pi}{2}\right) \qquad (3.162)$$

Thus, we see that a quantum treatment of the variables θ and ϕ asymptotically corresponds to quantizing the vibrational and rotational motion, whereas the rotational projection is treated classically. The propagation of wave functions in two or three dimensions is, as mentioned, straightforward on modern computers using for example, grid methods. Table 3.10 shows results obtained propagating a 3D wave packet in the variables ρ, θ, and ϕ, treating the fourth variable γ as a classical quantity. The agreement with exact numbers is reasonable, but not as good as for the inelastic collisions. This indicates that some quantum-classical correlation between γ and the other variables still persists. The extension of the method to treat electronic multi-surface problems is straightforward. The method scales about linearly with the number of surfaces [187].

Table 3.10: Vibrationally resolved reaction cross sections for the reaction $D + H_2(v = 0, j = 0) \rightarrow HD(v') + H$ compared with exact values. Data from ref. [99].

v'	Energy/eV	Cross section/Å^2	"Exact" ref. [100]
0	0.600	0.0333	0.0400
	0.650	0.0993	0.209
	0.700	0.226	0.440
	0.780	0.623	0.689
	0.850	0.718	0.885
	0.930	0.934	1.08
	1.086	1.22	1.31
	1.250	1.45	1.44
	1.350	1.59	1.50
1	0.900	0.0203	0.0064
	0.930	0.0290	0.0243
	1.086	0.103	0.0924
	1.250	0.201	0.180
	1.350	0.280	0.236
2	1.250	0.00485	0.00319
	1.350	0.0300	0.0182

B. State-to-all rates

If we are interested in cumulative cross sections and rates, it is possible to introduce considerable simplifications. In such cases, it is sufficient just to calculate the flux over a barrier which separates the reactants from the products. Consider the collision between two diatomic molecules AB and CD. The position of the molecules is defined in a space-fixed Jacobi coordinate frame, where r_1 and r_2 are the bond distances, and $\mathbf{r}_3$ is the vector connecting the centers of mass of the two molecules, respectively.

For such a system, the hamiltonian can be written as:

$$\hat{H} = -\sum_{i=1}^{3} \frac{\hbar^2}{2\mu_i} \left[\frac{\partial^2}{\partial r_i^2} + \frac{2}{r_i} \frac{\partial}{\partial r_i} + \frac{1}{r_i^2} \left(\frac{\partial^2}{\partial \theta_i^2} + \cot \theta_i \frac{\partial}{\partial \theta_i} + \frac{1}{\sin^2 \theta_i} \frac{\partial^2}{\partial \phi_i^2} \right) \right]$$
$$+ V(r_i, \theta_i, \phi_i) \tag{3.163}$$

where the Jacobi vector $\mathbf{r}_i$ is expressed in terms of its spherical polar components $r_i, \theta_i,$ and ϕ_i and

$$\mu_1 = \frac{m_A m_B}{m_A + m_B} \tag{3.164}$$

$$\mu_2 = \frac{m_C m_D}{m_C + m_D} \tag{3.165}$$

$$\mu_3 = \frac{(m_A + m_B)(m_C + m_D)}{(m_A + m_B + m_C + m_D)} \tag{3.166}$$

By introducing a wave function transformation in order to eliminate the first derivatives in r_i in eq. (3.163), we can obtain the following hamiltonian:

$$\hat{H}' = -\sum_{i=1}^{3} \frac{\hbar^2}{2\mu_i} \left[\frac{\partial^2}{\partial r_i^2} + \frac{1}{r_i^2} \left(\frac{\partial^2}{\partial \theta_i^2} + \cot \theta_i \frac{\partial}{\partial \theta_i} + \frac{1}{\sin^2 \theta_i} \frac{\partial^2}{\partial \phi_i^2} \right) \right]$$
$$+ V(r_i, \theta_i, \phi_i). \tag{3.167}$$

The connection between the two wave functions is

$$\psi_{new} = r_1 r_2 r_3 \psi_{old} \tag{3.168}$$

and the volume element for normalization of the new wave function

$$d\tau = \prod_{i=1}^{3} \sin \theta_i dr_i d\theta_i d\phi_i \tag{3.169}$$

We can now introduce different quantum-classical schemes:

I: r_1, r_2, and r_3 are quantized. The rotational motions are treated classically [84].

II: Only r_1 and r_2, that is, the vibrational motions are quantized.

In order to do this, we replace in the hamiltonian (3.167) the momentum operators of the variables we wish to treat classically with their classical counterparts, such that, in the latter case we have

$$\hat{H}'_{sc} = -\sum_{i=1}^{2} \frac{\hbar^2}{2\mu_i} \frac{\partial^2}{\partial r_i^2} + \sum_{i=1}^{3} \frac{1}{2\mu_i r_i^2} \left(p_{\theta_i}^2 + \frac{1}{\sin^2 \theta_i} p_{\phi_i}^2 \right) + \frac{p_{r_3}^2}{2\mu_3} + V(r_i, \theta_i, \phi_i)$$

(3.170)

The coupling between classical and quantal degrees of freedom is accomplished by means of an effective hamiltonian which is defined as the expectation value of the quantum-classical hamiltonian of eq. (3.170):

$$H_{sc}^{eff} = \frac{\langle \psi | \hat{H}'_{sc} | \psi \rangle}{\langle \psi | \psi \rangle}$$

(3.171)

where the brackets denote integration over r_1 and r_2. The reason for normalizing with the norm $\langle \psi | \psi \rangle$ is that in grid methods, we usually absorb part of the wave packet with an imaginary potential [103]. Hence, the norm over the grid is not conserved to unity. For the effective hamiltonian, we obtain

$$H_{sc}^{eff} = T_q + \frac{p_{r_3}^2}{2\mu_3 r_3^2} \left(p_{\theta_3}^2 + \frac{1}{\sin^2 \theta_3} p_{\phi_3}^2 \right)$$
$$+ \sum_{i=1}^{2} \frac{1}{2\mu_i} \left(p_{\theta_i}^2 + \frac{1}{\sin^2 \theta_i} p_{\phi_i}^2 \right) \left\{ \frac{1}{r_i^2} \right\} + V^{eff}(r_3, \theta_i, \phi_i)$$

(3.172)

where T_q is the energy associated with the quantal part,

$$V^{eff}(r_3, \theta_i, \phi_i) = \frac{\langle \psi | V(r_i, \theta_i, \phi_i) | \psi \rangle}{\langle \psi | \psi \rangle}$$

(3.173)

and

$$\left\{ \frac{1}{r_i^2} \right\} = \frac{\langle \psi | 1/r_i^2 | \psi \rangle}{\langle \psi | \psi \rangle}$$

(3.174)

The dynamical treatment of the system requires the simultaneous propagation of the Hamilton equations for the classical variables and the solution of the time-dependent Schrödinger equation for the vibrations of the two diatoms. For the latter we can write:

$$i\hbar \frac{\partial \psi}{\partial t} = \left\{ \sum_{i=1}^{2} \left[-\frac{\hbar^2}{2\mu_i} \frac{\partial^2}{\partial r_i^2} + \frac{1}{2\mu_i r_i^2} \left(p_{\theta_i}^2 + \frac{1}{\sin^2 \theta_i} p_{\phi_i}^2 \right) \right] + V(r_1, r_2; t) \right\} \psi$$

(3.175)

where the time-dependence of the potential comes from the time-dependence of the classical variables. Equation (3.175) is solved by a grid method: ψ is represented on a two-dimensional discrete grid for the quantum variables r_1 and

r_2. The initial wave function is taken as the product of two Morse-oscillator wave functions representing the vibrations of the two molecules,

$$\psi(r_1, r_2, t = 0) = \phi_{v_1}(r_1)\phi_{v_2}(r_2) \tag{3.176}$$

The effect of the kinetic energy operators on the wave function can be evaluated using a fast Fourier transform (FFT) method. The time propagation of the wave function can be obtained using, for example, the Lanczos [139] or split [140] propagator methods (see also [138]).

In addition to the quantum-classical hamiltonian, we obtain the classical equations of motion, which require the computation of the derivatives of the effective potential. The computation using eq. (3.173) is a very time-consuming procedure, so it is convenient to approximate it by:

$$V^{eff}(r_3, \theta_i, \phi_i) \approx V(\{r_1\}, \{r_2\}, r_3, \theta_i, \phi_i) \tag{3.177}$$

where $\{r_i\}$ is the average value of the quantum variable r_i, calculated with the same procedure as in eq. (3.174). Note that this approximation is employed only in the integration of the classical equations of motion.

The classical trajectories are integrated using, for example, a predictor corrector method, starting with random values for the classical variables, where the distance r_3 must be sufficiently large in order for the interaction potential to become negligible.

The accuracy of the integration procedure can be verified by checking the conservation of the norm of the wave function, of total energy, and of total angular momentum for non-reactive trajectories, while when the trajectory is reactive (or partly reactive) because of the presence of the absorbing potential, only the total angular momentum is conserved.

Every trajectory obtained in the dynamical propagation of the system can either be non-reactive, reactive or, unlike pure classical trajectory methods, only part of the wave packet can react. From every type of trajectory, we obtain information on the cross sections corresponding to reactive and to inelastic scattering. However, since only r_1 and r_2 are treated quantally, only initially state-selected or total cross sections and rate constants can be calculated with this method—detailed state-to-state information could be obtained by increasing the number of quantal degrees of freedom.

In order to calculate reactive cross sections, we analyse the flux along a dividing line in the product region, placed at the value r_2^*, where r_2 denotes the bond which breaks when the reaction takes place. Thus, we have

$$J(r_1, t; r_2^*) = \frac{\hbar}{\mu_2} \text{Im} \left(\psi(r_1, r_2, t)^* \frac{\partial \psi(r_1, r_2, t)}{\partial r_2} \right)\Bigg|_{r_2 = r_2^*} \tag{3.178}$$

The reactive components of the wave function are eliminated by means of an absorbing potential positioned on the last points of the grid in r_2. In this way, the reactive problem is converted to an essentially non-reactive one. Trajectories can be terminated when either the norm of the wave function remaining on the grid is negligible (holding for completely reactive trajectories) or the value of the distance between the two centers of mass r_3 is larger than a minimum distance. Typical values are 5–7 Å. The 2D quantum (r_1, r_2) and 7D classical $(r_3, \theta_i, \phi_i, i = 1, 3)$ calculation has been used to treat, for instance, the HO+CO→CO$_2$+H reaction.

Figure 3.8 shows the six bond distances between the atoms in the reaction. The two oxygen atoms have been labeled by (1) if in CO and (2) if in HO. The propagation time needed for each trajectory depends on the initial energy and can amount to a few hundred fs for completely non-reactive trajectories up to 2–3 ps (for low translational energies) for trajectories reacting slowly. Figure 3.9 shows the reaction probability as a function of time for the same trajectory. We notice that in a completely classical treatment of the collision, the contribution of this trajectory to the reaction probability would be zero. In the quantum-classical treatment it is about 0.011.

Initial state-selected reactive cross sections can be calculated by the expression:

$$\sigma_{v_{1i},v_{2i},j_{1i},j_{2i}} = 2\pi \int_0^{b_{max}} P^R_{v_{1i},v_{2i},j_{1i},j_{2i}} b\,db \qquad (3.179)$$

where the reaction probability $P^R_{v_{1i},v_{2i},j_{1i},j_{2i}}$ is obtained by integrating the flux $J(r_1, t; r_2^*)$ over time and over the remaining quantal variable, and where b is the

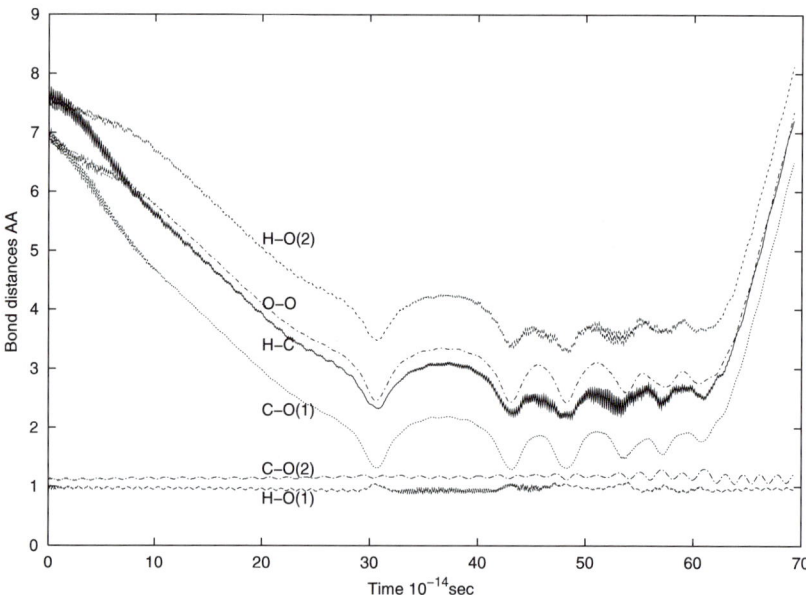

Fig. 3.8: The six bond distances as a function of time for the CO + OH collision. The initial kinetic energy is 13 kJ/mol, the rotational angular momenta are $j_1 = 15.7, j_2 = 2.3$, and the system is in its vibrational ground state $v_1, v_2 = 0, 0$. Although the trajectory is non-reactive, some quantum flux leaks to the reactive side $(r_2 > r_2^*)$, where $r_2^* = 2.9$ Å.

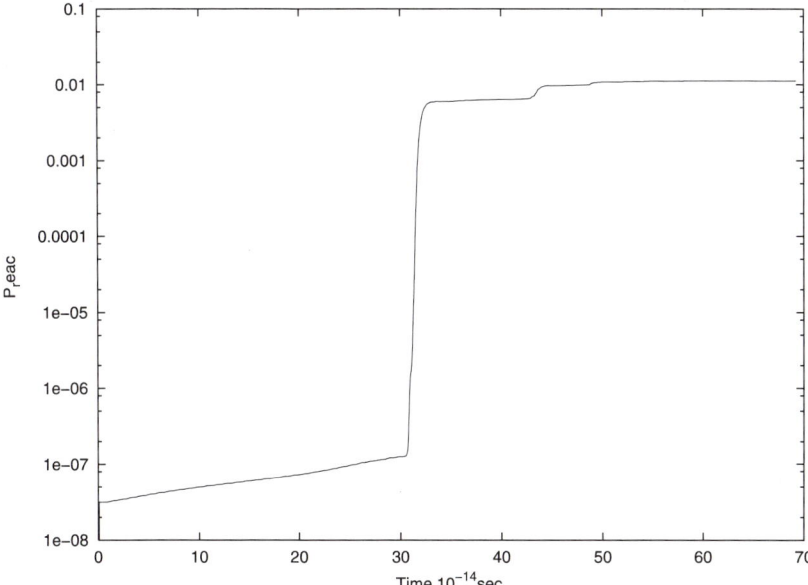

Fig. 3.9: The reaction flux for the trajectory shown in fig. 3.8. We notice an increase in the flux each time the C-O(1) distance hits a turning point.

impact parameter. For computing rates, it is more convenient to introduce the average cross section defined as:

$$\langle \sigma_{v_{1i},v_{2i}}(E_{cl}, T_0)\rangle = \frac{\pi\hbar^6}{8\mu_3(k_BT_0)^3 I_1 I_2}$$

$$\times \int_0^{l_{max}} dl(2l+1) \int_0^{j_{1max}} dj_1(2j_1+1)$$

$$\times \int_0^{j_{2max}} dj_2(2j_2+1)P^R_{v_{1i},v_{2i},j_{1i},j_{2i}} \qquad (3.180)$$

where k_B is the Boltzmann constant and T_0 is an arbitrary reference temperature (arbitrary because it cancels out when calculating the rate constants (see eq. 3.181)). T_0 has normally been taken as equal to 300 K; l_{max}, j_{1max}, and j_{2max} are the maximum values taken by l, j_1, and j_2 for a given total classical energy E_{cl}; I_1 and I_2 are the moments of inertia of the two reactants. The total classical energy E_{cl} is the sum of the kinetic energy coming from the contribution of translational and orbital motions plus the rotational energies of the diatoms. The integral in eq. (3.180) is evaluated with a standard Monte Carlo technique. From the cross section (3.180) we obtain the rate constant as:

$$k_{v_{1i},v_{2i}}(T) = \sqrt{\left(\frac{8k_BT}{\pi\mu_3}\right)}\left(\frac{T_0}{T}\right)^3 \int_0^\infty d(\beta E_{cl})e^{-\beta E_{cl}}\langle\sigma_{v_{1i},v_{2i}}(E_{cl}, T_0)\rangle \quad (3.181)$$

where $\beta = 1/(k_BT)$.

Note that from the wave function remaining on the grid after the collision, the probabilities for inelastic vibrational transitions could also be obtained by projecting the final wave function on the different asymptotic vibrational states:

$$P^{NR}_{v_{1i},v_{2i} \to v_{1f},v_{2f}} = \left| \int dr_1 \int dr_2 \psi(r_1, r_2, t) \phi_{v_{1f}}(r_1) \phi_{v_{2f}}(r_2) \right|^2 \qquad (3.182)$$

where the superscript NR stands for non-reactive. It is worth noting that if a sufficiently large number of final vibrational states is included, it is also possible to obtain the reaction probability from this procedure as:

$$P^R = 1 - \sum_{v_1, v_2} P^{NR}_{v_1, v_2} \qquad (3.183)$$

as long as other processes—for instance, collision induced dissociation—do not take place.

Figure 3.10 shows the rate constant for the reaction HO+CO→H+CO$_2$ obtained by calculating the average cross section at 10 values of the energy E_{cl} in the range 2.5 to 100 kJ/mol using 2–500 trajectories at each energy. The cross sections are compared with experimental data from [104]. We

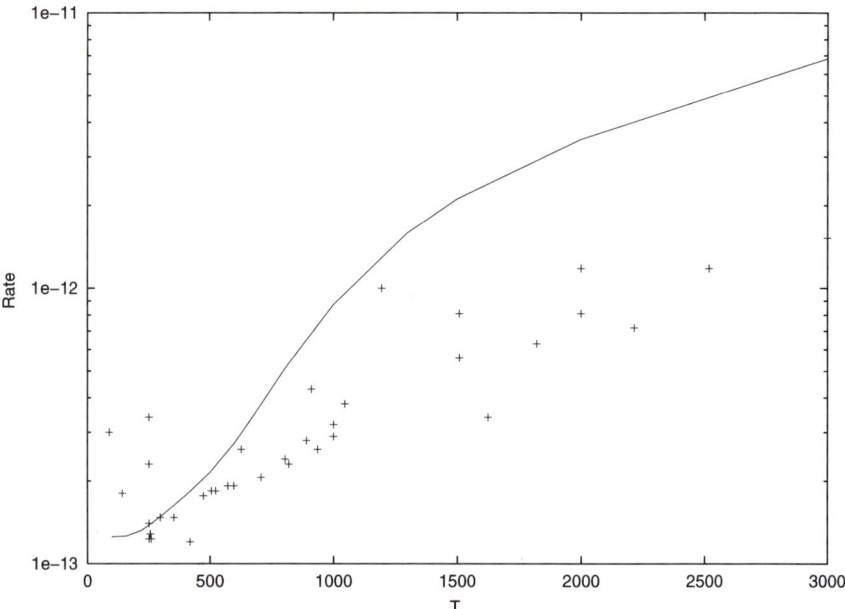

Fig. 3.10: The reaction rate constant k(T) for HO + CO → H + CO$_2$ as a function of temperature. The experimental data from [104] are indicated by (+), and the full line was obtained by a 2D quantum-classical model using a potential energy surface of Schatz et al. [106]. The rates have been multiplied with an electronic degeneracy factor of 0.5. The "inverse" temperature effect is due formation of collision complexes due to trapping in the attractive well.

notice that instead of using static grids in r_1, r_2, we can introduce the moving grid method as defined previously in the quantum dressed classical mechanics approach. This method has recently been used to treat the $H_2+CN \rightarrow HCN+H$ reaction [105]. In these calculations, the number of DVR points in the CN "spectator" bond could, in these calculations, be kept as small as 11.

3.2.7 Summary

In the last sections, we discussed the approximations behind the so-called classical path theory for inelastic and reactive scattering. The basic assumption is a separability assumption in which the degrees of freedom are divided in two groups: a classical group and a quantum group. The coupling between them is introduced as just the mean field limit. If the initial translational velocity is obtained so as to restore detailed balance. We have called the method "the symmetrized Ehrenfest approach" (see fig. 1.3). The method has been tested on a number of systems, varying the mass, energy range, and potentials. In general, good agreement has been obtained with either accurate calculations when available or with detailed experimental data. A representative selection of these comparisons has been given in this chapter.

It is possible to improve on the description by introducing a variational determination of the free parameters (as discussed in section 3.2). Thus, although the theory is approximate, the approximations are well understood and the theory can gradually be improved using the so-called Hermite corrected classical path approach [61, 125, 102]. In many cases of practical interest, it will be sufficiently accurate to use the CP method, and it has consequently been used to generate large tables of vibrationally resolved rate-constants. Such rates are crucial for the simulation of kinetic processes involving non-equilibrium vibrational distributions. As such, the method must be judged upon the results it has produced. To cite H. A. Lorentz:

In a theory which has given results like these, there must certainly be a great deal of the truth.

3.3 Non-adiabatic transitions

Non-adiabatic electronic transitions are important, not only in gas-phase chemistry, but also in condensed phases, liquids, and solids. An obvious approach here is to mix quantum and classical mechanics in such a way that the dynamics of the nuclei are treated classically whereas the electronic motion is quantized. A number of approaches have been formulated along these lines, and we shall mention several of these methods.

3.3.1 Pechukas theory

In 1969, Pechukas [108] published two papers dealing with the quantum-classical problem through the Feynman path-integral formulation of quantum dynamics [3]. We have already mentioned this formulation in chapter 2. The Feynman path-integral formulation of quantum dynamics has an appealing form, as far as the possibility of taking the classical limit is concerned. Thus, the classical limit $\hbar \to 0$ leads to the requirement that the variation of the action in the path integral

$$\int d[R(t)]\exp\left(\frac{i}{\hbar}S([R(t)])\right) \tag{3.184}$$

is set to zero $\delta S[R(t)] = 0$. This requirement gives the classical equations of motion. Pechukas [108] now considered the reduced propagator, defined as

$$K_{mn}(R_2, t_2; R_1, t_1) = \int d[R(t)]\exp\left(\frac{i}{\hbar}\int_{t_1}^{t_2} Ldt\right)\alpha_{mn}(R(t)) \tag{3.185}$$

where α_{mn} is the quantum mechanical amplitude for the r-system to go from state n to state m under the path $R(t)$. For non-adiabatic transitions, n and m denote two adiabatic electronic states. Note that in the path integral formulation, this path is any path which starts in R_1 at time t_1 and ends in R_2 at time t_2. L is the Lagrangian for the R-system

$$L = \frac{1}{2\mu}\dot{R}^2 - V_0(R) \tag{3.186}$$

The quantum states m are, as mentioned, adiabatic electronic states or, in an inelastic scattering problem, eigenstates of the hamiltonian H_0 for the isolated r-system, that is,

$$H_0\phi_m(r) = E_m\phi_m(r) \tag{3.187}$$

and the interaction potential is $V(R, r)$. By introducing

$$S = \int_{t_1}^{t_2} dt(L(R, \dot{R}, t) + \hbar \mathrm{Im}ln(\alpha_{mn}(R(t)))) \tag{3.188}$$

and defining the trajectory by the stationary phase condition, Pechukas obtained the following equation of motion for the path

$$\mu\ddot{R} = -\mathrm{Re}\left(\frac{\left\langle \psi_m(t_2, t)\left|\frac{\partial V}{\partial R}\right|\psi_n(t, t_1)\right\rangle}{\langle\psi_m(t_2, t)|\psi_n(t, t_1)\rangle}\right) \tag{3.189}$$

where the brackets denote integration over r, and where the wave function $\psi_n(t, t_1)$ evolves forward in time from t_1 to t with $\psi_n(t_1, t_1) = \phi_n$. $\psi_m(t_2, t)$, on the other hand, evolves backwards from t_2 to t with $\psi_m(t_2, t_2) = \phi_m$. Thus, the trajectory which determines the force on the quantum system is non-local, and it depends not only on the past and present (as in classical mechanics) but also upon future events. This problem makes it cumbersome to work with in practice, and convergence problems in the iterative scheme have been encountered [4, 8]. It has also been shown that occasions may occur where there is no solution (for real trajectories) and where there are many and sometimes even an infinite number of solutions also [9]. Also, interference effects between these solutions need to be accounted for. Aside from the fact that the trajectory has to be found by using iterative techniques [4, 8, 9, 109]—a technique which not always will converge—the propagator is not necessarily unitary when it is approximated with just the sum over the stationary phase terms. In order to obtain a unitary propagator, one needs to sum over all paths [110]. This problem is overcome in an approximate theory by Tully [111].

3.3.2 Tully's approach

As we have seen in the previous sections of this chapter, the quantum-classical theory is especially simple if the classical trajectories are governed by a mean field potential, constructed by the "Ehrenfest" average over the potential with the quantum wave function. However, if the system is such that it jumps from one potential energy surface to the other, it is expected that the mean field approximation is inadequate. The mean field approximation introduces, for a two-surface problem, the following effective Ehrenfest type potential

$$V_{eff} = |c_0|^2 W_{00}(R) + |c_1|^2 W_{11}(R) \qquad (3.190)$$

where W_{00} and W_{11} denote electronic adiabatic surfaces, R a nuclear coordinate, and c_i the quantum mechanical amplitude for being on surface i. The problems with mean field potentials have been discussed previously when treating inelastic collisions.

It is feasible to assume that the classical trajectory follows one or the other of the adiabatic surfaces and makes localized jumps from one to the other. At least it is in the spirit of classical mechanics to know what state the quantum system is in at a given time, and not just that it is in a given state with a certain probability.

This speculation led Tully and Preston to formulate the so-called surface hopping method in the seventies [114]. The method was actually introduced previously by Bjerre and Nikitin [115], who used it to study electronic quenching of excited sodium atoms in collisions with nitrogen molecules. In this theory, the classical trajectories are integrated using the adiabatic potential energy surfaces. At the avoided crossing or whenever the so-called Massey parameter

$$z_M = |\hbar \dot{\mathbf{R}} \cdot \mathbf{d}_{01}/(W_{11} - W_{00})| \qquad (3.191)$$

exceeds a critical value, the trajectories are allowed to jump to another potential energy surface. The Massey parameter [116] depends upon the velocity vector $\dot{\mathbf{R}}$, and the non-adiabatic coupling vector

$$\mathbf{d}_{01} = \langle \phi_0(\mathbf{r}; \mathbf{R}) | \nabla_{\mathbf{R}} \phi_1(\mathbf{r}; \mathbf{R}) \rangle \qquad (3.192)$$

where the bracket indicates integration over the electronic coordinates $\mathbf{r}$. However, whether the system jumps or not is decided by the transition probability, which often is estimated using simple formula taken from one-dimensional models. Such models are for instance the Landau-Zener model [112], or the improved

Nakamura models for non-adiabatic transitions [113]. A random number is generated and used to decide whether to make the jump or not. If the random number ξ is less than the transition probability, as, for instance, evaluated by the Landau-Zener expression

$$P_{LZ} = \exp(-2\pi\gamma/\hbar) \tag{3.193}$$

a transition is forced. The parameter γ is evaluated as

$$\gamma = \frac{1}{4}(W_{00} - W_{11})^2/|\dot{w}| \tag{3.194}$$

where the transition rate $\dot{w}$ in the simple one-dimensional L-Z model is given as the velocity v times the difference in slope of the diabatic curves in the crossing point, that is,

$$\dot{w} = v|F_1 - F_0| \tag{3.195}$$

where F_i $(i = 0, 1)$ denote the slopes. This expression can be generalized to the multi-dimensional situation as

$$\dot{w} = \sum_k \frac{\partial w}{\partial q_k}\dot{q}_k = \mathbf{n}\mathbf{M}^{-1}\mathbf{p}^{(2)} \tag{3.196}$$

where q_k denotes components of the nuclear position vector, $\mathbf{p}^{(2)}$ the momentum vector after the transition, $\mathbf{M}$ a (diagonal) matrix with the masses, and $\mathbf{n}$ a vector along which the surface of intersection behaves as a one-dimensional curve crossing [117]. After the jump, the trajectory continues on the upper(lower) surface according to classical mechanics. The momenta on the upper/lower surface are computed using the formula [117]

$$\mathbf{p}^{(2)} = \mathbf{p}^{(1)} - \mathbf{n}\frac{\mathbf{n}\mathbf{M}^{-1}\mathbf{p}^{(1)}}{\mathbf{n}\mathbf{M}^{-1}\mathbf{n}}\left(1 - \sqrt{1 - (W_{00} - W_{11})\frac{\mathbf{n}\mathbf{M}^{-1}\mathbf{n}}{(\mathbf{n}\mathbf{M}^{-1}\mathbf{p}^{(1)})^2}}\right) \tag{3.197}$$

where $\mathbf{p}^{(1)}$ denotes the momentum before the jump.

In a later model by Tully [118], the criterion for jumping is determined from the time-dependent expansion coefficients $c_j(t)$ in the expansion

$$\Psi(\mathbf{r}, \mathbf{R}, t) = \sum_j c_j(t)\phi_j(\mathbf{r}; \mathbf{R}) \tag{3.198}$$

of the total wave function in electronic basis functions ϕ_j. From the TDSE we obtain a set of coupled equations for the expansion coefficients

$$i\hbar\dot{c}_k(t) = \sum_j c_j(t)(V_{kj} - i\hbar\dot{\mathbf{R}} \cdot \mathbf{d}_{kj}) \tag{3.199}$$

where the last term represents the non-adiabatic coupling and where $V_{kj} = 0$ $(k \neq j)$ if the basis set refers to the adiabatic representation. Considering the density matrix elements $\rho_{jk} = c_j^* c_k$, we can obtain the values of $(d/dt)\rho_{kk}$, the change in population per time unit, from the above equation, and express it in terms of

$$\frac{d}{dt}\rho_{kk} = \sum_{j \neq k} b_{kj} \tag{3.200}$$

where $b_{kj} = -\dot{\rho}_{kj}$ is the rate for change from state k to j. A switch from state 0 to 1 is now made in a given time interval Δt if

$$\frac{\Delta t b_{01}}{\rho_{00}} > \xi \tag{3.201}$$

where ξ is a random number between zero and unity. The model may be called classical trajectories with quantum jumps. More recently, the model has been improved to the so-called "generalized surface hopping" method [123]. The trajectories are integrated using the force given as the mean field average over the electronic state ϕ_i, that is, from the Hellman-Feynmann theorem we obtain

$$\mu\ddot{\mathbf{R}} = -\langle\phi_i|\nabla_R H_0|\phi_i\rangle \tag{3.202}$$

where the brackets indicate integration over the electronic coordinates $\mathbf{r}$. The TDSE for the system is

$$i\hbar\frac{\partial\psi(\mathbf{r},\mathbf{R},t)}{\partial t} = H_0(\mathbf{r},\mathbf{R})\psi(\mathbf{r},\mathbf{R},t) \tag{3.203}$$

where, due to the classical treatment of $\mathbf{R}$, we have $\mathbf{R} = \mathbf{R}(t)$. As we have seen previously, the above equation is consistent with a narrow wave packet description of the nuclear motion and a product-type wave function. The correlation between the nuclear and electronic motion is, therefore, described approximately through the mean field description. In the generalized hopping method one expands the wave function as

$$\psi(\mathbf{r},\mathbf{R},t) = \sum_{j=1}^{N} c_j(t)\phi_j(\mathbf{r};\mathbf{R}) + \Phi(\mathbf{r},\mathbf{R},t) \tag{3.204}$$

where Φ denotes an additional wave function spanning the space not spanned by the N states ϕ_j. The state used to propagate the trajectory according to eq. (3.202) is either one of the functions ϕ_j or Φ. The mean field result is then obtained if $N = 0$ and the hopping method described above if $\Phi = 0$.

The hopping method obviously has some arbitrariness in it. One could argue that the system should switch in a continuous way from one surface to the other. This is actually fulfilled if the nuclear dynamics are allowed to follow the Pechukas force. Although problems with convergence of the Pechukas scheme have been noticed [4, 9, 8] the problem diminishes if the switch is made in a small time interval Δt, that is, the force is well defined in this time limit. Thus, this technique has been used, for instance, by Rossky and coworkers [109]. The advantage of the approach outlined here is that ordinary classical mechanics can be assumed for the bulk part of the system, and quantum mechanics introduced in those few degrees of freedom involved in the non-adiabatic (non-classical) events. The method is, therefore, a molecular dynamics method with localized quantum transitions (MDQT) and can be used to study quantum processes, as, for instance, electronic transitions or vibrational predissociation [124], as well as proton transfer [119] in solution and clusters. Methods which combine the use of a mean field or Ehrenfest potential for defining the trajectory and the surface hopping techniques have also been considered recently, see refs. [120, 121].

3.3.3 Spawning method

The so-called spawning method developed by Levine and coworkers [33] is essentially a generalization of the minimum error method mentioned in chapter 2 to multi-surface problems. In the spawning methods, more basis functions are added (spawned) when the dynamical coupling makes it necessary. The method uses the fact that for non-adiabatic transitions, the quantum process is localized in time and space to the region where the coupling between the electronic surfaces is strong. The underlying basis set used in this method is a set of frozen Gaussian wave packets propagated by classical trajectories. Thus, the initial nuclear wave function is expanded in a set of these wave packets. This expansion defines the initial position and momenta for the trajectories, which are used to propagate the wave packet in time and space. Hence, classical dynamics is used as a guide for the quantum basis set selection and truncation.

Aside from the classical trajectories, one has to solve equations for the time-dependent expansion coefficients. For a two-surface problem, we have

$$\Psi = \begin{bmatrix} \xi_1(\mathbf{R}, t) \\ \xi_2(\mathbf{R}, t) \end{bmatrix} \tag{3.205}$$

where the nuclear wave functions, as mentioned, are expanded in traveling Gaussians

$$\xi_l(\mathbf{R}, t) = \sum_k d_{lk}(t) \Phi_k^l(\mathbf{R}, \mathbf{R}_k^l(t), \mathbf{P}_k^l(t), t) \tag{3.206}$$

$\mathbf{R}_k^l(t)$ denotes a classical trajectory on surface l for nuclear mode k, and $\mathbf{P}_k^l(t)$ is a momentum parameter associated with the GWP. For multi-dimensional problems, the nuclear wave function is taken as a product between Gaussian wave packets in each dimension. The classical mechanical equations of motion for the trajectories and momenta are solved and used to propagate the wave packets on each electronic surface.

Transitions among the electronic states are induced by the non-adiabatic coupling elements and the transition matrices between the Gaussian basis sets on each surface. For a two-surface problem, the method yields the following set of equations for the expansion coefficients:

$$i\hbar \begin{bmatrix} \mathbf{S}_{11} & 0 \\ 0 & \mathbf{S}_{22} \end{bmatrix} \begin{bmatrix} \dot{\mathbf{d}}_1(t) \\ \dot{\mathbf{d}}_2(t) \end{bmatrix} = \begin{bmatrix} \mathbf{H}_{11} & \mathbf{H}_{12} \\ \mathbf{H}_{21} & \mathbf{H}_{22} \end{bmatrix} \begin{bmatrix} \mathbf{d}_1(t) \\ \mathbf{d}_2(t) \end{bmatrix}$$

$$- i\hbar \begin{bmatrix} \dot{\mathbf{S}}_{11} & 0 \\ 0 & \dot{\mathbf{S}}_{22} \end{bmatrix} \begin{bmatrix} \mathbf{d}_1(t) \\ \mathbf{d}_2(t) \end{bmatrix} \tag{3.207}$$

where the overlap matrices S_{11} and S_{22} contains matrix elements between the Gaussian wave packets on each potential surface, and where $d_i(t)$ is a vector of dimension equal to the number of wave packets included in the propagation. Coupling between the electronic states occurs in the diabatic representation through the matrix H_{12} with elements

$$H_{12}^{ij} = \langle \Phi_i^1 | \hat{H}_{el} | \Phi_j^2 \rangle \qquad (3.208)$$

where the brackets indicate integration over electronic and nuclear coordinates.

The spawning process adds more Gaussians when needed, that is, near an avoided crossing. Each new basis function needs an initialization of the momentum and position parameters at the time when they are created. For a discussion of the various possibilities, see, for example, ref. [35]. In order to decide whether to extend the basis set, the magnitude of the non-adiabatic coupling term for each term in the coupling matrix is estimated, and if it exceeds a certain value, additional basis functions are added in the region of non-adiabatic coupling. The so-called FMS (full multiple spawning), in which enough basis functions are used to account for coupling also away from the "localized" inter-state couplings, is in principle exact—but has, as the TDGH-DVR method, mentioned previously an option for treating a large fraction of the system approximately. In the IMS (independent multiple spawning) method, the Gaussians are propagated independently of each other by classical trajectories. Recently, the method has been extended by Ben-Nun and Martinez [34] to allow for spawning in connection with tunneling processes on a single potential energy surface.

3.3.4 The multi-trajectory and TDGH-DVR methods

As discussed in the previous sections, the problem with combining classical trajectory methods with quantum state expansion techniques has to do with the splitting of the trajectory. In the usual Born-Oppenheimer picture, the nuclear motion is governed by a given adiabatic potential energy surface. The information needed for classical mechanics to be uniquely defined is the specification of what surface the system is on. If the system evolves on more than one surface, we have a problem. In the mean field method, this problem is solved (or ignored) by letting the motion be governed by an average potential surface with quantum amplitudes defining how much the surfaces contribute. In the surface hopping method, the problem is solved by simply letting the trajectory switch from one surface to the other, that is, a given trajectory is propagated on one specific surface at a given time. All of these problems are, in fact, created by the insistence on taking a classical limit for some degrees of freedom. As soon as this limit is taken, the two parts do not any more "speak the same language" and there is no unique solution to the problem. In order to invoke a trajectory concept in a manner which in principle does not introduce any assumptions, we can use the FMS approach or the trajectory driven Gauss-Hermite basis set. However, since each electronic state requires a separate basis set, it is natural to let these basis sets evolve according to their own dynamics, that is, to incorporate two "trajectories," one for each surface (assuming just a two-surface problem). Thus for a two-surface 1D problem, we would have

$$i\hbar \frac{\partial}{\partial t} \begin{bmatrix} \psi_1(x,t) \\ \psi_2(x,t) \end{bmatrix} = \begin{bmatrix} V_{11}(x) & V_{12}(x) \\ V_{21}(x) & V_{22}(x) \end{bmatrix} \begin{bmatrix} \psi_1(x,t) \\ \psi_2(x,t) \end{bmatrix} \qquad (3.209)$$

where the wave functions are expanded as

$$\psi_l(x,t) = \sum_n c_{nl}(t)\Phi_{nl}(x,t) \qquad (3.210)$$

and where the G-H basis functions are centered around trajectories $x_l(t)$, that is,

$$\Phi_{nl}(x,t) = \exp\left(\frac{i}{\hbar}(\gamma_l(t) + p_l(t)(x - x_l(t)) + \mathrm{Re}A_l(t)(x - x_l(t))^2)\phi_n(\xi_l(t))\right) \qquad (3.211)$$

where

$$\phi_n(\xi_l(t)) = \frac{1}{\sqrt{n!2^n}} \exp(-\xi_l(t)^2/2)H_n(\xi_l(t)) \qquad (3.212)$$

$$\xi_l(t) = \sqrt{2\mathrm{Im}A_l(t)/\hbar}(x - x_l(t)) \qquad (3.213)$$

If we insert this expansion in the above TDSE for the problem, we obtain a set of coupled equations in the expansion coefficients c_{nl} and equations of motion for the trajectories $x_l(t)$. The coupling between the two surfaces is taken care of through the equations for c_{nl}, and the trajectories are propagated by their own effective potentials, that is, any coupling between the trajectories will come about through the dependence of the effective potential on c_{nl}. Thus, the trajectories need not jump or switch from one surface to the other. Each surface has its own trajectory [125]. Multi-trajectory methods have also been considered on less formal grounds in work by Sidis et al. and by Wang et al. [126], [127] using an MCTDSCF approach. In ref. [126], one trajectory was introduced for each vibrational manifold, as well as for intermanifold coupling terms.

The disadvantage of the method developed here, and also of the FMS method mentioned in the previous paragraph, is that integrals over the basis functions have to be evaluated. Usually rather approximate or crude methods have to be invoked in order not to be prohibitively expensive. But the problem can be avoided by switching to a DVR representation. This is possible through an extension to two or more surface problems of the TDGH-DVR method mentioned in section 2.8. In order to solve the problem, we have to ensure that the moving grid points are the same on each surface. In the case of the DVR representation of the problem, we then have

$$i\hbar \begin{pmatrix} \dot{\mathbf{d}}^{(1)} \\ \dot{\mathbf{d}}^{(2)} \end{pmatrix} = \begin{pmatrix} \mathbf{W}^{(11)} & \mathbf{W}^{(12)} \\ \mathbf{W}^{(21)} & \mathbf{W}^{(22)} \end{pmatrix} \begin{pmatrix} \mathbf{d}^{(1)} \\ \mathbf{d}^{(2)} \end{pmatrix}$$
$$+ \begin{pmatrix} \hbar\mathrm{Im}A_1/m\mathbf{T} & \mathbf{0} \\ \mathbf{0} & \hbar\mathrm{Im}A_1/m\mathbf{T} \end{pmatrix} \begin{pmatrix} \mathbf{d}^{(1)} \\ \mathbf{d}^{(2)} \end{pmatrix} \qquad (3.214)$$

where the kinetic energy matrices $\mathbf{T}$ couple elements within a given mode (here just a single mode) and the potential coupling matrices $\mathbf{W}^{(ij)}$ are diagonal in the grid representation. Thus the two states are coupled through the potential coupling by a 2×2 matrix at each grid point, provided that the grid points on surface 1 and 2 are the same! This can be obtained by letting the trajectory which carries the DVR points propagate on one of the surfaces only. Which surface should be chosen is determined by the one which has the turning point furthest in the scattering region. In principle, it does not matter which one is chosen, but if a surface with turning point for the trajectory far from the coupling region between the surfaces is chosen, we would need more grid points to cover the relevant space. Thus, the effective potentials W are defined by

$$W_{ij;i'j'}^{(nn)} = V_{ij;i'j'}^{(nn)} \delta_{ii'}\delta_{jj'} - V'_{eff}(x_i - x_n(t)) - \tfrac{1}{2}V''_{eff}(x_i - x_n(t))^2 \qquad (3.215)$$

where $x_n(t)$ is the trajectory evolving on the surface of choice, surface n. $W_{ij;i'j'}^{nm} = V_{ij;i'j'}^{(nm)}$ for $n \neq m$. Here the matrix, $V_{ij;i'j'}$ refers to diabatic potential surfaces. We notice that W is not an adiabatic surface but here denotes the diabatic reference potential as defined for the TDGH-DVR method. The trajectory is driven by the effective force which could, for instance, be taken as the derivative of the adiabatic lower potential. In the fixed-width approach we have $V''_{eff} = 4\mathrm{Im}A_n^2/m$. Figure 3.11 shows the adiabatic potential energy surfaces for the so-called dual avoided crossing model. In fig. 3.12 we have shown the result obtained by defining the moving grid points around a trajectory propagated on the lower adiabatic surface. Since the grid points follow the trajectory, we need to propagate just 20–30 grid points along a trajectory starting at -4 Å and ending at $+4$ Å on the lower adiabatic surface (see fig. 3.11). A Gaussian wave packet is used to initialize the coefficient $d_i^{(1)}(t_0)$ on the lower surface

$$d_i^{(1)}(t_0) = \phi_0(z_i)/\sqrt{A^{(i)}} \qquad (3.216)$$

where $A^{(i)} = \sum_{n=0}^{N-1} \phi_n(z_i)^2$ and where N is the number of grid points. The initial width parameter $\mathrm{Im}A(t_0)$ is set to 0.5 amu/τ ($\tau = 10^{-14}$ sec.) and the amplitudes $d_i^{(2)}(t_0)$ on the upper surface are initialized to zero. The classical trajectory $x(t)$ is integrated according to the equations

$$\dot{x}(t) = p_x(t)/m \qquad (3.217)$$

$$\dot{p}_x(t) = -\frac{\partial}{\partial x}V_{11}^{ad} \qquad (3.218)$$

where V_{11}^{ad} denotes the lower adiabatic surface. The above equations for the coefficients $d_i^{(k)}(t)$ are integrated using the split Lanczos method mentioned previously (section 2.8). When the trajectory has reached the asymptotic region,

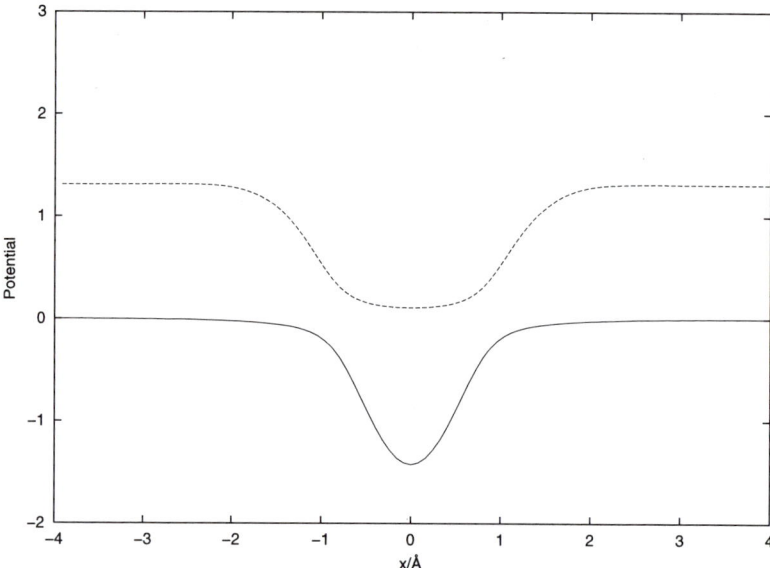

Fig. 3.11: The two adiabatic potentials for the dual avoided crossing model [118]. $V_{11} = 0, V_{22} = C\exp(-Dx^2)$ and $V_{22} = E_0 - A\exp(-Bx^2)$, where $E_0 = 1.3128$, $A = 2.6255$, $C = 0.2143$, and $B = 1\text{ Å}^{-2}$, $D = 0.2143\text{ Å}^{-2}$. Energy unit is 100 kJ/mol.

the wave packet is projected on plane outgoing waves on the two surfaces, that is, the S-matrix elements are obtained from

$$\Psi^{(1)}(x, t_0) = \exp\left(\frac{i}{\hbar}(p_x(t_0)(x - x(t_0))(2\text{Im}A(t_0)/\pi\hbar)^{1/4}\right.$$

$$\left. \times \exp(-\text{Im}A(t_0)(x - x(t_0))^2/\hbar)\right)$$

$$= \sum_{k_0} c_{k_0}\exp(ik_0 x) \tag{3.219}$$

$$\Psi^{(j)}(x, t) = \exp\left(\frac{i}{\hbar}p_x(t)(x - x(t))\right)\sum_i d_i^{(j)}(t)\phi_i(x, t)$$

$$= \sum_k S_{00}\exp(ikx) + S_{01}\sqrt{k/k_1}\exp(ik_1 x) \tag{3.220}$$

where the $j = 1, 2$ and where the connection between the initial wave number k_0 and the final is

$$E = \frac{\hbar^2 k_0^2}{2m} = \frac{\hbar^2 k^2}{2m} = \frac{\hbar^2 k_1^2}{2m} + E_0 \tag{3.221}$$

E_0 is the asymptotic energy difference between the two surfaces. Figure 3.12 shows $|S_{00}|^2$ for transmission on the lower surface and $|S_{01}|^2$ for transmission on the upper.

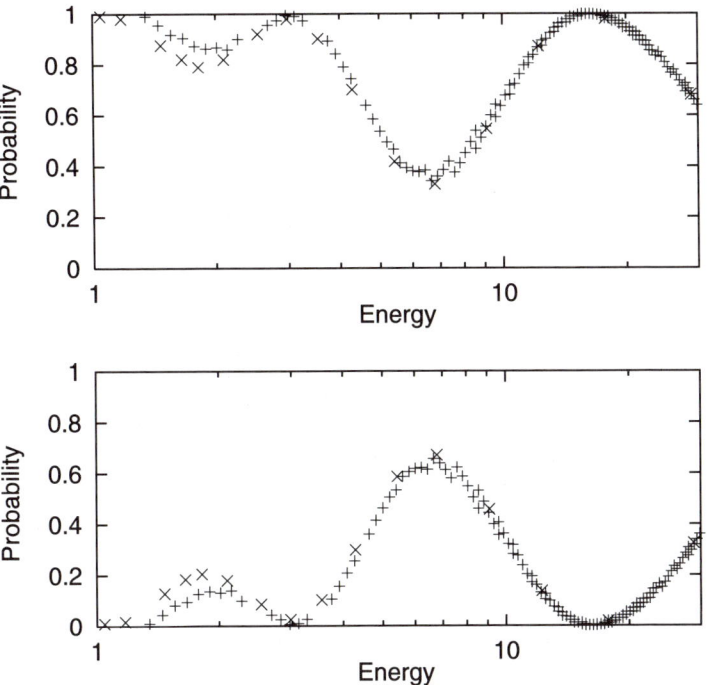

Fig. 3.12: Comparison of probabilities obtained using an stationary grid method ($\times$) and numbers obtained using the TDGH-DVR method in which the grid points follow a trajectory propagated on the lower adiabatic basis.

The average probability can be obtained in a much more simple manner, namely by "grid summation," that is,

$$\langle P_{in} \rangle = \int dk_0 \frac{\text{Incoming flux}}{k_0} = \int dk_0 c_{k_0}^2 = 1 \tag{3.222}$$

and likewise, for the outgoing average probability

$$\langle P_{out} \rangle = \langle P_{0n}(k_0) \rangle = \int dk \frac{\text{Outgoing flux}}{k}$$

$$= \int dk |S_{0n}|^2 f(k, k_0)(k_0/k) = \sum_i |d_i^{(n+1)}|^2 \tag{3.223}$$

where the momentum distribution is

$$f(k, k_0) = |c_{k_0}|^2 = \Delta x \sqrt{2/\pi} \exp(-2(\Delta x(k - k_0))^2) \tag{3.224}$$

with the width of the distribution $\Delta x = 0.5\sqrt{\hbar/\text{Im}A(t_0)}$. We have, furthermore, used the fact that the initial and final wave functions are superpositions of plane waves, that is, for the initial we have

$$\Psi = \sum_{k_0} c_{k_0} \exp(ik_0 x) \tag{3.225}$$

and for the final wave function

$$\sum_{k_0} c_{k_0} \sqrt{k_0/k} \exp(ikx) S_{0k} \qquad (3.226)$$

In ref. [122] it was shown that the average probability was in good agreement with the exact numbers for the model systems studied by Tully [118]. The average probabilities could be obtained with just 2–5 grid points on each surface, whereas the energy resolution required more grid points (about 30). Hence, it could be important to be able to obtain the exact probability by an inversion technique, that is, from known values of $\langle P_{0n}(k_0) \rangle$ to obtain $P_{0n} = |S_{0n}|^2$. Figure 3.13 shows the squared amplitudes $|d_i(t)|^2$ as a function of time, using just three grid points on each surface. The average kinetic energy is 181.6 kJ/mol. The probability for non-adiabatic transitions is well determined by this number of grid points, except in the threshold region, where a projection on plane waves is necessary. The probability build up on the upper surface at the first avoided crossing is depleted when the next is reached (see fig. 3.13). At this energy, the probability for excitation to the upper surface is 0.1961 (the exact number is 0.205 [118]). See also table 3.11 for a more extensive comparison.

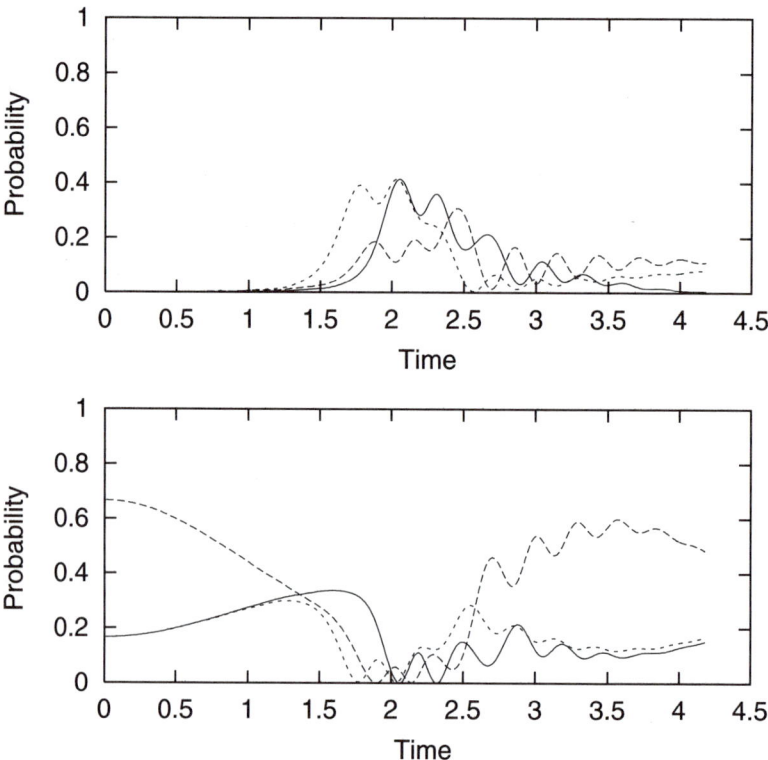

Fig. 3.13:   The amplitudes $|d_i(t)|^2$ for grid points on the lower surface (lower panel) and upper surface (upper panel) for the dual avoided crossing model. Three grid points are used, giving $P_{01} = 0.1961$. The exact number from [118] is 0.205.

Table 3.11: Average probabilities $\langle P_{01}(k_0) \rangle$ for transmission on the upper surface for a simple (system 1) and a dual (system 2) avoided crossing model. The exact numbers are obtained from the graphic representations in ref. [118]. The number of grid points on each surface is $N = 2, 3$ and 5. The initial width parameter $\alpha_0 = 0.5$ amu/τ and the trajectory governing the grid points is propagated on the lower adiabatic surface.

	System 1					System 2			
$k_0/\text{Å}^{-1}$	$N = 2$	$N = 3$	$N = 5$	"Exact"	$k_0/\text{Å}^{-1}$	$N = 2$	$N = 3$	$N = 5$	"Exact"
18.27	0.154	0.153	0.183	0.15	29.89	0.183	0.217	0.207	0.18
21.16	0.200	0.200	0.224	0.20	33.62	0.089	0.152	0.191	0.18
30.24	0.363	0.374	0.369	0.36	48.13	0.355	0.346	0.313	0.30
42.52	0.564	0.574	0.566	0.56	98.06	0.020	0.022	0.022	0.018
60.47	0.743	0.747	0.743	0.77	124.8	0.326	0.325	0.323	0.33

Chapter 4

Second quantization

Second quantization (SQ) concepts were introduced in chapter 2 as a general tool to treat excitations in molecular collisions for which the dynamics were described in cartesian coordinates. This SQ-formulation, which was derived from the TDGH representation of the wave function, could be introduced if the potential was expanded locally to second order around the position defined by a trajectory. It is, however, possible to use the SQ approach in a number of other dynamical situations, as for instance when dealing with the vibrational excitation of diatomic and polyatomic molecules, or with energy transfer to solids and chemical reactions in the socalled reaction path formulation (see table 4.1).

Table 4.1: Dynamical problems in which an SQ formulation is convenient. The first- and second-order forces are defined as the terms with linear or quadratic boson operators.

Situation	Linear forces	Quadratic forces
SQ-version of molecular dynamics in the Gauss-Hermite expansion	Difference between classical and quantum forces.	Local off-diagonal second derivatives of potential.
Energy transfer to molecules	First derivatives of potential. Centrifugal stretch coupling.	Second derivatives of potential. Coriolis coupling.
External fields (classical field)	Dipole coupling term.	
Energy transfer to solids	First derivative of molecule-solid interaction.	Electron-hole pair excitation[a].
Reaction path treatment of chemical reactions	Linear coupling of reaction path motion to perpendicular vibrations. Rot-vib coupling.	Quadratic coupling of reaction path motion to vibrations. Coriolis coupling.

[a]Fermion operators.

Since the formal expressions in the operators are the same, irrespective of the system or dynamical situation, the algebraic manipulations are also identical, and, hence, the formal solution the same. But the dynamical input to the scheme is of course different from case to case (see table 4.1).

In the second quantization formulation of the dynamical problems, one solves the operator algebraic equations formally. Once the formal solution is obtained, we can compute the dynamical quantities which enter the expressions. The advantage over state or grid expansion methods is significant since (at least for bosons) the number of dynamical operators is much less than the number of states. In order to solve the problem to infinite order, that is, also the TDSE for the system, the operators have to form a closed set with respect to commutations. This makes it necessary to drop some two-quantum operators (see below). Historically, the $M = 1$ quantum problem, namely that of a linearly forced harmonic oscillator, was solved using the operator algebraic approach by Pechukas and Light in 1966 [131]. In 1972, Kelley [128] solved the two-oscillator ($M = 2$) problem and the author solved the $M = 3$ and the general problem in 1978 [129] and 1980 [147], respectively. The general case was solved using graph theory designed for the problem and it will not be repeated here. But the formulas are given in this chapter and in the appendices B and C.

4.1   Diatom-diatom collisions

In order to illustrate the SQ approach, we consider the collision of two diatomic molecules in the classical path framework, that is, the rotational motion of both diatoms and the center-of-mass motion are treated classically. The vibrational motion of both molecules will be treated quantally. The intermolecular potential is expanded around the oscillator equilibrium distances r_i^{eq} as

$$V_{int} = V_0(R, c1, c2, sc1, sc2, ss1, ss2) + \sum_{i=1,2} \left[V_i^{(1)}(r_i - r_i^{eq}) \right.$$
$$\left. + \frac{1}{2} V_{ii}^{(2)}(r_i - r_i^{eq})^2 \right] + V_{12}^{(2)}(r_1 - r_1^{eq})(r_2 - r_2^{eq}) \tag{4.1}$$

where the notation from section 3.2.5 has been adopted. We now introduce the creation and annihilation operators, that is,

$$r_k - r_k^{(eq)} = b_k(a_k + a_k^+) \tag{4.2}$$
$$\hat{p}_k = i\sqrt{\hbar\omega_k m_k/2}(a_k^+ - a_k) \tag{4.3}$$

where $b_k = \sqrt{\hbar/2m_k\omega_k}$. In the SQ formulation we then have for the hamiltonian of the isolated molecules

$$H_0 = \sum_{k=1}^{2} \hbar\omega_k \left(a_k^+ a_k + \tfrac{1}{2}\right) \tag{4.4}$$

and, hence, the quantum part of the hamiltonian can be expressed as

$$H_q = \sum_k \hbar\tilde{\omega}_k \left(a_k^+ a_k + \tfrac{1}{2}\right) + \sum_k f_k(a_k^+ + a_k) + f_{12}(a_1^+ a_2 + a_1 a_2^+) \tag{4.5}$$

where the linear force is (see eq. (3.109))

$$f_k = b_k \left(V_k^{(1)} + \frac{2i}{m_k (r_k^{(eq)})^2 \omega_k} j_k \frac{dj_k}{dt} \right) \tag{4.6}$$

The last term is the contribution to the linear force arising from coupling between rotation and vibration ("the centrifugal stretch coupling"). The quantities $j_k(t)$ are the classical rotational angular momenta of the two molecules.

The modulated frequency is defined by

$$\tilde{\omega}_k = \omega_k + \frac{1}{2m_k \omega_k} V_{kk}^{(2)} \tag{4.7}$$

and the quadratic force as

$$f_{12} = b_1 b_2 V_{12}^{(2)} \tag{4.8}$$

where the double quantum operators a_k^2 and $(a_k^+)^2$ have been neglected and $[a_k, a_k^+] = 1$ has been used. If the double quantum operators are omitted, we can solve the operator algebra formally.

However, this does not imply that double quantum transitions are also neglected, but that they occur through successive use of single quantum operators (a process, which is included in an infinite order solution of the problem).

The evolution operator under the hamiltonian H_q is $U(t, t_0)$, such that

$$i\hbar \frac{d}{dt} U(t, t_0) = H_q U(t, t_0) \tag{4.9}$$

It is now convenient to switch to an interaction representation, that is, to express $U(t, t_0)$ as

$$U(t, t_0) = \exp \left(i \int \tilde{H}_0 dt / \hbar \right) U_I(t, t_0) \tag{4.10}$$

where $\tilde{H}_0 = \sum_k \hbar \tilde{\omega}_k (a_k^+ a_k + 1/2)$. The equation, which has to be solved for U_I is

$$i\hbar \frac{d}{dt} U_I(t, t_0) = \left[\sum_k (F_k^+ a_k^+ + F_k^- a_k) + F_{12}^+ a_1^+ a_2 + F_{12}^- a_1 a_2^+ \right] U_I(t, t_0) \tag{4.11}$$

where

$$F_k^{\pm} = \exp(\pm i\theta_k(t)) f_k \tag{4.12}$$
$$F_{12}^{\pm} = b_1 b_2 V_{12} \exp(\pm i(\theta_1 - \theta_2)) \tag{4.13}$$

and

$$\theta_k = \int dt \tilde{\omega}_k(t) \tag{4.14}$$

The linear terms in the operators are usually denoted VT terms, and the quadratic ones as VV terms. This notation comes from the early treatment of energy transfer in diatomic molecules using this approach [128]. Thus VT denotes vibration-translation and VV vibration-vibration coupling terms. The VV operators conserve the number of vibrational quanta, whereas the VT operators create or destroy vibrational quanta.

4.2 General hamiltonians

The concept developed above can now be extended to arbitrarily sized systems relevant for treating energy transfer in polyatomic molecules and solids (see chapter 5). However, as we have seen previously, the operators may also be introduced in a cartesian coordinate formulation of molecular dynamics, and in this case the operators denote excitations in local harmonic oscillators with time-dependent frequencies. The general hamiltonian for which we may solve the time-dependent Schrödinger equation algebraically is of the type

$$H_q = \sum_{k=1}^{M} \hbar\omega_k \left(a_k^+ a_k + \frac{1}{2} \right) + \sum_{k=1}^{M} f_k(a_k^+ + a_k) + \sum_{kl} f_{kl} a_k^+ a_l \tag{4.15}$$

where a_k are Boson or Fermion operators, and where f_k are linear and f_{kl} quadratic forces. We notice immediately that the diagonal part of the last term can be introduced in the first by defining the so-called modulated frequencies

$$\tilde{\omega}_k = \omega_k + f_{kk}(t) \tag{4.16}$$

The first term can be taken care of by introducing the "interaction" representation as illustrated above (see eq. (4.10)). This leaves us with the equation

$$i\hbar\frac{d}{dt}U_I(t,t_0) = \tilde{H}_q U_I(t,t_0) \tag{4.17}$$

where

$$\tilde{H}_q = \sum_{k}(F_k^+ a_k^+ + F_k^- a_k) + \sum_{k<l}(F_{kl}^+ a_k^+ a_l + F_{kl}^- a_k a_l^+) \tag{4.18}$$

and

$$F_k^+ = f_k \exp(i\theta_k(t)) \tag{4.19}$$

$$F_{kl}^+ = f_{kl} \exp[i(\theta_k(t) - \theta_l(t))] \tag{4.20}$$

with $F_k^- = (F_k^+)^*$, $F_{kl}^- = (F_{kl}^+)^*$.

In order to solve the TDSE for the evolution operator, we can conveniently split it in a product form as

$$U_I(t,t_0) = U_{VV} U_{VT} \tag{4.21}$$

and, hence, the eq. (4.17) can be divided in two equations

$$i\hbar\frac{d}{dt}U_{VV} = H_{VV} U_{VV} \tag{4.22}$$

and

$$i\hbar\frac{d}{dt}U_{VT} = [U_{VV}^+ H_{VT} U_{VV}]U_{VT} \tag{4.23}$$

where

$$H_{VT} = \sum_{k}(F_k^+ a_k^+ + F_k^- a_k) \tag{4.24}$$

and as the generalization of the simple diatom-diatom case above we have

$$H_{VV} = \sum_{k<l}(F_{kl}^+ a_k^+ a_l + F_{kl}^- a_k a_l^+). \tag{4.25}$$

The two equations for the VT and the VV problem may now be solved formally to give [147]

$$U_{VT} = \prod_{k=1}^{M} U_{VT}^{(k)} \tag{4.26}$$

where

$$U_{VT}^{(k)} = \exp(i\beta_k)\exp(i(\alpha_k^+ a_k^+ + \alpha_k^- a_k)) \tag{4.27}$$

and

$$\alpha_k^{\pm}(t, t_0) = -\frac{1}{\hbar}\int_{t_0}^{t} dt' W_k^{\pm}(t') \tag{4.28}$$

$$\beta_k(t, t_0) = -\frac{i}{\hbar^2}\int_{t_0}^{t} dt'' \int_{t_0}^{t''} dt'[W_k^-(t')W_k^+(t'') - W_k^+(t')W_k^-(t'')] \tag{4.29}$$

We notice that β_k is real and, hence, enters the evolution operator just as a phase factor. The quantities $W_k^{\pm}$ are related to the linear forces through the equations

$$W_k^+ = F_k^+ + \sum_{j=1}^{M} F_j^+ Q_{jk}^* \tag{4.30}$$

$$W_k^- = F_k^- + \sum_{j=1}^{M} F_j^- Q_{jk} \tag{4.31}$$

where $*$ denotes a complex conjugate, and where the functions Q_{jk} are defined through

$$Q_{jk} = R_{jk} - \delta_{jk} \tag{4.32}$$

In order to determine the functions Q_{ij}, we need to solve a set of equations of the type

$$i\hbar\frac{d}{dt}\mathbf{R} = \mathbf{BR} \tag{4.33}$$

which is an $M \times M$ matrix equation with the elements of the hermitian matrix $\mathbf{B}$ given as $B_{kl} = F_{kl}^+$ for $k > l$ and $B_{lk} = F_{kl}^-$. The evolution operator for the VV-problem can also be obtained

$$U_{VV} = \prod_{i=1}^{M}\exp(\alpha_{ii}O_{ii})\prod_{i=1}^{M}\prod_{j\neq i}\exp(\alpha_{ji}O_{ji}) \tag{4.34}$$

where we have introduced the operator notation

$$O_{ij} = a_i^+ a_j \tag{4.35}$$

and some new functions α_{ij}. These latter functions are also related to the solution of equation (4.33) by solving the so-called "inversion problem" (see appendix C). It is now possible to obtain transition amplitudes for a quantum transition $\{m\}$ to $\{n\}$ where

$$|\{m\}\rangle = |m_1 m_2, \dots, m_M\rangle = |m_1\rangle|m_2\rangle, \dots, |m_M\rangle \tag{4.36}$$

Thus, we have

$$\langle\{n\}|U_I|\{m\}\rangle = \sum_{\{k\}} \langle\{n\}|U_{VV}|\{k\}\rangle\langle\{k\}|U_{VT}|\{m\}\rangle \tag{4.37}$$

where the last term can be expressed as

$$\langle\{k\}|U_{VT}|\{m\}\rangle = \prod_{j=1}^{M} \left\langle k_j|U_{VT}^{(j)}|m_j\right\rangle \tag{4.38}$$

The matrix elements over the VT operator can easily be evaluated [131, 138]. The result is

$$\left\langle k_j|U_{VT}^{(j)}|m_j\right\rangle = \exp\left(i\beta_j - \frac{1}{2}\rho_j\right)\sqrt{k_j!/m_j!}$$
$$\times (i\alpha_j^+)^{k_j-m_j} \sum_{p=0}^{m_j} \frac{(-1)^p}{p!} \frac{\rho_j^p}{(k_j - m_j + p)!(m_j - p)!} \tag{4.39}$$

where $\rho_j = \alpha_j^+\alpha_j^-$ and the summation is carried out over terms with non-negative factorials. The VT problem constitutes the solution for the linearly forced harmonic oscillator and, as mentioned, it was solved using operator algebra by Pechukas and Light [131]. However, the solution was obtained previously by Kerner and Treanor [132], who identified the value of $\hbar\omega_j\rho_j$ as the classical energy transfer to a harmonic oscillator initially at rest and subjected to a linear force.

The matrix elements over the VV operator U_{VV} can also be evaluated (see appendices B and C). Thus, by using the expressions given in appendix B, we can evaluate state-resolved transition probabilities from eq. (4.37). Due to the $(M-1)^2$ summations involved when solving for the VV matrix elements, the method has so far not been used for more than 5–6 dimensional systems, but this case covers, for instance, vibrational energy transfer in collisions between two triatomic molecules. Various approximate schemes, as, for instance, a perturbational treatment, can be proposed for solving larger systems.

The SQ formulation of energy transfer assumes a second-order expansion of both intra- and inter-molecular interactions. Thus the method can be used in cases where such a description suffices. Often the intra/intermolecular forces for a heat bath, a solid, or a rigid part of a polyatomic molecule cannot be characterized better than through first- and second-derivative terms. Hence, the SQ formulation is convenient in these cases for that part of the system. Other techniques may then be introduced for the remaining part. They could be either quantum (say, grid methods) or classical mechanical treatment of some motions.

Table 4.2: Splitting of degrees of freedom in three groups and the appropriate dynamical treatment.

Few degrees of freedom $(< 10)^a$	Rigorous quantum treatments
	grid or state expansion techniques
Rigid degrees of freedom $(<1000–10000)$	SQ-approach
Non-rigid degrees of freedom (< 100000)	Classical mechanics

^{a}The number depends of course on available computer facilities.

The quantum methods could be used with advantage for reactive, dissociative, or tunneling degrees of freedom, and the classical methods for "large amplitude" motions as that of solution molecules, torsional motions of methyl groups or reactive modes of heavy atoms, and so on. Thus, the degrees of freedom are divided into three parts according to the scheme in table 4.2. At present the treatment of up to about 7 quantum degrees of freedom is possible, but certainly not practical. Consequently, we have set the upper limit to about 10. If the quadratic operators are included, the SQ treatment involves the solution of a matrix problem of dimension M. Since we have to propagate the solution of the equations (4.33), we have set the upper limit of the matrix size to about 10^3–10^4. The solution can be effectively obtained using for instance the Billing-Baer propagator method [133].

    Table 4.2 shows that within the quantum-classical scheme we do have techniques for solving dynamical problems involving from diatomics and polyatomics to biomolecular systems. The only question is then, whether such MD calculations on large systems are capable of revealing any information of interest to the experimentalist, or answering any questions which could not have been answered, for instance, by experimental means. The restrictions on the number of classical equations of motion is set to 100000 which is certainly somewhat conservative. The splitting of the DOFs into three sets is illustrated in fig. 4.1. The quantum DOFs treated exactly, for instance, using a grid representation, are solved in an effective potential obtained from the SQ degrees of freedom and subject to classical forces. The SQ system is also forced classically, and provides in turn an effective potential to the classical DOFs.

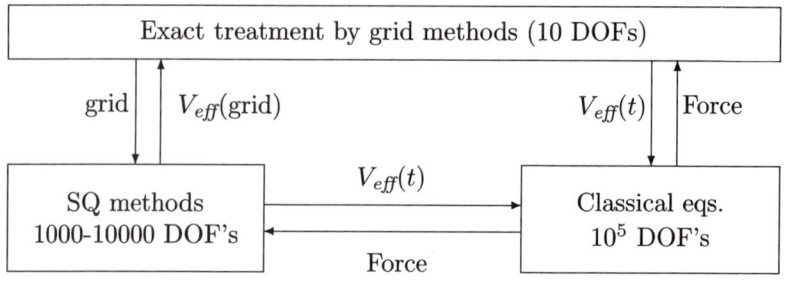

Fig. 4.1: Flow diagram for a triangular splitting of the DOFs.

Chapter 5

More complex systems

By more complex systems we mean systems containing on the order of hundreds or thousands of atoms, or molecules with less atoms but with "complicated" motions, the latter being the case when considering collisions between poly-atomic molecules. In the present chapter we deal with quantum-classical methods for treating energy transfer in collisions involving polyatomic molecules, molecule surface scattering, reactions in polyatomic systems and solution. We will assume that it is possible to construct realistical potential energy surfaces for the systems. Obviously, these surfaces will be of empirical or semi-empirical nature. In some of the methods, as for instance the reaction path method, one tries to minimize the information needed on potential energy surfaces.

5.1 Potential energy surfaces

Chemical reactions and energy transfer processes in the gas phase are often studied using just a single adiabatic Born-Oppenheimer potential energy surface. However non-adiabatic effects, that is, coupling between different electronic states, is an important aspect in chemistry. If the coupling between the various electronic states can be neglected, the "electronic" effect reduces to that of a statistical degeneracy factor g_e [180]. In order to incorporate non-adiabatic effects, the starting point is the non-relativistic Schrödinger equation $H\Phi = E\Phi$, where the hamiltonian can be written as

$$H = -\sum_a \frac{\hbar^2}{2M_a}\nabla_a^2 - \sum_j \frac{\hbar^2}{2m_e}\nabla_j^2 - \sum_a\sum_j \frac{Z_a e^2}{R_{aj}} + \sum_i\sum_{j<i} \frac{e^2}{R_{ij}}$$

$$+ \sum_a\sum_{b<a} \frac{Z_a Z_b e^2}{R_{ab}} \tag{5.1}$$

where a and b denote nuclei, i and j electrons. The masses of the nuclei are M_a and that of the electrons m_e. Z_a is the charge of the nuclei, whereas R_{ab}, R_{aj}, and R_{ij} denote the distances between the particles. In order to facilitate the solution of the problem one invokes the so-called Born-Oppenheimer separation of the electronic and nuclear degrees of freedom. Thus, in order to obtain an approximate solution to

$$H\Phi(\mathbf{x}, \mathbf{R}) = E\Phi(\mathbf{x}, \mathbf{R}) \tag{5.2}$$

where the electronic and nuclear coordinates are given by $\mathbf{x}$ and $\mathbf{R}$, respectively, we may utilize a Born-Oppenheimer procedure that involves the following steps:

1. We assume that the nuclei are held in fixed positions.
2. Solve the electronic structure problem and obtain the electronic wave functions and corresponding energies for the given nuclear configuration.
3. Assume that the total wave function for a given electronic state can be written as a product of an electronic and nuclear wave function.
4. Substitute the product wave function into eq. (5.2).
5. Neglect the effect of the nuclear kinetic energy operator on the electronic wave function.

Although it is certainly desirable to deal with just a single electronic potential energy surface, and although this has been justified by claiming the weakness of non-adiabatic coupling terms, the appearance of conical intersections between electronic states complicates matters somewhat. It turns out that even if the actual coupling can be neglected, the solution of the nuclear problem should include topological phase factors in the presence of a conical intersection [184]. The phase factor should take care of the sign change in the nuclear wave function when encircling the conical intersection. For triatomic systems, the phase factor is especially easy to incorporate when using hyper-spherical variables [185]. Here the phase effect can be incorporated in the angle ϕ which when increased by 2π makes the system encircle the intersection by a pseudo rotation. Thus, if the encircling of the intersection can be attributed to a certain angle, the basis functions can simply be constructed so as to incorporate the sign change.

The effect of the geometric phase can also be incorporated by adding a vector potential to the original hamiltonian [183] and, hence, the nuclear dynamics can be treated in the ways outlined in this book—just in an effective potential, which consists of the ordinary Born-Oppenheimer potential with the vector potential added to it. As a matter of fact, this methodology has been used to incorporate the geometric phase effect in classical trajectory calculations [191]. Recently the effect of topological phases has been considered by Baer [192]. Baer's extended B-O method is capable of treating topological effects also in cases where several conical intersections are present.

5.1.1   The Born-Oppenheimer separation

The wave function for the electron-nuclei problem is, according to ansatz 3 above, expressed as

$$\Phi(\mathbf{x}, \mathbf{R}) = \xi(\mathbf{x}; \mathbf{R})\psi(\mathbf{R}) \tag{5.3}$$

where $\mathbf{x}$ and $\mathbf{R}$ denote the set of electron and nuclear cartesian coordinates, respectively. We now imagine (ansatz 2) that we could solve the electronic problem first for fixed nuclei configuration. Thus the wave function $\xi(\mathbf{x}; \mathbf{R})$ is obtained as the solution to

$$\left(-\sum_i \frac{\hbar^2}{2m_e}\nabla_i^2 + V_{eN} + V_{ee}\right)\xi(\mathbf{x}; \mathbf{R}) = E_{el}(\mathbf{R})\xi(\mathbf{x}; \mathbf{R}) \tag{5.4}$$

where V_{eN} and V_{ee} denote the third and fourth terms in eq. (5.1). The electronic surface $E_{el}(\mathbf{R})$ is, of course, a function of the chosen nuclear configuration. Since the equation above has several solutions corresponding to different electronic states, the surface, as well as the wave function ξ, should be labeled by an electronic state number, which for simplicity is omitted here.

If we insert the product form eq. (5.3) into the Schrödinger equation, we get

$$\left(-\sum_a \frac{\hbar^2}{2M_a}\nabla_a^2 + \sum_a\sum_{b<a}\frac{Z_aZ_b}{R_{ab}} + E_{el}(\mathbf{R}) \right)\psi(\mathbf{R})\xi(\mathbf{x};\mathbf{R}) = E\psi(\mathbf{R})\xi(\mathbf{x};\mathbf{R})$$

(5.5)

Multiplying from the left with $\xi^*(\mathbf{x};\mathbf{R})$ and integrating over the electronic coordinates, we obtain

$$\left(-\sum_a \frac{\hbar^2}{2M_a}\nabla_a^2 + V(\mathbf{R}) \right)\psi(\mathbf{R}) = E\psi(\mathbf{R})$$

(5.6)

where we have neglected the non-adiabatic coupling elements

$$-\sum_a \frac{\hbar^2}{2M_a}(2\nabla_a\psi\nabla_a\xi + \psi\nabla_a^2\xi)$$

(5.7)

The magnitude of these terms is a measure of the rate of non-adiabatic transition between the different electronic levels.

Within the B-O approximation, the problem is reduced to the solution of the nuclear motion on a single potential energy surface

$$V(\mathbf{R}) = E_{el}(\mathbf{R}) + \sum_{a<b}\frac{Z_aZ_b}{R_{ab}}$$

(5.8)

The surface $V(\mathbf{R})$ is available from electronic structure calculations, that is, by solving eq. (5.4). However, in many cases of practical interest it has to be obtained by some semi-empirical or empirical method.

Returning to eq. (5.4), we may write it as

$$H_{elec}\xi(\mathbf{x};\mathbf{R}) = E_{el}(\mathbf{R})\xi(\mathbf{x};\mathbf{R})$$

(5.9)

where

$$H_{elec} = -\sum_i \frac{\hbar^2}{2m_e}\nabla_i^2 + V_{eN} + V_{ee}$$

(5.10)

We then multiply with $\xi^*(\mathbf{x};\mathbf{R})$ from the left and integrate over the electronic coordinates to obtain

$$E_{el}(\mathbf{R}) = \int \xi^*(\mathbf{x};\mathbf{R})H_{elec}\xi(\mathbf{x};\mathbf{R})d\mathbf{x}$$

(5.11)

We can now evaluate the derivative of $E_{el}(\mathbf{R})$ with respect to a nuclear coordinate R_i

$$\frac{\partial E_{el}(\mathbf{R})}{\partial R_i} = \frac{\partial}{\partial R_i} \int \xi^*(\mathbf{x};\mathbf{R}) H_{elec} \xi(\mathbf{x};\mathbf{R}) d\mathbf{x}$$

$$= \int \frac{\partial \xi^*(\mathbf{x};\mathbf{R})}{\partial R_i} H_{elec} \xi(\mathbf{x};\mathbf{R}) d\mathbf{x} + \int \xi^*(\mathbf{x};\mathbf{R}) \frac{\partial H_{elec}}{\partial R_i} \xi(\mathbf{x};\mathbf{R}) d\mathbf{x}$$

$$+ \int \xi^*(\mathbf{x};\mathbf{R}) H_{elec} \frac{\partial \xi(\mathbf{x};\mathbf{R})}{\partial R_i} d\mathbf{x} \tag{5.12}$$

Since the sum of the first and third integrals is zero we find that

$$\frac{\partial E_{el}(\mathbf{R})}{\partial R_i} = \int \xi^*(\mathbf{x};\mathbf{R}) \frac{\partial H_{elec}}{\partial R_i} \xi(\mathbf{x};\mathbf{R}) d\mathbf{x} \tag{5.13}$$

which is known as the Hellmann-Feynman theorem. It states that within the Born-Oppenheimer approximation, the forces on the nuclei are given as the derivative of H_{elec} averaged over the electronic wave function. In order to use the reaction path method (see section 5.4) for chemical reactions, one also needs the hessian matrix, that is,

$$H_{ij} = \frac{\partial^2 E_{el}(\mathbf{R})}{\partial R_i \partial R_j} \tag{5.14}$$

which may be obtained in a similar manner. Thus, if the electronic structure problem is solved and the electronic wave function known, only a little additional work is required to obtain the gradient and hessian.

In many dynamical problems of chemical significance, the non-adiabatic coupling terms cannot be ignored. Hence, also the terms $\langle \xi_\alpha | \nabla_a \xi_\beta \rangle$, where α, β denote two different solutions to eq. (5.9), are needed from electronic structure calculations.

The information on gradients and hessians is, as we shall see below, what we need in order to determine the reaction path. But it is also the information, which enters the hamiltonian for the nuclear motion, that is, it determines the nuclear dynamics.

5.1.2   Approximate interaction potentials

The intermolecular interaction between atoms or molecules and a polyatomic molecule is conveniently split into a short- and a long-range part. The short-range potential is usually modeled by a sum of pairwise exponential terms

$$V_{sr} = \sum_{ij} A_{ij} \exp(-\alpha_{ij} R_{ij}) \tag{5.15}$$

where R_{ij} is the distance between atom i in one molecule and atom j in the other. The reason for choosing this expression has to do with the general lack of data for potential surface dependence on the intramolecular bond distances. The atom-atom expression provides a guess of this dependence in terms of few parameters which can be obtained by other means, as, for instance, by fitting to data on second virial coefficients, transport properties, molecular beam data, and so on, or ab initio data with the molecules in their equilibrium configurations. Once the coefficients A_{ij} and α_{ij} are determined with the atoms in their equilibrium positions in each molecule, we introduce the vibrational coordinate dependence by using the fact that

$$\mathbf{R}_{ij} = \mathbf{R} + \mathbf{r}_j - \mathbf{r}_i \tag{5.16}$$

where $\mathbf{R}$ is the center-of-mass distance between the molecules. $\mathbf{r}_i$ is the position vector of atom i in a space-fixed coordinate system placed in the center of mass of molecule (2) and $\mathbf{r}_j$ the position vector of atom j measured in a space-fixed coordinate system placed in the center of mass of molecule (1). The position vectors can now be expressed as

$$\mathbf{r}_j = \mathbf{a}_j + \frac{1}{\sqrt{m_j}} \sum_k \mathbf{l}_{kj} Q_k^{(1)} \tag{5.17}$$

$$\mathbf{r}_i = \mathbf{a}_i + \frac{1}{\sqrt{m_i}} \sum_k \mathbf{l}_{ik} Q_k^{(2)} \tag{5.18}$$

where $Q_k^{(i)}$ are normal mode coordinates in molecule i. The $\mathbf{a}_i$s are reference vectors specifying the equilibrium positions of the atoms in the two molecules and $\mathbf{l}_{ik}$ a normal mode eigenvector. Neglecting terms of order R^{-2} and higher, we obtain

$$R_{ij} \sim R + \frac{\mathbf{R} \cdot (\mathbf{r}_j - \mathbf{r}_i)}{R} \tag{5.19}$$

which can be introduced in eq. (5.15).

For the interaction of an atom with a diatomic molecule or two diatomic molecules, the short-range potential is often expressed in a simple fashion, using the dumbell expression. For two identical homonuclear diatomic molecules, we have

$$V_{sr} = 4C \exp(-\alpha R) \cosh(\delta \alpha r_1 \cos \gamma_1) \cosh(\delta \alpha r_2 \cos \gamma_2) \tag{5.20}$$

where R is the center-of-mass distance, r_i $(i = 1, 2)$ the two bond distances and γ_i the angle between r_i and R. We notice that there are just three parameters C, α and δ. Thus, δ has been taken as a parameter rather than being set to a mass ratio ($1/2$ for homonuclear diatoms). Also the steepness parameter α can be given more flexibility by using

$$\alpha = \alpha_0 + \alpha_1 R + \alpha_2 R^2 \tag{5.21}$$

The parameters $C, \alpha_0, \alpha_1, \alpha_2$, and δ are fitted to information on the short-range potential, either experimental information or information obtained through ab initio calculations, or both.

To the short-range potential we add two types of long-range interactions, namely the dispersion part and the electrostatic part. The dispersion potential (or van der Waals potential) depends upon R as

$$V_{disp} = -C_6 R^{-6} - C_8 R^{-8} \ldots \tag{5.22}$$

The dispersion potential is often multiplied by a switching function $f(R)$ with the property that $f(R) \to 0$ for small values of R and $f(R) \to 1$ for large values. A popular choice for $f(R)$ is

$$f(R) = \begin{cases} 1 & R \leq R_0 \\ \exp(-a(R/R_0 - 1)^2) & R > R_0 \end{cases} \tag{5.23}$$

The values of a and R_0 can be determined by information on the van der Waals well, the second virial coefficient and elastic scattering data. Sometimes a switching function is designed for each individual term in the dispersion series, that is,

$$f_n(R) = 1 - \exp(-\alpha R) \sum_{k=0}^{n} \frac{(\alpha R)^k}{k!} \quad n = 6, 8, \ldots \tag{5.24}$$

as proposed by Tang and Toennies [155]. We notice that for an ion-molecule interaction, the "dispersion" part starts with an R^{-4} term [157].

For molecule-molecule or ion-molecule interactions, additional long-range forces are present through, dipole, quadrupole, octopole, and higher order terms interacting with either the ionic charge or the multipole on the other molecule. The multipole moments depend on the interatomic distances in the molecules— a dependence which is crucial to incorporate for describing vibrational energy transfer induced by the long-range forces.

For diatomic molecules the intra-molecular potential is often approximated by a Morse-potential.

However, if highly lying vibrational levels are considered, it is necessary to improve the potential and introduce an expression with more parameters. Such an expression is, for instance, the extended Rydberg formula

$$V(r) = -D_e(1 + a_1(r - r_e) + a_2(r - r_e)^2 + \cdots) \exp(-\alpha(r - r_e)) \tag{5.25}$$

where r_e is the equilibrium bond distance.

For polyatomic molecules, information on the intramolecular potential is available from, for instance, spectroscopic measurements, and the potential is often parametrized in terms of force field expressions of the type

$$2V_{intra} = \sum_i K_i(\Delta r)_i^2 + \sum_{ij} H_{ij} r_{ki} r_{kj} \Delta \alpha_{ijk}^2 + \sum_{ij} F_{ij}(\Delta q_{ij})^2$$
$$+ \sum_i P_i(\Delta \pi_i)^2 + \sum_i Y_i(\Delta t_i)^2 + \cdots \tag{5.26}$$

where the first term includes force constants for bond stretching, the next for bond bending, repulsive, out-of-plane bending, torsional motions, and so on, some of which may, of course, be absent in specific cases. Extensive information on force-field parameters is available in the literature [159]. Thus, it appears that as far as energy transfer to polyatomic molecules is concerned, the situation is rather good. Both potential parameters for inter- and intramolecular degrees of freedom are available. Also, since the dynamical theory is in good shape as we shall see later, information on mode specific rates and energy transfer may readily be calculated.

The above outline of procedures for constructing the potential energy surface has to be modified if chemical reactions can occur. Here, we may use the fact that most reactions, although consisting of many atoms, are still three-center reactions, that is, one bond is formed and another broken simultaneously. The natural starting point for constructing an analytical potential is then the (London, Eyring, Polanyi, Sato) LEPS surface [181], which has the property that it gives the correct asymptotic fragment potentials. Thus the LEPS surface is a good starting point for a potential of the three-center part of the system. How, the potential is constructed from here has at present no accepted best route and it is outside the scope of the present book to discuss the possibilities. For a recent review, see, for instance, [182].

It is important that the surfaces used for molecular scattering calculations have the correct global structure or character, the reason being that the trajectories probe a large global fraction of the surfaces and that rates or cross sections depend on a weighted average over these trajectories. Thus surfaces which, for a dimer as $(HF)_2$, for instance, are of high quality from a spectroscopic point of view may not be adequate for treating the collision dynamics between two HF molecules, and vice versa.

5.2   Polyatomic molecules

In large polyatomic molecules or in solids, the most important degrees of freedom are the $3N-6$ vibrational modes, and energy transfer to these modes constitutes the dominating energy loss mechanism in collisions. In the previous chapter, we saw how energy transfer to a set of harmonic oscillators could be described within the second quantization formalism. Thus, large systems are not necessarily the most complicated to treat, the reason being that the vibrational modes are often well treated in the harmonic approximation. In other words, the information content we either need, or can obtain from a large class of systems, stays at the harmonic level.

For some solids or molecules with electron band structure, electronic excitation processes also need to be considered in the energy transfer calculations. Here, also, the equations can be brought into a convenient form using second quantization concepts. The lack of detailed experimental information on large systems makes it justifiable for us to perform calculations which contain the essential dynamical features. Actually, the most difficult cases in energy transfer problems are constituted by collisions between, say, two triatomic molecules. In these problems, neither rotational motions, Coriolis coupling, Fermi resonances, nor other anharmonic couplings can be neglected.

In the present chapter, we formulate a theory in which the harmonic solution is the zeroth-order solution on which anharmonic corrections can be added. The method has been used to treat energy transfer from triatomics to molecules with several hundreds of atoms. The method is again based on the classical path approach to the problem.

For the collision between an atom and a polyatomic molecule, we consider the following approximate hamiltonian

$$H = \hat{H}_0 + v_{anh} + H_{co} + V_{int} + H^0_{rot} + T_{kin} \tag{5.27}$$

If we treat the dynamics within the quantum-classical theory, we would typically consider a quantum treatment of the vibrations, or perhaps only the high frequency vibrational modes, and a classical treatment of the translational motion and the rotational motion of the polyatomic molecule. Thus, the kinetic energy is

$$T_{kin} = \frac{1}{2\mu}(P_X^2 + P_Y^2 + P_Z^2) \tag{5.28}$$

where $\mu = m_A m_B/(m_A + m_B)$ is the reduced mass for collision between the molecules A and B. The zeroth-order hamiltonian for the vibrational modes is $\hat{H}_0$,—with eigenfunctions $\phi_{\{m_k\}} = \Pi_{k=1}^M \phi_{m_k}(Q_k)$, that is, a product of harmonic oscillator states in the normal mode coordinates with $M = 3N - 6$ or $3N - 5$ for linear molecules.

Thus, the motion of the polyatomic molecule is separated into a rotational motion of a fixed frame, and the oscillatory displacements from it, that is, for atom i we have

$$\mathbf{r}_i = \mathbf{A}\mathbf{a}_i^0 + \rho_i \tag{5.29}$$

where $\mathbf{a}_i^0$ is the reference position of the molecule with the atoms in their equilibrium positions. The rotation of the molecule is described through the Euler angles (ξ, θ, ϕ) and the rotation matrix is defined as [101]

$$\mathbf{A} = \begin{bmatrix} \cos\xi\cos\phi - \cos\theta\sin\phi\sin\xi & -\sin\xi\cos\phi - \cos\theta\sin\phi\cos\xi & \sin\theta\sin\phi \\ \cos\xi\sin\phi + \cos\theta\cos\phi\sin\xi & -\sin\xi\sin\phi + \cos\theta\cos\phi\cos\xi & -\sin\theta\cos\phi \\ \sin\theta\sin\xi & \sin\theta\cos\xi & \cos\theta \end{bmatrix} \tag{5.30}$$

The displacement vector ρ_i is then expressed in terms of the normal mode coordinates Q_k [156], that is, with $\rho_i = m_i^{-1/2}(\xi_i, \eta_i, \zeta_i)$ where

$$\xi_i = \sum_k \ell_{ik}^{(x)} Q_k \tag{5.31}$$

$$\eta_i = \sum_k \ell_{ik}^{(y)} Q_k \tag{5.32}$$

and

$$\zeta_i = \sum_k \ell_{ik}^{(z)} Q_k \tag{5.33}$$

The notation is, due to degeneracies in the bending mode, slightly different in the case of linear polyatomic molecules [156]. Thus, for linear molecules we need to make two modifications of the approach outlined here. Firstly, the perturbational treatment of the anharmonic coupling terms has to be carried out using the "Fermi-transformed" basis [143, 130], that is, by using perturbation theory for degenerate systems. Secondly, the vibrational angular momentum coupling in the bending mode can be transformed out of the problem such that the $M = 4$ quantum problem for a linear triatomic molecule is converted to an $M = 2$ problem for the vibrational angular momentum coupling in the bending mode and an $M = 3$ problem for the three normal modes (see [144] for details). These additional complications will not be dealt with in any detail here.

The anharmonic coupling terms are collected in the term v_{anh}, and they are included perturbationally (see below). In the quantum-classical framework, the Coriolis coupling depends on both classical and quantum variables, that is,

$$H_{co} = -(p_x P_x(t)/I_{xx}^e + p_y P_y(t)/I_{yy}^e + p_z P_z(t)/I_{zz}^e) \qquad (5.34)$$

where the rotational angular momenta $P_x(t)$ and so on, depend upon time through the trajectory and the so-called vibrational angular momenta are given as [146]

$$p_\alpha = \sum_k \sum_{k'} \sum_i \ell_{ik}^{(\beta)} \ell_{ik'}^{(\gamma)} (Q_k p_{k'} - Q_{k'} p_k) \qquad (5.35)$$

where α, β and γ denote x, y, z such that $\alpha \neq \beta \neq \gamma$. The momenta conjugate to Q_k, the normal mode coordinates, are p_k. The zeroth-order rotational hamiltonian is given as

$$H_{rot}^0 = \tfrac{1}{2}(P_x(t)^2/I_{xx}^e + P_y(t)^2/I_{yy}^e + P_z(t)^2/I_{zz}^e) \qquad (5.36)$$

where I_{xx}^e, I_{yy}^e and I_{zz}^e denote the principal moments of inertia for the molecule. Since the momenta P_x, P_y, P_z have no conjugate coordinates, we need to introduce Euler angles and, hence, the connection between the momenta P_x, P_y, P_z in order to achieve Hamilton equations of motion, and the momenta conjugate to the Euler angles is also needed. For the rotation of a rigid molecule, it is

$$P_x = p_\theta \sin\xi - p_\phi \cos\xi \csc\theta + p_\xi \cos\xi \cot\theta \qquad (5.37)$$
$$P_y = p_\theta \cos\xi + p_\phi \sin\xi \csc\theta - p_\xi \sin\xi \cot\theta \qquad (5.38)$$
$$P_z = p_\xi \qquad (5.39)$$

The intermolecular potential is expanded to second order in the normal coordinates as

$$V_{int} = V_{int}^0 + \sum_k V_k^{(1)} Q_k + \frac{1}{2}\sum_{kk'} V_{kk'}^{(2)} Q_k Q_{k'} + \cdots \qquad (5.40)$$

For many practical purposes we may neglect the anharmonic terms of the intermolecular potential, that is, the cubic and higher order terms in eq. (5.40). However, the anharmonic terms in the molecular hamiltonian are included in v_{anh}. The eigenstates of the molecular hamiltonian are now obtained through the equation

$$[\hat{H}_0 + v_{anh} + H_{co}(t)]\psi_{\{m_k\}}(t) = E_{\{m_k\}}\psi_{\{m_k\}}(t) \qquad (5.41)$$

where all terms not vanishing asymptotically (static terms) have been included. We have also indicated that the states depend on time through the time-dependence of the classical variables in the Coriolis coupling term. If the anharmonic perturbation is small, we can proceed using perturbation theory, that is, to first order we have

$$\psi_{\{m_k\}}(Q_k, t) = \phi_{\{m_k\}} + \sum_{\{n_k\}} \frac{\langle \phi_{\{n_k\}} | v_{anh} + H_{co} | \phi_{\{m_k\}} \rangle}{E^0_{\{n_k\}} - E^0_{\{m_k\}}} \phi_{\{n_k\}} \tag{5.42}$$

where $\phi_{\{n_k\}}$ are the harmonic oscillator states and the brackets $\langle \ \rangle$ indicate integration over Q_k, that is, we have

$$\hat{H}_0 \phi_{\{n_k\}} = E^0_{\{n_k\}} \phi_{\{n_k\}} \tag{5.43}$$

In order to solve the quantum mechanical part of the problem, we should expand the total wave function in the perturbed basis set, that is,

$$\Psi(t) = \sum_{\{m_k\}} b_{\{m_k\}}(t) \psi_{\{m_k\}}(t) \tag{5.44}$$

From the TDSE, we obtain a set of coupled equations in the expansion coefficients

$$i\hbar \dot{b}_{\{m_k\}}(t) = (E_{\{m_k\}} + H^{(0)}_{rot} + T_{kin}) b_{\{m_k\}}(t)$$
$$+ \sum_{\{n_k\}} b_{\{n_k\}}(t) \left[\langle \psi_{\{m_k\}} | V_{int} | \psi_{\{n_k\}} \rangle - i\hbar \left\langle \psi_{\{m_k\}} \left| \frac{d}{dt} \psi_{\{m_k\}} \right\rangle \right] \right.$$
$$\tag{5.45}$$

where the last term to first order in the perturbation v_{anh} and H_{co} can be written as

$$\sum_{\{n_k\}} b_{\{n_k\}} [\langle \phi_{\{m_k\}} | V_{int} + H^d_{co} | \phi_{\{n_k\}} \rangle] \tag{5.46}$$

The coupling through the Coriolis term can, furthermore, be expressed in terms of creation/annihilation operators of the vibrational degrees of freedom. We have called this term the dynamical Coriolis term and it has the form

$$H^d_{co} = -i\hbar \frac{dH_{co}}{dt} \frac{1}{E^0_{\{n_k\}} - E_{\{m_k\}}}$$
$$= \frac{\hbar}{2} \sum_\alpha \frac{\dot{P}_\alpha}{I^e_{\alpha\alpha}} \sum_{k<k'} \zeta^{(\alpha)}_{kk'} \frac{\omega^+}{\omega_{k'} - \omega_k} (a^+_k a_{k'} + a^+_{k'} a_k) \tag{5.47}$$

where we have used eq. (5.35), from which it follows that

$$Q_k = b_k(a_k + a^+_k) \tag{5.48}$$
$$p_k = i\sqrt{\hbar\omega_k/2}(a^+_k - a_k) \tag{5.49}$$

with $b_k = \sqrt{\hbar/2\omega_k}$ and where we have introduced the notations $\omega^+ = (\omega_k + \omega_{k'})/\sqrt{\omega_k \omega_{k'}}$ and

$$\zeta_{kk'}^{(\alpha)} = \sum_l (\ell_{ik}^{(\beta)} \ell_{ik'}^{(\gamma)} - \ell_{ik}^{(\gamma)} \ell_{ik'}^{(\beta)}) \tag{5.50}$$

Thus, the dynamical Coriolis coupling term contains the so-called VV operators $a_k a_l^+$.

It is now convenient to introduce an interaction representation with respect to the energies $E_{\{n_k\}}$ and to divide the intermolecular potential in a harmonic and an anharmonic part, that is, the harmonic part of V_{int} is defined by

$$V_h = \sum_k V_k^{(1)} b_k (a_k + a_k^+) + \sum_{k<k'} V_{kk'}^{(2)} (a_k a_{k'}^+ + a_k^+ a_{k'}) \tag{5.51}$$

where we have introduced the operator representation of the normal mode coordinates. The anharmonic part includes, aside from the cubic and higher order terms in the normal modes, the double quantum operators $a_k a_{k'}$ and $a_k^+ a_{k'}^+$ ($k \neq k'$) missing from V_h, and the part of the dynamical Coriolis coupling containing similar operators.

Introducing the matrix representation $\mathbf{U}$ of the evolution operator under the linear (VT) and quadratic (VV) operators, we have, in the interaction representation,

$$i\hbar \frac{d}{dt} \mathbf{U} = \exp(i\theta_0)[\mathbf{V}_h + i\hbar \mathbf{H}_{co}^d] \exp(-i\theta_0)\mathbf{U} \tag{5.52}$$

where we have included the harmonic and part of the dynamical Coriolis coupling. The θ_0 matrix is diagonal with elements given as

$$\theta_{\{n_k\}}^0 (t) = \frac{1}{\hbar} \int dt E_{\{n_k\}}^0 \tag{5.53}$$

where

$$E_{\{n_k\}}^0 = \sum_k \hbar \tilde{\omega}_k (n_k + 1/2) \tag{5.54}$$

and $\tilde{\omega}_k = \omega_k + 1/2 b_k^2 V_{kk}^{(2)}$ is the modulated vibrational frequency.

Introducing now the fact that the first-order treatment of the static anharmonic terms gives a basis set

$$\psi = \phi + \mathbf{h}\phi \tag{5.55}$$

where the $\mathbf{h}$ matrix is defined through eq. (5.42), we express the evolution operator as β in this basis and obtain the coupled equations (in matrix notation) as

$$i\hbar \frac{d}{dt} \beta = \exp(i\theta)[\mathbf{V}_h + i\hbar \mathbf{H}_{co}^d + \mathbf{V}_a + \mathbf{h}^* \mathbf{V}_h - \mathbf{V}_h \mathbf{h}] \exp(-i\theta)\beta \tag{5.56}$$

Thus, we obtain $\beta = \mathbf{UT}$ where

$$i\hbar\frac{d}{dt}\mathbf{T} = \mathbf{U}^+ \exp(i\theta)[\mathbf{V}_a + \mathbf{h}^*\mathbf{V}_h - \mathbf{V}_h\mathbf{h}]\exp(-i\theta)\mathbf{UT} \tag{5.57}$$

where $\mathbf{T}(t_0) = \mathbf{I}$ is the unit operator. The advantage of this splitting into a $\mathbf{U}$ operator and the correction $\mathbf{T}$ is that the first problem can be solved without resorting to a state-expansion technique.

5.2.1 Second quantization solution for U

In the previous chapter we showed that the SQ method could be used for solution of the TDSE with a general hamiltonian expressed as:

$$H = \sum_k \hbar\omega_k(a_k^+ a_k + \tfrac{1}{2}) + \sum_k f_k(t)(a_k + a_k^+) + \sum_{kl} f_{kl}(t)a_k^+ a_l \tag{5.58}$$

This hamiltonian is a model hamiltonian for a polyatomic molecule perturbed by linear and quadratic forces:

$$f_k = b_k \left.\frac{\partial V_{int}}{\partial Q_k}\right|_{eq} \tag{5.59}$$

$$f_{kl} = b_k b_l \frac{1}{2}\left.\frac{\partial^2 V_{int}}{\partial Q_k \partial Q_l}\right|_{eq} + \frac{\sqrt{\omega_k/\omega_l} + \sqrt{\omega_l/\omega_k}}{(\omega_l - \omega_k)}\frac{\hbar}{2}\sum_\alpha \dot{P}_\alpha \zeta_{kl}^{(\alpha)}/I_{\alpha\alpha}^{(e)} \tag{5.60}$$

where V_{int} is the intermolecular potential, and the coordinates Q_k are normal mode coordinates. If a classical path is assumed for the relative motion of two colliding molecules, the forces become time-dependent through the classical equations of motion. Following chapter 4, we have for the evolution operator

$$U(t,t_0) = \exp(-iH_0 t/\hbar)U_I(t,t_0) \tag{5.61}$$

where $H_0 = \sum_k \hbar\omega_k(a_k^+ a_k + \tfrac{1}{2})$. Hence, we have

$$i\hbar\frac{dU_I}{dt} = \left[\sum_k (F_k^+ a_k^+ + F_k^- a_k) + \sum_{kl}(F_{kl}^+ a_k^+ a_l + F_{kl}^- a_k a_l^+)\right]U_I \tag{5.62}$$

where

$$F_k^\pm = f_k \exp(\pm i\omega_k t) \tag{5.63}$$

$$F_{kl}^\pm = f_{kl}\exp(\pm i(\omega_k - \omega_l)t) \tag{5.64}$$

The solution is partitioned into a VT and a VV part

$$U_I = U_{VV}U_{VT} \tag{5.65}$$

where U_{VT} can be factorized as $U_{VT} = \Pi_k U_{VT}^{(k)}$ and $U_{VT}^{(k)}$ must satisfy the equation:

$$i\hbar\frac{dU_{VT}^{(k)}}{dt} = U_{VV}^+(F_k^+ a_k^+ + F_k^- a_k)U_{VV}U_{VT}^{(k)} \tag{5.66}$$

The solution U_{VV} to eq. (4.23) can be expressed in terms of the operators $O_{kl} = a_k^+ a_l$, that is,

$$U_{VV} = \Pi_{ij} \exp(\alpha_{ij}(t, t_0) O_{ij}) \tag{5.67}$$

In order to solve the problem posed by the eq. (5.66), we have to evaluate the quantities

$$U_{VV}^+ (a_k^+ \text{ or } a_k) U_{VV} \tag{5.68}$$

This can be done using operator algebra. The result is [145]

$$U_{VV}^+ a_k U_{VV} = a_k + \sum_{j=1}^{M} Q_{kj} a_j \tag{5.69}$$

$$U_{VV}^+ a_k^+ U_{VV} = a_k^+ + \sum_{j=1}^{M} Q_{kj}^* a_j^+ \tag{5.70}$$

where $Q_{kj} = R_{kj} - \delta_{kj}$ (see chapter 4 for a definition of the **R** matrix). The terms "VV" and "VT" come (as mentioned) from molecular collision theory [130]. The VV-operators induce transition between two vibrational modes, such that the number of vibrational quanta are conserved. For the remaining processes, we use the term VT processes. The VT operators excites a specific mode through vibrational-translational/rotational coupling.

The operator approach for the vibrational degrees of freedom is conveniently used in the classical path method, where the relative translational motion of the two molecules is treated classically, that is, the classical equations of motion are solved simultaneously with the eq. (4.34) (for the VV problem) and the calculation of the integrals (4.29), (4.30) (for the VT problem). Once the formal solution has been obtained in the operator space, we can, in order to solve the eq. (5.57), switch to a matrix representation in the harmonic oscillator basis set.

The various technical aspects of the quantum-classical SQ theory for energy transfer in polyatomic have been illustrated in the flow diagram 5.1.

5.2.2   The classical degrees of freedom

The quantum and classical subsystems can be coupled self-consistently (see also chapter 3) through an effective hamiltonian written as

$$H_{eff} = \langle \Psi_0 | U_{VT} U_{VV} | H | U_{VV} U_{VT} | \Psi_0 \rangle \tag{5.71}$$

where Ψ_0 is the initial wave function, that is,

$$\Psi(t) = \exp(-iH_0 t/\hbar) U_I(t, t_0) \Psi_0 \tag{5.72}$$

and the integration is over the "quantum-coordinates." Thus, the effective hamiltonian H_{eff} will depend upon time either explicitly or through the classical

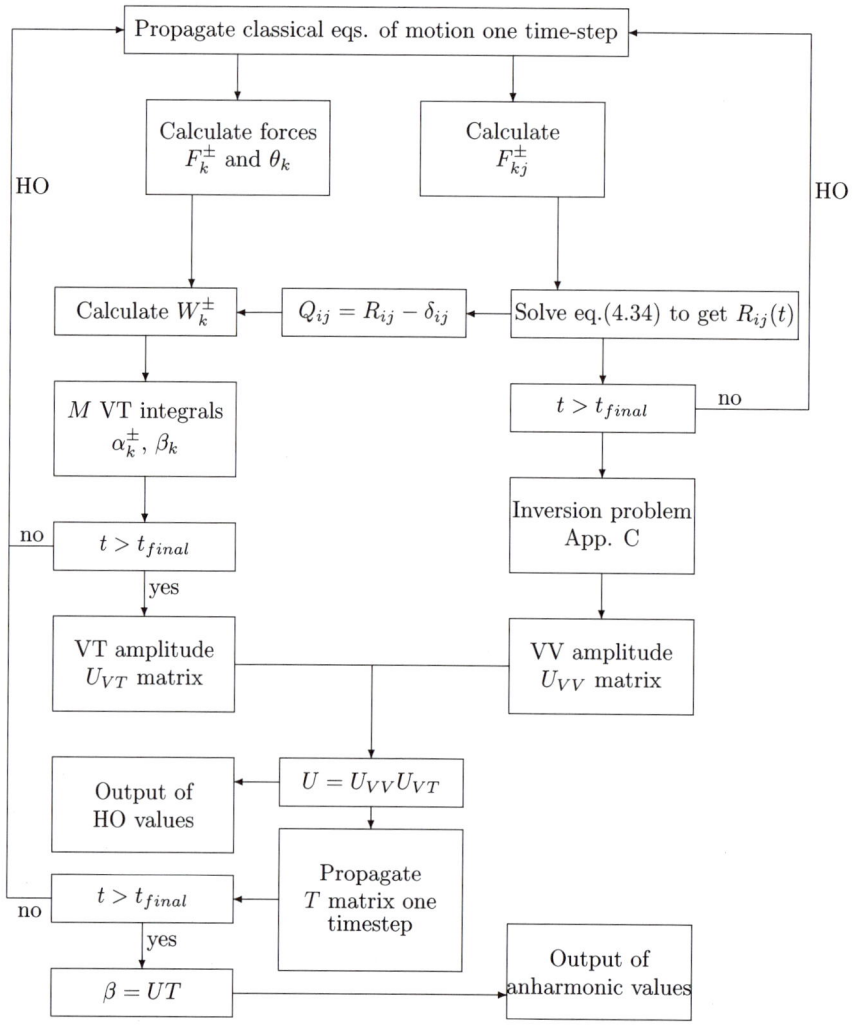

Fig. 5.1: Flow diagram for a quantum-classical approach to energy transfer in polyatomic molecules.

coordinates and momenta. The effective hamiltonian can be expressed in terms of the functions $\alpha_k^{\pm}$ and Q_{kj}, using the eqs. (5.69), (5.70) and

$$a_k \exp(i\alpha_k^+ a_k^+) = \exp(i\alpha_k^+ a_k^+)(a_k + i\alpha_k^+) \tag{5.73}$$

$$a_k^+ \exp(i\alpha_k^- a_k) = \exp(i\alpha_k^- a_k)(a_k^+ - i\alpha_k^-) \tag{5.74}$$

The equations of motion for the classical degrees of freedom are obtained using Hamilton's principle on H_{eff}, that is,

$$\frac{dH_{eff}}{dt} = \frac{\partial H_{eff}}{\partial t} + \frac{\partial H_{eff}}{\partial R}\dot{R} + \frac{\partial H_{eff}}{\partial P_R}\dot{P}_R = 0 \tag{5.75}$$

Unless the operators U_{VV} and U_{VT} depend explicitly upon some of the classical variables, then $\partial H_{eff}/\partial t = 0$ and we have

$$\dot{X} = \frac{\partial H_{eff}}{\partial P_X} \tag{5.76}$$

$$\dot{P}_X = -\frac{\partial H_{eff}}{\partial X} \tag{5.77}$$

with corresponding equations for the Y and Z components. For the rotational motion of the polyatomic molecule, we introduce the Euler angles $(\theta, \phi$ and $\xi)$ and corresponding momenta. Thus, we obtain the additional classical equations of motion

$$\begin{pmatrix} \dot{\theta} \\ \dot{\phi} \\ \dot{\xi} \end{pmatrix} = \begin{pmatrix} \sin\xi & \cos\xi & 0 \\ -\cos\xi\csc\theta & \sin\xi\csc\theta & 0 \\ \cos\xi\cot\theta & -\sin\xi\cot\theta & 1 \end{pmatrix} \begin{pmatrix} P_x/I^e_{xx} \\ P_y/I^e_{yy} \\ P_z/I^e_{zz} \end{pmatrix} \tag{5.78}$$

and

$$\dot{p}_\theta = -\frac{\partial H_{eff}}{\partial \theta} - (p_\theta \cos\theta - p_\phi)[(P_x/I^e_{xx})\cos\xi - (P_y/I^e_{yy})\sin\xi]\sin^{-2}\theta \tag{5.79}$$

$$\dot{p}_\phi = -\frac{\partial H_{eff}}{\partial \phi} \tag{5.80}$$

$$\dot{p}_\xi = -\frac{\partial H_{eff}}{\partial \xi} - (p_\phi\csc\theta - p_\xi\cot\theta[(P_x/I^e_{xx})\sin\xi + (P_y/I^e_{yy})\cos\xi$$
$$- p_\theta[(P_x/I^e_{xx})\cos\xi - (P_y/I^e_{yy})\sin\xi] \tag{5.81}$$

The classical equations of motion are integrated using a standard Predictor-Corrector method together with the equations for the operators. We have seen that in order to include the anharmonic terms, we need to use a state-expansion approach. The zeroth order solution—the harmonic—solution can, however, be obtained with much less effort by using the SQ methodology.

If the anharmonic terms from the intermolecular potential can be neglected, we can proceed in a different manner. Here we expand the wave function in the harmonic basis set and solve the dynamical problem in the SQ approximation. When the collision is over, we can then calculate transition amplitudes in the harmonic basis $a^0_{\{m_k\}}$ set, and in order to obtain them in the anharmonic, we reexpand the wave function in this basis set and, hence, obtain

$$b_{\{n_k\}}(t) = \sum_{\{m_k\}} a^0_{\{m_k\}} \exp(i(E_{\{n_k\}} - E^0_{\{m_k\}})t)\langle\psi_{\{n_k\}}|\phi_{\{m_k\}}\rangle \tag{5.82}$$

where the overlap integral is between anharmonic and harmonic basis functions [153].

The methodology presented above is extremely powerful for treating energy transfer in polyatomic molecules, retaining a quantum description of the vibrational modes.

In order to obtain the solution, we have used the fact that the time-dependent Schrödinger equation is algebraically solvable if the operators in the hamiltonian form a closed set with respect to commutations. The closed set of operators constitutes a Lie group, and, hence, group theoretical methods can be used (see, for example, [154]) when solving the equations.

The expression for the transition amplitude from a state $|n_1 \ldots n_M\rangle$ to $|n'_1 \ldots n'_M\rangle$, induced by the operator U_{VV}, is given in the appendix B. The VT transition amplitudes for transitions induced by the operators $U_{VT}^{(k)}$ can be obtained using eq. (4.40). Thus, the numerical effort is reduced to the solution of the coupled equations (4.34), M integrals of the type (4.29) and (4.30), and equations of motion for the classical variables.

5.2.3 Rate-constants

If, for instance, we consider a collision between an atom and a triatomic (non-linear) molecule, we could imagine that the three vibrational modes are quantized, whereas the relative motion and the rotational motions of the tri-atom are treated classically. Introducing the energy $U = E_{rot} + E_{kin} = E - E_{vib}$ we obtain the rate-constant for a vibrational transition in the molecule as

$$k_{nn'}(T) = \sqrt{8k_B T/\pi\mu}(T_0/T)^{p/2} \int d(\beta U) \exp(-\beta U)\langle\sigma_{nn'}(U,T_0)\rangle \qquad (5.83)$$

where T_0 is a reference temperature, μ the reduced mass for the relative motion and $\langle\sigma_{nn'}(U,T_0)\rangle$ an average cross section for transition from the vibrational quantum state $n = n_1, n_2, n_3$ to $n' = n'_1, n'_2, n'_3$. The power p is 5 for an A+BCD system, but is, in general, (for systems where only the vibrations are quantized) given as 2+(the number of non-zero moments of inertia). The average cross sections can often be fitted to an expression of the type

$$\langle\sigma\rangle \sim AU^\gamma \qquad (5.84)$$

and, hence, the rate-constant for the "exothermic" transition takes the form

$$k_{nn'}(T) = A\sqrt{8k_B T/\pi\mu}(T_0/T)^{p/2}(2\gamma)\exp(\Delta E/2k_B T)$$

$$\times (\Delta E/4)^\gamma K_\gamma(\Delta E/2k_B T) \qquad (5.85)$$

where $\Delta E = E_{n'} - E_n > 0$ for an exothermic transition, and where K is a Bessel function. The average cross section has nothing to do with the ordinary total cross section, except that it has the same unit Å^2. The average cross section is obtained as an average over those degrees of freedom not treated quantally, that is, in the present case over rotational and translational degrees of freedom.

Thus, for an A+BCD system we have [158]

$$\langle\sigma_{nn'}(U,T_0)\rangle = Q_{rot}(T_0)^{-1}(\pi\hbar^2/2\mu k T_0)h^{-3}\int_0^\infty dl(2l+1)$$

$$\times \int_0^\pi d\theta \int_0^{2\pi} d\phi \int_0^{2\pi} d\xi P_{nn'}(U,l,\theta,\phi,\xi,p_\theta,p_\xi,p_\phi) \qquad (5.86)$$

where the rotational partion function at T_0 is

$$Q_{rot}(T_0) = \sqrt{\pi}(T_0^3/\theta_{xx}\theta_{yy}\theta_{zz})^{\frac{1}{2}} \qquad (5.87)$$

and the rotational temperature is related to the principal moments of inertia $I_{\alpha\alpha}^e$ by $\theta_{\alpha\alpha} = \hbar^2/2I_{\alpha\alpha}^e k_B$. The multi-dimensional integral is over the orbital angular momentum (l), and contains an average over the Euler angles (θ, ϕ, ξ) and the conjugate momenta. In the present case, the seven-dimensional integral is conveniently evaluated using Monte Carlo sampling. Thus, the Euler angles are picked randomly between zero and π or 2π and the momenta so that P_x is between zero and P_x^{max}, where $P_x^{max} = \sqrt{2U'/I_{xx}^e}$. Likewise, for P_y with $P_y^{max} = \sqrt{2(U' - P_x^2/(2I_{xx}^e)/I_{yy}^e)}$, and P_z is obtained by energy conservation such that $E_{rot} = U' = U - E_{kin}$. The momenta conjugate to the Euler angles can now be obtained from eqs. (5.37)–(5.39). This statistical sampling technique is used to define the initial variables for the classical trajectories, and the average cross section typically converges to 10–20% accuracy with a few hundred trajectories. Table 5.1 shows some average cross sections at two energies for excitation of H_2O colliding with He atoms. We notice that the small cross sections especially are sensitive to inclusion of enough coupling terms, and that the harmonic approximation is useful for at least a first estimate of the important transitions. Aside from the allowed dynamical approximations, the construction of a sufficiently reliable potential energy surface poses problems

Table 5.1: The sensitivity of average cross sections for vibrational excitation of H_2O colliding with He to various dynamical approximations. The energy U is 10 kJ/mol (upper number) and 20 kJ/mol (lower number). $T_0 = 300$ K. Data from [158]. HO means harmonic oscillator model, cubic, quartic that the cubic/quartic anharmonic couplings are included. Coriolis means that the coriolis coupling is also added. 4.1(−4) means 4.1×10^{-4}.

Transition	Cross section $\mathring{A}^2$			
000-010	4.1(−4)	3.4(−4)	3.4(−4)	3.3(−4)
	0.10	0.078	0.076	0.077
000-020	3.4(−8)	2.6(−8)	2.7(−8)	2.4(−8)
	3.2(−4)	1.9(−4)	1.5(−4)	1.7(−4)
000-100	3.2(−10)	6.7(−9)	1.3(−8)	2.8(−8)
	2.4(−8)	5.0(−7)	2.1(−6)	2.1(−6)
010-020	8.1(−4)	6.4(−4)	6.8(−4)	6.3(−4)
	0.20	0.18	0.17	0.17
010-001	1.8(−5)	1.5(−5)	1.8(−5)	1.8(−5)
	6.0(−5)	7.6(−5)	7.0(−5)	5.6(−4)
100-001	6.5(−2)	7.1(−2)	7.1(−2)	7.1(−2)
	0.26	0.59	0.59	2.0
Approximation	HO	cubic	coriolis	quartic

for more complicated molecules. This is especially true as far as the vibrational coordinate dependence of the short-range potential is concerned.

5.2.4 Some case studies

Energy transfer to polyatomic molecules is a fundamental part of the theory for unimolecular reactions, and reliable estimates can be obtained using the theory outlined above. State-resolved rates are, furthermore, necessary for modeling many non-equilibrium situations, and mode selective excitation is important for understanding energy pathways of importance for reactivity. For solids and proteins the site specificity is sometimes of great significance.

The quantum-classical theory is ideal for treating all of these processes. Only the vibrational degrees of freedom need to be quantized, and in the harmonic approximation the operator approach can handle even large systems. Consequently, the method has been used to calculate energy transfer in a number of systems $Ar+SF_6$ [160], Ar or He+glyoxal [161], He+α-Helix [163], and others. Some general trends from these calculations will be mentioned here.

Table 5.2 shows the energy transfer to the low frequency modes of glyoxal in collisions with He and Ar. The table shows general trends, namely that kinetic energy is primarily transferred to the low frequency modes, and that He atoms are more efficient for inducing vibrational excitation than Ar atoms. This trend can be explained by considering simple theories for vibrational excitation and by noticing that it is the velocity which enters (see eq. (3.12)) the expression for the probability. Hence, at a given kinetic energy, the He atom will have a higher impact velocity. However, for rotational excitation the opposite is true. Here the Ar-atoms are more efficient. For a molecule such as glyoxal, with principal moments of inertia which are still small (19.9, 82.6, and 136 amu $Å^2$), we see that

Table 5.2: Energy transfer ΔE_i to the three lowest vibrational modes of glyoxal as a function of kinetic energy in collisions with He and Ar atoms. Energies in kJ/mol. The total energy transfer to vibrational and rotational modes are also shown. The low frequency vibrational modes are $\omega_i = 237$, 414 and 536 cm^{-1}, respectively.

E_{kin}	ΔE_1	ΔE_2	ΔE_3	ΔE_{vib}	ΔE_{rot}	
3.67	0.011	0.0006	$< 10^{-4}$	0.012	0.32	He
	0.005	0.0003	0.00003	0.005	0.69	Ar
4.59	0.023	0.0014	0.0002	0.024	0.33	He
	0.0089	0.0006	0.0001	0.010	0.95	Ar
6.52	0.051	0.011	0.0014	0.064	0.80	He
	0.026	0.0022	0.0004	0.029	1.21	Ar
11.4	0.154	0.068	0.012	0.24	1.43	He
	0.090	0.0099	0.0029	0.11	1.93	Ar
21.4	0.51	0.31	0.063	0.96	2.21	He
	0.35	0.092	0.023	0.47	3.16	Ar

Table 5.3: Energy transfer in kJ/mol to CF_4 by proton collisions as a function of mode and impact parameter range. The kinetic energy is 1000 kJ/mol. Data from ref. [153].

Mode	ω_i Impact parameter	0–1 Å	1–2 Å	2–3 Å	3–4 Å	4–5 Å
2	434.5 cm^{-1}	166	55	8	2	1
4	631.2 cm^{-1}	94	23	5	1	0
1	908.4 cm^{-1}	4	4	0	0	0
3	1283 cm^{-1}	50	85	64	11	2
	ΔE_{rot}	11	9	0	0	0

the rotational degrees of freedom play a significant role in the energy exchange process. The above rules for vibrational excitation can be violated if, for instance, the long-range potential induces some mode specificity, as is the case of ion-molecule collisions (see table 5.3).

As the kinetic energy increases, the role of energy transfer to the rotational degrees of freedom also becomes less important. In this limit, the collision becomes more sudden, and due to the higher energy more vibrational modes are accesible for energy transfer processes.

For large polyatomics, the overall rotational motion may be neglected. But on the other hand, such molecules contain groups such as -CH$_3$ methyl groups, which should be treated as hindered rotors rather than vibrators [162]. A hybrid theory which quantizes some of the vibrational modes and treats energy transfer to these rotors classically has been used for energy transfer to an α-helix of alanine [163]. Energy transfer to a polyatomic molecule may be negative if the site of impact appears hot compared to the incoming kinetic energy. Thus the high frequency modes can deliver energy to an impinging atom or molecule at the same time as the low frequency modes receives energy (see fig. 5.2).

Figure 5.3 shows the collision between CO_2 and $O(^3P)$. The total energy of the classical degrees of freedom (rotation and translation) is 10 kJ/mol, and the CO_2 molecule is rotationally quenched by the oxygen atom. In order to obtain the average cross sections, the initial total angular momentum and the rotational angular momenta, as well as the orientation of the CO_2 molecule, are chosen randomly. Thus, the average cross section for a vibrational transition is obtained as

$$\langle \sigma_{nn'}(U, T_0) \rangle = \frac{\pi \hbar^4}{4\mu I (k_B T_0)^2} \sum_{J=0}^{J_{max}} \sum_{j=0}^{j_{max}} \sum_{\ell=|J-j|}^{J+j} (2J+1) P_{nn'} \tag{5.88}$$

where U is the classical energy, that is, the sum of rotational and translational energy. n is a collective quantum number $n = n_1 n_{21} n_{22} n_3$ for CO_2. The transition probabilities $P_{nn'}$ are obtained by solving the classical equations of motion for the relative translational and rotational motion, together with the coupled equations for the expansion coefficients (5.56). Thus $P_{nn'}$ depends on the initial rotational and orbital angular momenta as well as molecular orientation angles and angles conjugate to j and ℓ.

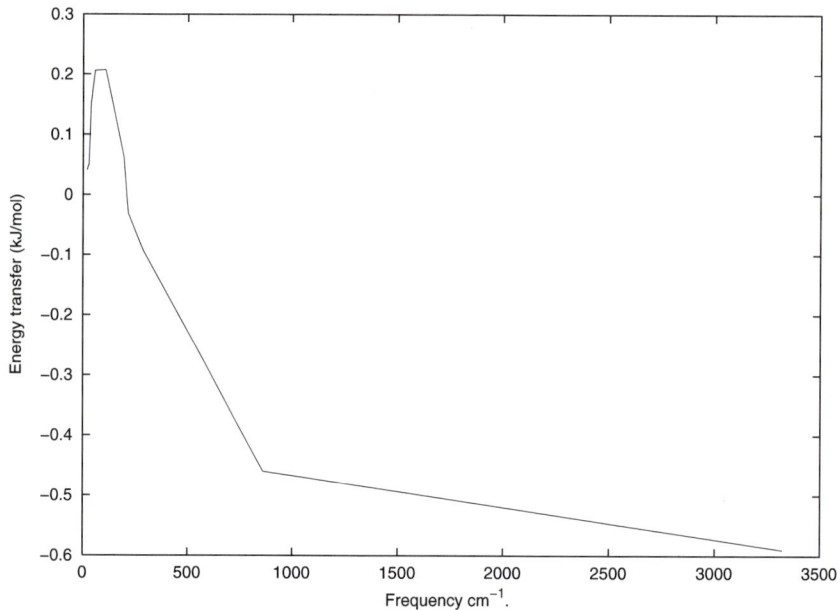

Fig. 5.2: Accumulative energy transfer to the vibrational modes of an α-helix of alanine as a function of the mode frequencies in cm^{-1}. Thus $\Delta E = \sum_{k'=7}^{k} \Delta E_k$ where $k = 7$ denotes the lowest frequency and ΔE_k the energy transfer to mode k. The helix consist of 20 petide units (133 atoms) and is hit by a He atom with kinetic energy $5\,kJ/mol$. The helix temperature is $300\,K$.

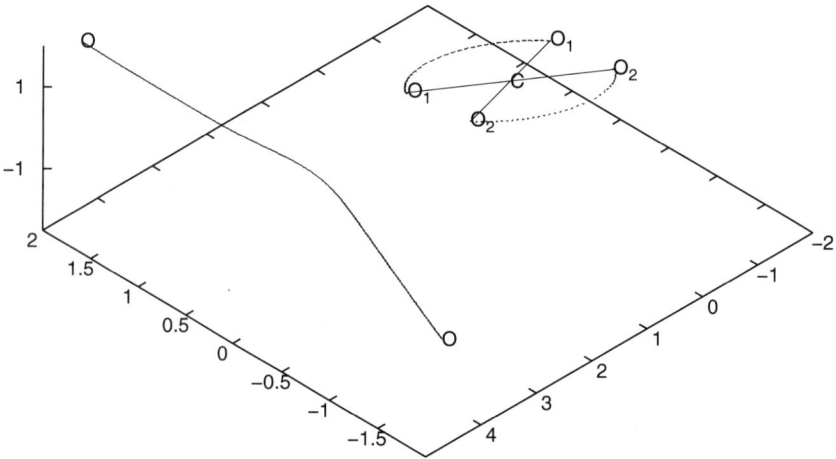

Fig. 5.3: An oxygen atom collides with CO_2. The initial center of mass energy is 6.64 kJ/mol. The rotational energy is 3.36 kJ/mol and is converted into translational energy by this collision. The intermolecular potential is from Redmon and Schatz [164].

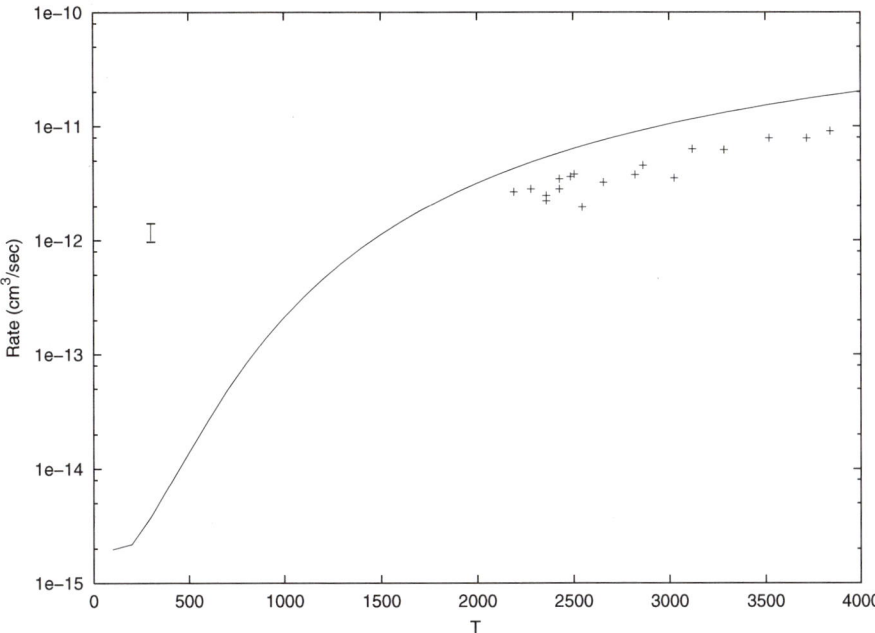

Fig. 5.4: Rate-constant for deactivation of the bending mode of CO_2 by collisions with $O(^3P)$ as a function of temperature. The high temperature data are in good agreement with experimental shock tube measurements [165]. But the room temperature result is too low compared with recent measurements [166], which estimate the rate at 300 K to $1.2 \pm 0.2 \times 10^{-12}$ cm^3/sec. This discrepancy indicates relaxation through a vibronic coupling model in which the spin-orbit coupled states are included [167, 166].

From the average cross sections, the rate constant is obtained as

$$k_{nn'}(T) = \sqrt{8k_B T/\pi\mu}(T_0/T)^2 \int d(\beta\epsilon) \exp(-\beta\epsilon)\langle\sigma_{nn'}(U, T_0)\rangle \qquad (5.89)$$

where $\epsilon = E - E_n$ and E is the effective energy given by eq. (3.26). Figure 5.4 shows the rate constant for vibrational relaxation of the bending mode of CO_2 in collisions with an oxygen atom. The agreement with experimental shock tube measurements is good at higher temperatures. At room temperature they are underestimated. The reason is believed to be that the various spin states of $O(^3P)$ can couple to nuclear motion through spin-orbit coupling terms. Hence, the vibrational relaxation will be significantly enhanced [167].

5.3 Chemical processes at surfaces

The understanding of surface processes (that is, energy transfer by collisions with and reactions at the surface, as well as diffusion processes along the surface) has gained importance over the last decades. The molecular dynamics is modified, as compared to that in the gas phase, by the shape, structure, and nature of the solid. Although the qualitative influence, "the catalytic effect," of solids is in principle known, many details at the molecular level are at present incompletely understood. For any theory to be of significance in this respect, it must include the effect of all the electron/nuclear dynamics relevant for the system. I believe that the quantum-classical theory again is the method of choice. The reason is that it combines the convenience of using classical dynamics for the reacting gas-phase molecules with a quantum boson/fermion treatment of the solid dynamics.

In the present section we give the equations for the classical path approach to the problem. As can be recalled from chapter 3, this involves a separability approximation between the two subsystems, and a detailed balance correction. It has also turned out to be possible to introduce the latter in energy transfer to solids [148]. The separability approximation is good if the coupling between the gas-phase molecules and the solid is weak. We notice that this condition may be fulfilled even if the energy transfer to the solid is large. The reason for this is that the energy transfer is obtained as a sum of contributions from many phonon modes, and that in a normal mode (phonon picture) it is the coupling to each individual mode which has to be weak. However, we have also seen in chapter 3, that the detailed balance correction gives reliable numbers for the excitation of a single oscillator. If the solid is treated as a sum of such oscillators, that is, in the phonon normal mode picture, we would expect this "symmetrized Ehrenfest" approach to be valid under these circumstances.

In order to use this methodology, it is necessary to determine the transformation matrix, which brings us from cartesian to normal mode coordinates. However, this matrix can easily be calculated for a finite-size crystal of about 1000 to 2000 atoms with present computer technologies. If the periodicity of the surface is incorporated, the problem can be reduced to a much simpler slab calculation [148]. However, we prefer the solid to be treated as being a finite-size crystal. Many of the quantities we are interested in "converge" with crystal sizes of a few hundred atoms arranged in three to four layers of atoms. The advantage of this approach is that any surface anomalies, corrugation, and so on, can also be treated within the same framework. Furthermore, three-dimensional rigid structures, zeolites, helix-backbones, and so on, can be dealt with without any major modifications.

As mentioned, the energy transfer to solids is conveniently treated within the second quantization approach, the reason being that the collective modes of the solid, which are excited by the collision, are the normal modes and the electron-hole pair excitations. Both of these modes can be described in SQ language using operators, namely those for bosons and fermions [148].

As the potential is usually expressed in terms of the interatomic distances (atom-atom potentials), in order to apply the SQ scheme we must introduce the normal mode coordinates of the solid, and in this respect we need the transformation between displacement $(\eta_{\alpha\gamma})$ and normal mode (Q_r) coordinates

$$\eta_{\alpha\gamma} = \sum_r T_{\alpha\gamma;r} Q_r \tag{5.90}$$

where α denotes an atom in the solid, γ takes the values 1, 2, 3 denoting the x, y, z directions and Q_r is the r'th normal mode. $T_{\alpha\gamma;r}$ denotes an element of the $\mathbf{T}$ matrix defined in such a way that it diagonalizes the force-constant matrix $\mathbf{A}$ of the solid, with elements given by:

$$A_{k\gamma;l\delta} = \frac{1}{\sqrt{M_k M_l}} \frac{\partial^2 V}{\partial r_{k\gamma} \partial r_{l\delta}} \tag{5.91}$$

where k and l label the atoms from 1 to N and where $\gamma, \delta = 1, 2, 3$. Thus the hamiltonian (H_c) for the solid includes first- and second-order terms in the displacement from the equilibrium positions

$$H_c = \epsilon_0 + \sum_{r=1}^{M} H_r \tag{5.92}$$

where ϵ_0 is the zero-point energy and

$$H_r = \tfrac{1}{2}(\dot{Q}_r^2 + \omega_r^2 Q_r^2) \tag{5.93}$$

The $M = 3N - 6$ non-zero frequencies are obtained by diagonalizing the matrix consisting of second derivatives of the potential for the solid. We can now quantize the vibrational modes r and write the normal mode coordinates in terms of the boson creation/annihilation operators

$$Q_r = b_r(a_r^+ + a_r) \tag{5.94}$$

where $b_r = (\hbar/2\omega_r)^{\frac{1}{2}}$.

Transitions among the phonon quantum states may now be induced by anharmonic terms in the solid or through coupling to the electronic motion. Transitions induced by gas-surface collisions are, however, the processes of interest in the present section.

In order to introduce the proper framework for the description, we introduce $R_{a\alpha}$ as the distance from the gas phase atom a to the surface atom α, that is,

$$R_{a\alpha} = \sqrt{\sum_{i=1,2,3} (\bar{X}_i - X_{\alpha i})^2} \tag{5.95}$$

with the gas atom positions given by $(\bar{X}_1, \bar{X}_2, \bar{X}_3)$ and the positions $X_{\alpha i}$ of the solid atoms. We can now expand the interaction potential in normal mode coordinates as

$$V_I = V_I^{(0)} + \sum_r V_r^{(1)} Q_r + \frac{1}{2} \sum_{rr'} V_{rr'}^{(2)} Q_r Q_{r'} \tag{5.96}$$

where $V_I^{(0)}$ is the interaction potential with the lattice atoms at their equilibrium position $X_{\alpha i}^{eq}$ and

$$V_r^{(1)} = \sum_\alpha \frac{\partial V_{a\alpha}}{\partial R_{a\alpha}}\bigg|_{eq} \frac{\partial R_{a\alpha}}{\partial Q_r} \tag{5.97}$$

is the first derivative of the interaction potential with respect to phonon coordinate Q_r. The second derivative is denoted by $V_{rr'}^{(2)}$ and given by

$$V_{rr'}^{(2)} = \sum_\alpha \frac{\partial V_{a\alpha}}{\partial R_{a\alpha}}\bigg|_{eq} \frac{\partial^2 R_{a\alpha}}{\partial Q_r \partial Q_{r'}} + \sum_{\alpha\beta} \frac{\partial^2 V_{a\alpha}}{\partial R_{a\alpha}\partial R_{a\beta}}\bigg|_{eq} \frac{\partial R_{a\alpha}}{\partial Q_r} \frac{\partial R_{a\beta}}{\partial Q_{r'}} \tag{5.98}$$

In order to calculate $\partial R_{a\alpha}/\partial Q_r$, we use eqs. (5.90), (5.95), and

$$\eta_{\alpha\gamma} = m_\alpha^{\frac{1}{2}}(X_{\alpha\gamma} - X_{\alpha\gamma}^{eq}) \tag{5.99}$$

that is,

$$\frac{\partial R_{a\alpha}}{\partial Q_r} = -m_\alpha^{-\frac{1}{2}} \sum_{i=1}^{3} T_{\alpha i;r}(\bar{X}_i - X_{\alpha i}^{eq})/R_{a\alpha}^{eq} \tag{5.100}$$

where $T_{\alpha i;r}$ is an element of the matrix which diagonalizes the force constant or second derivative matrix.

The second derivatives can be expressed as

$$\frac{\partial^2 R_{a\alpha}}{\partial Q_r \partial Q_{r'}} = \frac{1}{m_\alpha R_{a\alpha}^3} \sum_{ij} T_{\alpha i;r} T_{\alpha j;r'}(\bar{X}_i - X_{\alpha i}^{eq})(\bar{X}_j - X_{\alpha j}^{eq})(\delta_{ij} - 1) \tag{5.101}$$

The terms $V_r^{(1)}$ and $V_{rr'}^{(2)}$ in eq. (5.96) are those responsible for collision-induced phonon excitation. These terms depend on time through the time-dependence of the trajectory. If only the linear terms $V_r^{(1)}$ are included, the oscillators are called "linearly forced." Omitting the zero-point energy of the solid (ϵ_0), we can write the total hamiltonian for an atom colliding with the surface as

$$H = \frac{1}{2}\sum_{r=1}^{3N-6} (\dot{Q}_r^2 + \omega_r^2 Q_r^2) + V_I^{(0)} + \sum_r V_r^{(1)} Q_r +$$
$$+ \frac{1}{2}\sum_{rr'} V_{rr'}^{(2)} Q_r Q_{r'} + \frac{1}{2M} P_R^2 \tag{5.102}$$

where the last term is the kinetic energy of the atom and M its mass. The evolution operator for the phonons perturbed linearly and quadratically is known, as mentioned above, from the theory of collision-induced energy transfer to polyatomic molecules. It may be expressed in terms of the boson creation/annihilation operators. In the classical path treatment, the dynamics of the gas-phase atoms is treated classically. However, a partial quantization of the gas-phase degrees of freedom is sometimes necessary. This is the case if problems with zero-point vibrational energy occur or if tunneling phenomena need to be treated properly (see also below).

The effective hamiltonian which governs this motion is then conveniently approximated by

$$H_{\text{eff}} = \langle\Psi|H|\Psi\rangle = \langle\Psi_0|U^+ H U|\Psi_0\rangle \tag{5.103}$$

where $U(t, t_0)$ is the evolution operator for the phonons and $|\Psi_0\rangle$ the initial wave function, that is,

$$|\Psi_0\rangle = \prod_{k=1}^{M} |n_k^0\rangle \tag{5.104}$$

If only linear terms in Q_{lk} and $\alpha_k^{\pm}$ (see chapter 4) are included, the final expression for the effective hamiltonian is

$$H_{\text{eff}} = \sum_k \hbar\omega_k(\rho_k + n_k^0) + V_I^{(0)} + \sum_k \omega_k^{-1}\epsilon_k(t) + \frac{P_R^2}{2M}$$

$$+ \frac{1}{2} \sum_{kl} b_k b_l V_{kl}^{(2)} \{(2n_l^0 + 1)\delta_{lk} + \exp(i(\omega_k - \omega_l)t)[n_k^0 Q_{lk} + n_l^0 Q_{kl}^*]$$

$$+ \exp(i(\omega_l - \omega_k)t)[Q_{lk}^*(n_k^0 + 1) + Q_{kl}(n_l^0 + 1)]\} + O(Q_{kl}^2) \tag{5.105}$$

In the derivation, we have used the fact that the initial wave function is taken to be a "pure" phonon state, and the relation [147]

$$U_{VT}^+ U_{VV}^+ a_k U_{VV} U_{VT} = a_k + i\alpha_k^+ + \sum_j Q_{kj}(a_j + i\alpha_j^+) \tag{5.106}$$

$$U_{VT}^+ U_{VV}^+ a_k^+ U_{VV} U_{VT} = a_k^+ - i\alpha_k^- + \sum_j Q_{kj}^*(a_j^+ - i\alpha_j^-) \tag{5.107}$$

We have, furthermore, neglected terms of second or higher order in Q_{kl} and $\alpha_k^{\pm}$. We shall consider, below, the case where a Boltzmann distribution of initial phonon states is introduced.

The functions $Q_{lk}(t)$ are given to first order by

$$Q_{kl}(t) = Q_{lk}(t) = \frac{b_k b_l}{2i\hbar} \int_{t_0}^{t} dt' V_{kl}^{(2)}(t') \exp(i(\omega_k - \omega_l)t) \tag{5.108}$$

where the trajectory for the incoming atom is assumed to start at $t = t_0$. In eq. (5.105) ρ_k is the so-called "excitation strength" for oscillator k, that is, a measure for how much this particular phonon mode is excited. Initially, we then have $\rho_k(t_0) = 0$ and finally, that is, for $t \to \infty$, where the atom has left the surface, the only remaining terms in eq. (5.105) are with $n_k^0 = 0$

$$E_{int} = \sum_{k=1}^{3N-6} \hbar\omega_k \rho_k \tag{5.109}$$

and the kinetic energy term. Thus E_{int} is the energy transfer to the solid. The quantity $\epsilon_k(t)$ in eq. (5.105) is defined as

$$\epsilon_k(t) = V_k^{(1)}(R(t)) \int_{t_0}^{t} dt' V_k^{(1)}(R(t')) \sin(\omega_k(t' - t)) \tag{5.110}$$

This term arises from the expectation value of the linear $V_k^{(1)}Q_k$ term in eq. (5.102). These terms are called phonon VT terms because they exchange one quantum of vibrational energy when forced by the translational motion of the incoming atom. The second derivative terms in eq. (5.102) contain the operators $a_r^+ a_{r'}$ and $a_r a_{r'}^+$, that is, these terms create a quantum in one mode and destroy one in another. It turns out, however, that these so-called VV terms are of minor importance for energy transfer to solids. These terms may, therefore, be neglected [148]. But their effect can easily be incorporated using the methodology outlined in chapter 4.

Equation (5.109) shows that $E_{int} > 0$ irrespective of the initial state ofthe solid, and thereby also of the surface temperature. The explanation for this artifact is not to be found in the harmonic approximation for the oscillators, but is rather due to the self-consistent field (SCF) approximation (see chapter 3) made when coupling the two subsystems, the atom and the solid. In order to improve this situation, corrections must be introduced (see section 3.2). If the detailed balance corrections described in section 3.2 are introduced, we may derive an effective hamiltonian which depends on surface temperature (for details, see ref. [148]).

If we were to neglect the VV processes, we would get the following effective hamiltonian from eq. (5.105)

$$H_{eff} = E_{int} + V_I^{(0)}(R) + \sum_k \omega_k^{-1} \epsilon_k(R, t) + \frac{P_R^2}{2M} + \sum_k \hbar \omega_k n_k^0 \qquad (5.111)$$

The equations of motion are obtained using Hamilton's principle, that is,

$$\frac{dH_{eff}}{dt} = \frac{\partial H_{eff}}{\partial t} + \frac{\partial H_{eff}}{\partial R}\dot{R} + \frac{\partial H_{eff}}{\partial P_R}\dot{P}_R \qquad (5.112)$$

yield

$$\dot{R} = \frac{\partial H_{eff}}{\partial P_R} = \frac{P_R}{M} \qquad (5.113)$$

and

$$\dot{P}_R = -\frac{\partial H_{eff}}{\partial R} = -\frac{\partial V_I^{(0)}}{\partial R} + \sum_k \omega_k^{-1} \frac{\partial \epsilon_k(R, t)}{\partial R} \qquad (5.114)$$

Thus, if we require $(d/dt)H_{eff} = 0$, we get:

$$\frac{\partial H_{eff}}{\partial t} = \sum_{k=1}^{3N-6} \hbar \omega_k \dot{\rho}_k + \sum_k \omega_k^{-1} \frac{\partial \epsilon_k}{\partial t} = 0 \qquad (5.115)$$

or

$$\frac{\partial \epsilon_k}{\partial t} = -\hbar \omega_k^2 \dot{\rho}_k = -\omega_k \frac{d}{dt}\Delta E_k \qquad (5.116)$$

where $\Delta E_k = \hbar \omega_k \rho_k$. We now notice that eq. (5.116) may be used to evaluate ϵ_k in the effective potential

$$\langle V^{(1)} \rangle = \sum_k \omega_k^{-1} \epsilon_k(R, t) \qquad (5.117)$$

from ΔE_k. This observation has been used to incorporate a simple correction to the SCF-treatment of atom/molecule surface scattering and, hence, for obtaining an initial state dependence of the effective potential and the energy transfer [149]. This is essential for making it possible to include surface temperature effects on the energy transfer. Thus the function ϵ_k is replaced by

$$\epsilon_k(R,t) \to V_k^{(1)}(R(t))\omega_k \eta_k(t,T_s) \tag{5.118}$$

where $\eta_k(t,T_s)$ depends on the surface temperature T_s

$$\eta_k(t,T_s) = -\frac{1}{\hbar\omega_k} \int dt' \frac{d}{d\rho_k}[\Delta E_k^+(T_s) + \Delta E_k^-(T_s)]$$
$$\times [I_{ck}(t')\cos(\omega_k t') + I_{sk}(t')\sin(\omega_k t')] \tag{5.119}$$

where $\Delta E_k^\pm(T_s)$ is the energy transfer to the phonons in connection with excitation $(+)$ and deexcitation $(-)$ processes and

$$I_{ck}(t) = \int dt' V_k^{(1)} \cos\omega_k t \tag{5.120}$$

$$I_{sk}(t) = \int dt' V_k^{(1)} \sin\omega_k t \tag{5.121}$$

The reason for this splitting in the two processes has to do with a "detailed balance" correction introduced such that the value for ρ_k for the transition m_k to n_k in mode k is

$$\rho_k \to \rho_k \exp(\mp a_k |m_k - n_k|\omega_k^2) \tag{5.122}$$

where the parameter a_k can be derived using the simple collision model (see also eq. (3.12)). We prefer to express a_k as [150]

$$a_k = \frac{1}{2}\hbar\omega_k \frac{d\ln\rho_k}{dE_0} \tag{5.123}$$

where E_0 is the kinetic energy of the incoming molecule. ρ_k and, thereby, a_k can be estimated using a simple first-order collision model as

$$\rho_k = C\frac{\cosh(\omega\tau/a)^2}{\sinh(\omega\pi/a)^2} \tag{5.124}$$

where the constant C is independent of E_0, $a = \alpha v_0 \cos\theta/2$, $E_0 = \frac{1}{2}mv_0^2$, m the mass of the incoming atom/molecule, θ the approach angle and $\cot\tau = -\sqrt{E_m/E_0}$. E_m is the well depth. Thus, the parameters α and E_m can be obtained from the molecule-surface interaction potential.

However, it should be emphasized that although we know in principle how to make the detailed balance correction for systems where the classical path approximation is applicable, it is difficult to apply these methods in situations like the present, where each transition in the solid defines a best classical path. Hence, the introduction of the above approximate approach. But without this detailed balance correction, there would be no surface temperature dependence of the effective potential!

Aside from phonons, the electron-hole pair excitation (where the collision induces electronic transitions of electrons from below to above the Fermi level) may also be important for a realistic description of the energy loss to the solid. It is, therefore, important that also these processes can be described within the SQ approach, involving fermion operators rather than boson operators [149]. The excitation processes due to electron-hole pair excitations in solids also influence the dynamics of the incoming atom/molecule. The influence of these excitation processes on the incoming atom/molecule is again described through a mean-field potential, obtained as the expectation value of the molecule-electron interaction. This adds an additional time-dependent effective potential [148]. Hence the influence of both phonon and electron-hole pair excitation on the dynamics of the gas-phase atoms can be approximately accounted for by adding an effective potential to the hamiltonian for the molecule-surface interaction.

Thus we have

$$H_{eff} = T_{kin} + V_I^0 + V_{eff}^{ph} + V_{eff}^{elho} \tag{5.125}$$

where T_{kin} is the kinetic energy of the atoms in the gas phase. The next term is the interaction potential for the interaction between the atom/molecule and the surface with the solid atoms in their equilibrium positions. In the case of a molecule, V_I^0 also includes the intramolecular potential. The next two terms are then the mean field approximation to the interaction between the gas atoms and the solid, due to phonon and electron-hole pair excitation, respectively. It is important that both terms can be obtained within the SQ formulation, that is, without calculating the wave function explicitly. The first term can be written as

$$V_{eff}^{ph} = \sum_k V_k^{(1)} \eta_k(t, T_s) \tag{5.126}$$

and an approximate expression for the latter has been derived (see for, instance, ref. [148]) to be

$$V_{eff}^{elho} = 2\mathrm{Re} \sum_{k=1}^{F} \sum_{l=F+1}^{N} j_0\left(2\sqrt{A_{kk}^{(2)}}\right) \frac{H_{kl}}{\hbar} \int_0^t dt' H_{kl}^* \sin(\omega_{kl}(t - t')) \tag{5.127}$$

where $j_0(2x) = \sin(2x)/2x$, and where the index k runs over energy levels below the Fermi level F and l over levels above. H_{kl} is a matrix element for excitation of an electron from level k to level l (a hopping integral) and $\hbar\omega_{kl}$ the energy separation between the two levels.

Table 5.4 shows results obtained using the quantum classical theory for molecule surface scattering of hydrogen and deuterium from a Cu(100) crystal with 130 atoms arranged in 4 layers. The molecule approaches the surface with zero incident angle to the surface normal and the surface temperature is 300 K. We notice that the sticking probability, defined as the sum of the adsorption and the dissociation probability at the surface, is higher for the heavy molecule. This is due to stronger phonon coupling and, hence, increased energy transfer to the surface E_{int}. The rotational excitation is not much affected by the inclusion of coupling to the electron-hole pair mechanism.

Table 5.4: Energy transfer to a Cu(001) surface as a function of the coupling, that is, with and without electron-hole pair mechanism for deuterium scattered from the surface. The initial kinetic energy is 49.6 kJ/mol and the initial vibrational/rotation state is (0,0). The zero-point energies are: 25.9 and 18.5 kJ/mol for H_2 and D_2, respectively.

Quantity	Molecule	With phonon coupling	With phonon and electron/hole coupling
P_{stick}	D_2	0.58	0.46
	H_2	0.30	0.12
E_{int}	D_2	11.6	12.7
	H_2	6.8	7.4
E_{rot}	D_2	5.8	5.7
	H_2	7.2	9.9
E_{vib}	D_2	15.8	17.3
	H_2	20.9	27.7

Average over 100 collisions. Error bars about 5%.

Table 5.4 also shows that classical mechanics has a problem in conserving the zero-point vibrational energy—a common problem in so-called quasi-classical calculations in which the molecule is started in the ground vibrational state with the correct zero-point energy. Due to strong coupling to the substrate, some or in principle all of it is available for the other degrees of freedom (translation and rotation), that is, also for overcoming the barrier to reaction at the surface. Hence, low energy results of sticking probabilities obtained using classical mechanics should be considered with some reservation.

Introducing the center-of-mass coordinates of the molecule as (X, Y, Z) and (x, y, z) for the diatomic molecule with bond length r and orientation θ, ϕ we have

$$x = r \sin\theta \, \cos\phi \qquad (5.128)$$

$$y = r \sin\theta \, \sin\phi \qquad (5.129)$$

$$z = r \cos\theta \qquad (5.130)$$

In order to avoid the problem with the zero point energy, we need to quantize at least the vibrational degree of freedom r. Thus, we could use a quantum grid method, in for instance, the coordinates r and Z, where r is the bond distance and Z the perpendicular distance from the center of mass of the molecule and the surface. In this way, tunneling in the Z coordinate is also treated correctly. The remaining four degrees of freedom (X, Y, θ, ϕ) could still be treated classically [151]. Although the influence of electron-hole mechanism on molecule dynamics from table 5.4 appears to be modest, we can see a clear difference if more detailed information is considered. Figure 5.5 shows the so-called excitation strength for excitation of individual phonon modes. Since the electron-hole coupling has a tendency to enhance the lifetime at the surface, the

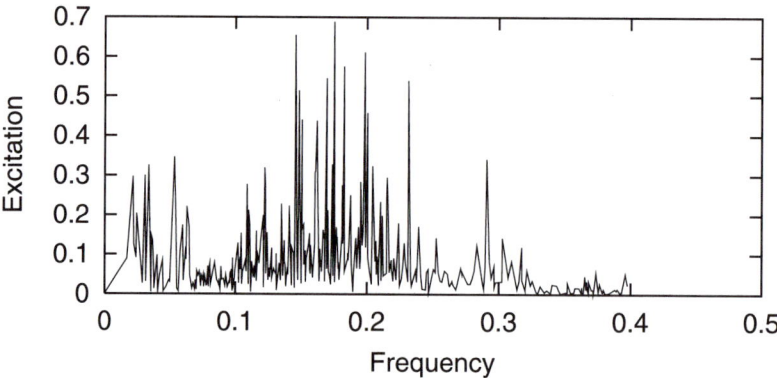

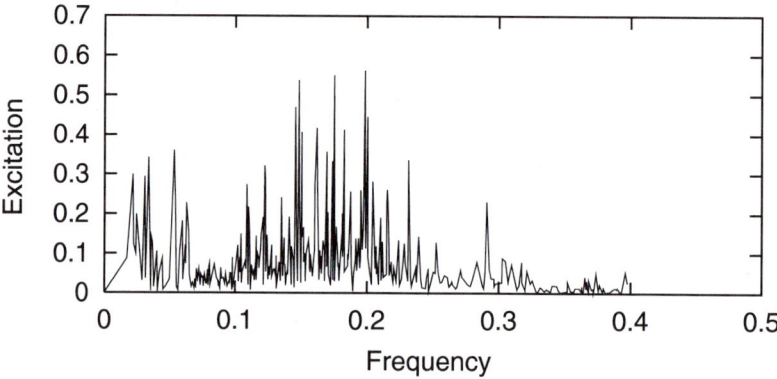

Fig. 5.5: The excitation ρ_k of phonon modes as a function of frequency ω_k in τ^{-1} for deuterium scattered from a Cu(001) surface. The numbers are obtained as averages over 100 collisions with perpendicular impact and initial kinetic energy 49.6 kJ/mol. The initial vibrational/rotational state is (0,0) and the surface temperature 300 K. The upper panel shows the excitation if electron-hole pair coupling is included, the lower is without electron-hole pair coupling.

interaction with the surface is stronger, and hence the increase in ρ_k for some modes.

The most elaborate calculations have included a quantum treatment of the six molecule coordinates through the TDGH-DVR method described in section 2.8 using from 1–3 million grid points and solving the dynamics in an effective potential including both phonon and electron-hole pair coupling [152].

5.4 Reaction path methods

The reaction path theory offers an approximate methodology for obtaining state-resolved reaction rate constants with a minimum of information about the potential energy surface, namely gradients and hessians along the reaction path.

The theory is applicable for systems with simple isolated transition states, that is, for systems having only one imaginary eigenvalue of the hessian (second derivative matrix of the potential) at the transition state. It is possible to handle reactions with several such transition states. However, if the reaction path bifurcates, or if the saddle point is of higher order, it is necessary to turn to other approaches, such as reaction surface or volume theories. The latter method is discussed in section 5.5.

As mentioned, the advantage of the reaction path (RP) method is that it requires only the potential, gradients, and hessians as a function of the nuclear coordinates along the reaction path. Thus, the theory avoids the often tedious construction of the full $3N - 6$ dimensional potential energy surface and is able to use the information given by standard ab initio program packages directly. It may, therefore, form the dynamical basis for a "direct" ab initio theory to dynamics of complicated systems [168]. In this respect, it is similar to traditional transition state theory but the advantage is that state-resolved rates can be obtained and not just total reaction rates. It should be mentioned, however, that vibrationally resolved rates are available from the somewhat more complicated generalized transition state theory [200].

Contrary to the ordinary transition state theory, the reaction path theory is a dynamical theory which incorporates all degrees of freedom. Thus the reaction path theory provides a hamiltonian, and the dynamics may be solved using classical, quantum-classical, or quantum mechanical descriptions. However, the hamiltonian is an approximate one and the results are therefore also approximate.

The fact that the state-resolved rates can be obtained is important. State-resolved rates are of importance for the simulation of many non-equilibrium systems in which the reactants do not necessarily have an equilibrium distribution of the quantum states. Especially the vibrational states have long relaxation times (on the time-scale of molecular reactions). Hence, vibrational state resolved rates are necessary in order to understand and simulate the kinetics of many important non-equilibrium systems.

The reaction path methodology has been developed over the last 20–30 years. Early theories involving natural collision coordinates are due to Marcus [169], Hofacker, and Levine [170]. Miller et al. [171] derived a reaction path hamiltonian by using a methodology known in spectroscopical approaches to the semirigid bender molecule [172]. Billing derived [173] the reaction path hamiltonian for $J \neq 0$ using an approach known from standard spectroscopic methodology [174].

The reaction path method has recently been reviewed by Kraka [178] and Quapp [198].

5.4.1   Surface characteristics

According to eq. (5.8), the potential energy surface is given as $V(\mathbf{R}) = E_{el}(\mathbf{R}) + V_{NN}(\mathbf{R})$. For a (non-linear) molecule composed of N atoms, the potential is a function of $3N - 6$ internuclear internal coordinates, that is, 3 variables for a triatomic, 6 for a four atomic system, and so on. The calculation and characterization of this multi-dimensional function presents a major challenge in theoretical chemistry. Although the potential is a global function of the coordinates, some features and regions are more important than others. Regions of interest for spectroscopy and for the study of chemical reactions are, for example, minima and saddle points of first order. Minima correspond to stable forms of the molecule: the global minimum to the most stable and local minima to stable forms of higher energy. A saddle point of first order is the point of highest energy along the minimum energy path between two minima. A saddle point of second order has two imaginary eigenvalues of the hessian. In such cases, the simple reaction path hamiltonian is not expected to be adequate for treating the dynamics, but reaction surface and volume theories will be applicable in these cases. In the following, we shall assume that the saddle point is of first order. At the minima we have that the gradient vanishes, that is,

$$g_k = \frac{\partial V}{\partial R_k} = 0 \tag{5.131}$$

where $k = 1, 2, \ldots, 3N$ and where R_k denotes the cartesian nuclear coordinates.

At the critical point, there is no force acting on the atoms in the molecule. For further characterization, we also need the hessian matrix as defined above, that is, the second derivative or force constant matrix. At a minimum, the eigenvalues of this hessian matrix are all positive. At a saddle point of first order, one of the eigenvalues is negative, that is, there is a maximum in one dimension and a minimum in $3N - 7$ dimensions. Saddle points of higher order will then be characterized by more negative eigenvalues of the hessian.

In order to find the critical points, one can use the so-called Newton-Raphson method, in which the potential is expanded at a point close to the critical point R_c

$$V(\mathbf{R}) = V(\mathbf{R}_0) + \sum_{k=1}^{3N-6} g_k(\mathbf{R}_0)\Delta R_k + \frac{1}{2}\sum_k \sum_l H_{kl}(\mathbf{R}_0)\Delta R_k \Delta R_l \tag{5.132}$$

where

$$\Delta R_k = R_k - R_{0k} \tag{5.133}$$

In matrix/vector notation, we write

$$V(\mathbf{R}) = V(\mathbf{R}_0) + \mathbf{g}^T \Delta \mathbf{R} + \tfrac{1}{2}\Delta \mathbf{R}^T \mathbf{H} \Delta \mathbf{R} \tag{5.134}$$

Or for the gradient near the critical point:

$$\mathbf{g}(\mathbf{R}) = \mathbf{g}(\mathbf{R}_0) + \mathbf{H}(\mathbf{R}_0)\Delta \mathbf{R} \tag{5.135}$$

and, hence, by the definition of a stationary point we get:

$$\Delta \mathbf{R} = -\mathbf{H}(\mathbf{R}_0)^{-1}\mathbf{g}(\mathbf{R}_0) \tag{5.136}$$

This equation defines the Newton-Raphson step in the direction of the minimum. Only if the potential is an exact quadratic function over the entire region $\mathbf{R}_0$

to $\mathbf{R}_c$ is it possible to reach the critical point in one step. This is usually not the case, hence, the derivative and the hessian have to be calculated at the new point and the procedure repeated until convergence.

As mentioned above, the potential energy surface (PES) is a function of $3N - 6$ internal coordinates describing the relative positions of the nuclei. Thus, at present, the complete surface is known to "chemical accuracy" only for few triatomic systems. By chemical accuracy, we mean the accuracy needed at room temperature to predict the rate constant with an accuracy better than 30–50%. This requires about 1 kJ/mol (0.2 kcal/mol) estimates of the barrier height. For the prediction of differential cross sections and scattering resonances in agreement with the best experimental resolution, even higher accuracy is needed, and also not just in the saddle point region. Although significant progress has been made in the "exact," that is, quantum mechanical treatment of reactive scattering processes for 3 and 4 atomic reactions, the requirements mentioned above give unfavorable odds for pursuing the ab initio quantum chemical + dynamical treatment of other than a few benchmark systems.

The quantum mechanical solution of the nuclear dynamical problem is inherently more difficult than the electronic structure problem. There are at least two reasons for this. The first is that the potential is unknown, contrary to the situation for the electronic structure problem. It may be determined by solving the electronic problem many times, using however a variety of methods, the reason being that an ab initio method, which works for some nuclear configurations, may not be accurate enough for others. The second reason is that the variational methods, which have been so successful for solving the electronic problem, although in principle also applicable for the nuclear dynamical one give absolute accuracies for the reaction probabilities or cross sections. This is not sufficient for calculation of many dynamical "properties." The reaction rate constant for instance varies as a function of temperature by many orders of magnitude and we sometimes have to determine very small reaction probabilities, where a relative rather than an absolute accuracy is required.

Because of these problems, we are forced to use methods which somehow minimize the requirement to the information on the PES and which allow for an efficient numerical solution of the many-body problem, and at the same time methods which are able to incorporate quantum effects when present (that is, to be able to account for tunneling and to incorporate a proper treatment of zero-point vibrational energy, and so on). The reaction path method is one such approach.

Although it is, in principle, possible to use the reaction path method in a direct dynamics set up—that is, to couple the method to an ab initio program package, most applications of the RP theory have still been carried out using analytical potential energy surfaces in which the parameters have been fitted so as to reproduce, as much as possible, all known properties. As a model system we have chosen H_2+CN. Table 5.5 shows the saddle point geometry and eigenvalues of the hessian matrix. This information is, together with the reactant frequencies and geometries, all that is needed for obtaining the transition state rate constant (see fig. 5.14 below).

Table 5.5: Transition state geometry obtained by
Newton-Raphson search on the H_2-CN surface [193]. The
search is carried on until the sum of the squared gradients
is less than 10^{-5}. The eigenvalues of the hessian matrix at
the saddle point are also listed. We notice that the
structure is linear and that two frequencies are degenerate.

Atom	Position	$\omega^2 \, \tau^2$
H	-0.00196	-1.68555
H	0.76102	0.04627
C	2.42089	0.04624
N	3.58066	1.12020
		1.12044
		16.90073
		35.00330

$1\tau = 10^{-14}$ sec.

5.4.2   The reaction path method

In order to determine the reaction path, one starts at the saddle point, determined by the Newton-Raphson technique described in the previous section, and
follows the steepest descent path to the reactant and product valleys (see flow
diagram 5.13). Along the steepest descent path, one takes steps in the direction
which leads to a maximal decrease in potential energy. Thus we wish to maximize

$$\Delta V(\mathbf{R}) = \sum g_k \Delta R_k \tag{5.137}$$

subject to the constraint that $\Delta \mathbf{R}^2$ is fixed. In order to do so we introduce the
functional

$$I(\mathbf{R}) = \sum_k g_k \Delta R_k + \lambda \sum_k \Delta R_k^2 \tag{5.138}$$

where λ is a Lagrange multiplier. The extremum condition for ΔR_k then
becomes:

$$\Delta R_k = -\frac{1}{2\lambda} g_k \tag{5.139}$$

that is, the steepest descent path is proportional to the negative gradient of the
energy in that direction. By following the steepest descent path, we change the
coordinates of all the N atoms of the system. If the atoms were moving infinitely
slowly, the path would be mapped out by the classical equations of motion.
However, the displacement of the atoms should not be mass dependent. This
can be avoided by introducing mass-weighted coordinates in a similar manner
to what is done in the normal mode treatment of vibrational motion. The shift
to mass dependent coordinates lead to the definition of the so-called intrinsic
reaction path [197]. The intrinsic reaction path (IRP) is simply defined as the

steepest descent path in mass-weighted coordinates. Thus in the mass-weighted cartesian position coordinates we have changed the notation, such that

$$x_i = \sqrt{m_j} X_{j,\alpha} \tag{5.140}$$

where $i = 1, 2, 3$ for $j = 1$ and $\alpha = x, y, z$, etc. $X_{j,\alpha}$ is the position coordinate of atom j in direction α.

Along the IRP we have

$$\frac{d\mathbf{x}(s)}{ds} = -\frac{\nabla V}{|\nabla V|} \tag{5.141}$$

where s is the IRP arc length along the path, that is, it is an integral over displacements ds defined as:

$$ds = \pm \sqrt{\sum_{i=1}^{3N} (dx_i)^2} \tag{5.142}$$

where the $(-)$ sign by convention refers to the path from the saddle point to reactants and the $(+)$ sign refer to the path to products. Note that the unit of s is $\sqrt{\text{amu}}\text{Å}$.

We notice that the IRP can be defined by the equation:

$$\frac{d\mathbf{x}(t)}{dt} = -\nabla V(\mathbf{x}) \tag{5.143}$$

where t is a parameter. We now introduce a local quadratic expansion around the point $\mathbf{x}_0$, that is,

$$V(\mathbf{x}) = V(\mathbf{x}_0) + \nabla V(\mathbf{x}_0) \cdot (\mathbf{x} - \mathbf{x}_0) + \tfrac{1}{2}(\mathbf{x} - \mathbf{x}_0)^T \cdot \mathbf{H}(\mathbf{x}_0) \cdot (\mathbf{x} - \mathbf{x}_0) \tag{5.144}$$

From the above two equations we have

$$\frac{d}{dt}(\mathbf{x} - \mathbf{x}_0) = -\nabla V(\mathbf{x}_0) - \mathbf{H}(\mathbf{x}_0)(\mathbf{x} - \mathbf{x}_0) \tag{5.145}$$

This equation has the solution

$$\mathbf{x}(t) = \mathbf{x}_0 - \sum_{i=1}^{3N} \mathbf{T}_i^T \cdot \nabla V(\mathbf{x}_0)[(\exp(\lambda_i(t - t_0)) - 1)/\lambda_i]\mathbf{T}_i \tag{5.146}$$

where

$$-\mathbf{H}\mathbf{T}_i = \lambda_i \mathbf{T}_i \tag{5.147}$$

that is, the $\mathbf{T}_i$s are eigenvectors and λ_i the corresponding eigenvalues to the hessian matrix evaluated in mass-weighted coordinates. Thus the IRP is found by starting at the saddle point and using the above scheme in the direction of products and reactants. Each atom of the composite system has its own path, independent of the choice of coordinate system [198]. At each point we need to calculate the gradient and the hessian. The value of s is updated along the path, using eq. (5.142). Note that the above procedure fails at the saddle point itself (the gradient being zero). Thus, the first small step is taken in the positive or negative direction of the vector corresponding to the imaginary frequency—found by diagonalizing the hessian at the saddle point. For large molecules as, for instance, proteins, special techniques for determining the global reaction path connecting minima arising from conformal changes have been devised [199].

Figure 5.6 shows the energy profile $V_0(s)$ for the H_2+CN system. The positions of the four nuclei are displaced according to the equation (5.146) using the potential surface from ref. [193].

5.4.3 Reaction path constraints

Consider a rigid body rotating about a fixed axis with a constant angular velocity, ω, in radians per second. This rotation can be described by a vector $\boldsymbol{\omega}$, with length ω, and a direction parallel to the axis of rotation. A point P not on the rotation axis will then have a linear velocity given by

$$\mathbf{v} = \boldsymbol{\omega} \times \mathbf{r} \tag{5.148}$$

where $\mathbf{r}$ is the radius vector from the point P to a fixed point O on the axis of rotation. If the point has a mass m, the momentum is

$$m\mathbf{v} = m(\boldsymbol{\omega} \times \mathbf{r}) \tag{5.149}$$

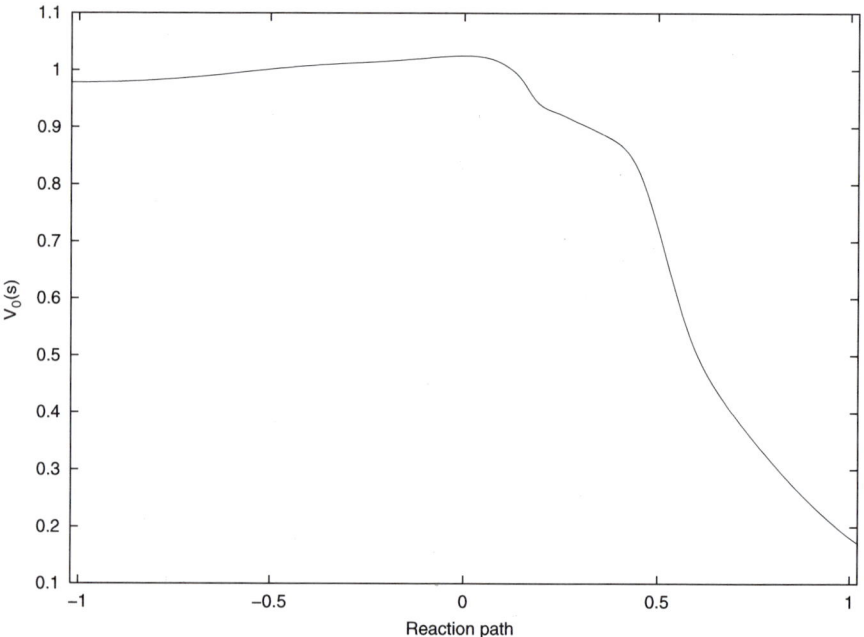

Fig. 5.6: The intrinsic reaction path for the reaction $H_2 + CN \rightarrow HCN + H$ using the method outlined in section 5.4.2.

and the angular momentum about the point O is

$$\mathbf{M} = \mathbf{r} \times m\mathbf{v} = m[\mathbf{r} \times (\omega \times \mathbf{r})] \tag{5.150}$$

Considering now the motion of a molecule consisting of N atoms, we introduce the position vector of atom i in a space-fixed coordinate system

$$\mathbf{r}_{0i} = \mathbf{r}' + \mathbf{r}_i \tag{5.151}$$

where the vector $\mathbf{r}'$ denotes the origin of the coordinate system $O'_{x'y'z'}$ located at the center of mass of the molecule. $\mathbf{r}_i$ is the position vector of atom i in the coordinate system $(x'y'z')$. The motion of an atom i relative to the center of mass can be thought of as consisting of two contributions: a rotational motion around the center of mass connected to the rotation of a rigid framework where the atoms have their equilibrium positions, and an oscillatory or displacement motion away from this position. Thus, the position vector $\mathbf{r}_i$ can be written as

$$\mathbf{r}_i = \mathbf{a}_i + \rho_i \tag{5.152}$$

where $\mathbf{a}_i$ denotes the equilibrium position vector and ρ_i the displacement vector. Considering now the velocity of an atom i we have, in the space-fixed coordinate system

$$\mathbf{v}_{0i} = \mathbf{v}' + (\omega \times \mathbf{r}_i) + \mathbf{v}_i \tag{5.153}$$

where the first term is the velocity of the center of mass, the second the linear velocity of the atom i due to the rotation of the molecule around the origin $O'_{x'y'z'}$, while the last term is the translational velocity of the atom relative to the $O'_{x'y'z'}$, that is, $\mathbf{v}_i = d\mathbf{r}_i/dt$. The motion is subject to six constraints, namely that

$$\sum_i m_i \mathbf{v}_i = \mathbf{0} \tag{5.154}$$

These three equations fix the center of mass at the origin of $O'_{x'y'z'}$. The other three constraints are due to Eckart and, are, therefore, called the Eckart condition [186]. It states that if atom i is in its equilibrium position, where $\mathbf{r}_i = \mathbf{a}_i$, then there should be no additional angular momentum relative to $O'_{x'y'z'}$.

Considering the angular momentum, we have

$$\mathbf{M} = \sum_i m_i[\mathbf{r}_i \times (\omega \times \mathbf{r}_i)] + \sum_i m_i \mathbf{r}_i \times \mathbf{v}_i \tag{5.155}$$

and setting $\mathbf{r}_i = \mathbf{a}_i$ we must (due to this constraint) require that:

$$\sum_i m_i \mathbf{a}_i \times \mathbf{v}_i = \mathbf{0} \tag{5.156}$$

since the first term is the angular momentum of a rigid body rotating around an axis through the origin $O'_{x'y'z'}$ with an angular velocity ω. These six constraints leave us with $3N - 6$ degrees of freedom for the displacement from the reference frame defined by the vectors $\mathbf{a}_i$. The complete motion of the 3N atoms has then been divided into three parts:

a)   Translational motion of the center of mass,
b)   rotational motion of a rigid frame around the center of mass, and
c)   displacement (vibrational) motion relative to the rotating frame.

This analysis is relevant whenever we are at a stationary point corresponding to a minimum. However, at a saddle point and along the IRP, the motion (in s) is not a small amplitude motion and, hence, must be treated differently. Thus we are left with not $3N - 6$ but only $3N - 7$ vibrational-like motions, and one translational motion along the reaction path. Formally, the s motion is projected out by introducing yet another constraint, namely that the vibrational displacement ρ_i must be perpendicular to a vector along the reaction path. The reference vectors $\mathbf{a}_i$ become dependent on the reaction path parameter s. The reaction path constraint is

$$\sum_i m_i \frac{d\mathbf{a}_i}{ds} \cdot \rho_i = 0 \tag{5.157}$$

where $d\mathbf{a}_i/ds$ is a vector along the reaction path (see fig. 5.7) for atom i and as we have seen above (see eq. (5.141)), this vector has the same direction as the force, which moves the atom along the path, that is,

$$\frac{d\mathbf{a}_i}{ds} = -const \left. \frac{\partial V}{\partial \mathbf{r}_i} \right|_{\mathbf{r}_i = \mathbf{a}_i} \tag{5.158}$$

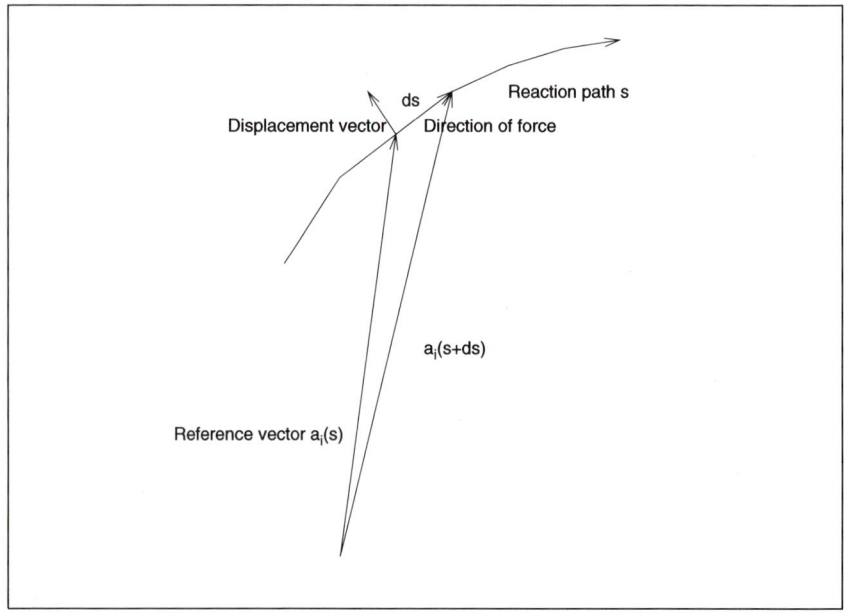

Fig. 5.7:   The reaction path constraint states that the "perpendicular" displacement from the reaction path is perpendicular to the force vector along the path.

By introducing this seventh constraint, it is possible to derive a reaction path hamiltonian [171], [173]. The result is after some lengthy algebra obtained as

$$H = \frac{p_s^2}{2\tilde{N}} + H_{rot} + H_{vib} + V_0(s) - \frac{p_s}{\tilde{N}} \sum_{kk'} P_k Q_{k'} B_{kk'} \tag{5.159}$$

where

$$\tilde{N} = 1 + 2 \sum_k B_{kF}(s) Q_k + \sum_{kk'} \bar{C}_{kk'}(s) Q_k Q_{k'} \tag{5.160}$$

and where Q_k are normal mode coordinates for the perpendicular vibrations $k = 1, 2, \dots, 3N - 7$. P_k is the momentum conjugate to Q_k and

$$\bar{C}_{kk'} = C_{kk'} - \sum_l B_{lk} B_{lk'} \tag{5.161}$$

The symbol F denotes the reaction path degree of freedom. B_{kF}, $B_{kk'}$ and $\bar{C}_{kk'}$ are Coriolis-like coupling terms for the coupling between the reaction path motion F and the vibrational mode k and two vibrational modes k and k', respectively. These coupling terms will be defined below in section 5.4.5. p_s is the momentum for the reaction path motion, H_{rot} the hamiltonian for the overall rotational motion of the molecule, $V_0(s)$ the potential along the steepest descent path, and

$$H_{vib} = \frac{1}{2} \sum_{k=1}^{3N-7} (P_k^2 + \omega_k(s)^2 Q_k^2) \tag{5.162}$$

By expanding the factor $\tilde{N}^{-1}$ to second order in the vibrational coordinates Q_k, it is possible to bring the hamiltonian into a form for which a quantum mechanical treatment, within the second quantization method, can be carried out. The reaction path methodology is, of course, not restricted to a second-order truncation in the normal mode coordinates. But this approximation makes the formulation of the method especially compact, and it is consistent with the derivation of eq. (5.159) which is correct up to second order in the normal mode coordinates and momenta.

Figure 5.8 shows the frequencies obtained by solving the constrained eigenvalue problem along the reaction path for the HHCN system. Since this system is linear at the transition state and along the RP, we have 6 rather than 7 independent constraints. For a treatment of the constrained eigenvalue problem, see appendix D. Thus the number of frequencies is $3N - 6 = 6$ in the present case. Two of these are degenerate at the transition state. Since the complex remains linear along the IRP, this degeneracy is not lifted.

The matrix elements responsible for coupling of the motion along the IRP to the perpendicular motion are B_{kF}. They are shown in fig. 5.9.

The coupling between the various perpendicular modes is induced by the Coriolis coupling terms $B_{kk'}$ and $C_{kk'}$ (see figs. 5.10 and 5.11).

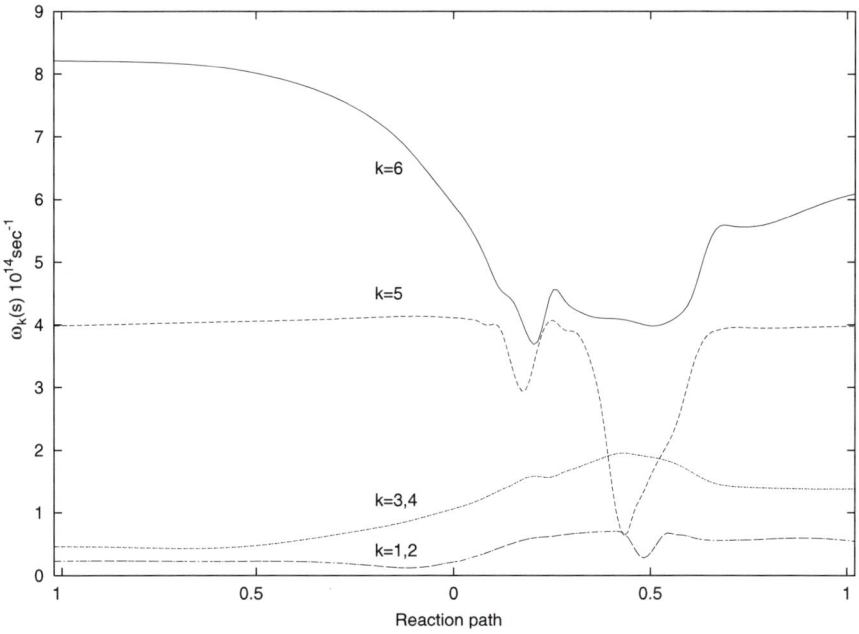

Fig. 5.8: The six vibrational frequencies $\omega_k(s)$ $(k = 1, \ldots, 6)$ for the HHCN complex.

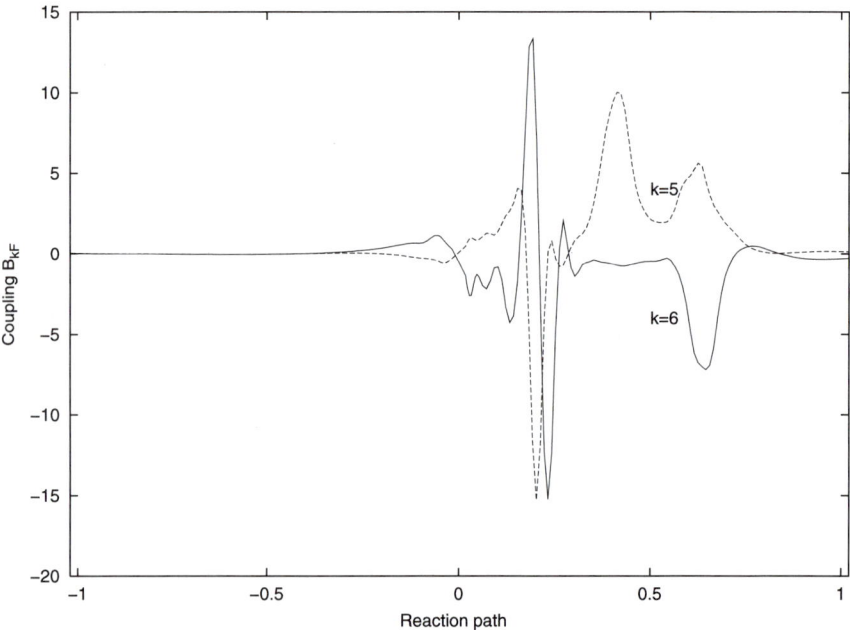

Fig. 5.9: The coupling term B_{kF} as a function of s for $k = 5$ and 6. We notice that $B_{kF} = 0$ for $k = 1, 2, 3, 4$.

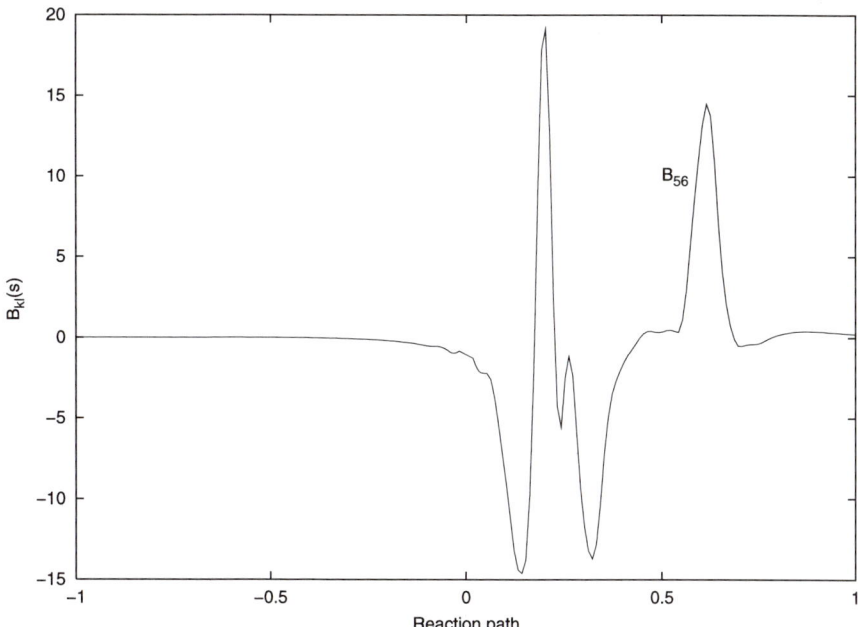

Fig. 5.10:   The coupling term B_{56} as a function of reaction path s.

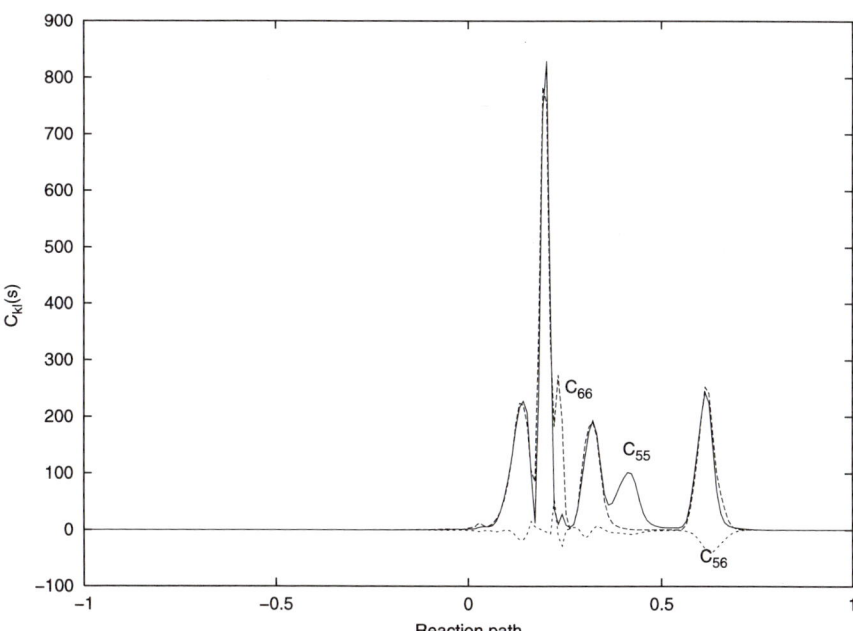

Fig. 5.11:   The coupling terms C_{55}, C_{56} and C_{66} as a function of the reaction path s.

5.4.4   Absolute rate-constants

The reaction path Hamiltonian given in the previous section describes the dynamics of a configuration (a certain molecular arrangement) obtained by following the IRC from the saddle point to the reactants, that is, from $s = 0$ to $-s_0$. The value of s_0 should be large enough for the coupling terms B_{kF}, $B_{kk'}$ and $C_{kk'}$ in (5.159) to vanish (see figs. 5.10 and 5.11), so that, at $-s_0$ we have

$$H = \frac{p_s^2}{2} + H_{vib}^0 + H_{rot}^0 \qquad (5.163)$$

This hamiltonian represents only a certain small part of the total configuration space available for the reactants in a normal collision. Thus, in order to use the reaction path hamiltonian theory in calculations of absolute cross sections, it is necessary to relate the original phase space of the reactants to that available for the system when governed by reaction path dynamics. The reaction path hamiltonian is the hamiltonian for a "supermolecule" consisting of the two reactant molecules forming a loose complex, the geometry of which is defined by following the reaction path from the transition state to the reactant channel. Since this geometry is only one particular of many possible collision encounters, it is necessary to introduce a phase space conversion factor. It has been suggested [189] that the conversion or kinematic factor, which should be introduced, is one which ensures that the reaction path approach, when the transition state approximation (no recrossing of a dividing surface) is introduced, gives the same result for the reaction rate as that obtained from transition state theory. The transition state limit is thereby contained in the reaction path approach. However, the difference is that reaction path theory defines a hamiltonian which can be used for a dynamical evaluation of the reaction rate–constant. Thus effects such as recrossing and tunneling can be incorporated, hence, the reaction path method retains the collisional or "dynamical point of view," whereas transition state theory adopts equilibrium or statistical concepts.

Considering, as an example, the collision of two diatomic molecules with initial vibrational quantum numbers n_1 and n_2, we have the following expression for a vibrationally state-resolved rate-constant

$$k_{n_1 n_2; n_1' n_2'}(T) = \sqrt{\frac{8k_B T}{\pi \mu}} \frac{1}{Q_{rot}} \frac{\pi \hbar^2}{2\mu k_B T} \sum_{l=0}(2l+1) \sum_{j_1=0}\sum_{j_2=0}(2j_1+1)(2j_2+1)$$

$$\times \int_0^\infty d(\beta E_{kin})\exp(-\beta E_{kin})\exp(-\beta E_{rot})P_{react}(E_{kin}, b, j_1 j_2)$$

$$(5.164)$$

where $\beta = 1/k_B T$, where l is the orbital angular momentum, μ the reduced mass $m_{AB}m_{CD}/(m_{AB} + m_{CD})$, Q_{rot} the rotational partition function

$$Q_{rot} = \frac{2I_1 k_B T}{\hbar^2} \frac{2I_2 k_B T}{\hbar^2} \qquad (5.165)$$

and

$$E_{rot} = \frac{\hbar^2}{2I_1}j_1(j_1+1) + \frac{\hbar^2}{2I_2}j_2(j_2+1) \tag{5.166}$$

where I_i $(i = 1, 2)$ are the moments of inertia. Consider the system at a given total energy $E = E_{kin} + E_{rot} + E_{vib}$, where $E_{vib} = E_{n_1 n_2}$ is the vibrational energy. The degeneracy factor is $g = (2l + 1)(2j_1 + 1)(2j_2 + 1)$.

If the system is treated using the reaction path hamiltonian, the four atoms enter a super-molecule with total angular momentum J. Assuming it to be a symmetric top, the rotational energy is

$$E_{rot} = AJ(J+1) + (B - C)K^2 \tag{5.167}$$

where K takes values $K = \pm J$ and A, B, C are rotational constants. Thus, the degeneracy factor for the complex is

$$g_J = (2J + 1)2\sigma_{symm} \tag{5.168}$$

where σ_{symm} is a symmetry factor which counts the number of ways we can make an identical complex. For a reaction between H_2 and OH or CN, for instance, $\sigma_{symm} = 2$ since the two hydrogen atoms can be interchanged in H_2. This symmetry factor is identical to the one entering transition state theory [190]. Introducing now the statistical factor

$$w_J = \frac{2(2J + 1)\sigma_{symm}}{(2l + 1)(2j_1 + 1)(2j_2 + 1)} \tag{5.169}$$

in the expression (5.165), we obtain the following expression for the rate-constant

$$k_{n_1 n_2 \to n_1' n_2'}(T) = \sqrt{\frac{8kT}{\pi\mu}} \frac{\pi\hbar^2}{2\mu k_B T} \frac{\sigma_{symm}}{Q_{rot}} \sum_J (2J + 1) \sum_{K=-J}^{J}$$
$$\times \int_0^\infty d(\beta E) \exp(-\beta(E - E_{n_1 n_2})) P_{react}(E, J, K) \tag{5.170}$$

where, as mentioned, we have assumed that the reaction path complex is a symmetric top with principal moments of inertia $I_{xx} \sim I_{yy}$. As a consequence, we have introduced the total angular momentum J and its projection on a body-fixed axis K. Q_{rot} is the rotational partition function of the reactants, J the total angular momentum, $E - E_{n_1 n_2}$ the energy available to reaction path motion and the rotation of the complex, and μ the reduced mass for the relative motion of the two diatomics. In the reaction path formulation, the reaction probability P_{react} is found by integrating the equations of motion numerically.

We can make a relation to transition state theory by noting that the factor

$$\sqrt{\frac{8k_B T}{\pi\mu}} \frac{\pi\hbar^2}{2\mu k_B T} \tag{5.171}$$

can be written as

$$\frac{k_B T}{h} \frac{1}{Q_{trans}} \tag{5.172}$$

where

$$Q_{trans} = \frac{(2\mu\pi k_B T)^{3/2}}{h^3} \tag{5.173}$$

is the translational partition function per unit volume [190]. The transition state result is obtained by assuming the following simple model for the reaction probability [190], [200]

$$P_{react} = 1 \quad \text{for } E - E_{n_1 n_2}(s) - H_{rot}(s) > E_0,$$
$$P_{react} = 0 \quad \text{otherwise}$$

Thus, we get

$$k_{n_1 n_2 \to n_1' n_2'}(T) = \frac{k_B T}{h} \frac{\sigma_{symm}}{Q_{trans} Q_{rot}} Q_{rot}^{\dagger} \exp(-\beta E_0) \exp(\beta(E_{n_1 n_2} - E_{n_1 n_2}(s))) \tag{5.174}$$

where E_0 is the activation energy (including zero-point energy difference between transition state and reactant vibrations), $E_{n_1 n_2}$ the energy of the reactant vibrational states and $E_{n_1 n_2}(s)$ the vibrational energy of the complex at the transition state. The rotational partition function of the complex at the transition state is

$$Q_{rot}^{\dagger}(s) = \sum_J (2J + 1) \sum_{K=-J}^{J} \exp(-\beta H_{rot}(s)) \tag{5.175}$$

For the total reaction rate, we obtain

$$k(T, s) = \frac{1}{Q_{vib}} \sum_{n_1 n_2 n_1' n_2'} \exp(-\beta E_{n_1 n_2}) k_{n_1 n_2 \to n_1' n_2'}(T) \tag{5.176}$$

where we have averaged over initial and summed over final vibrational states. The vibrational partition function is

$$Q_{vib} = \sum_{n_1 n_2} \exp(-\beta E_{n_1 n_2}) \tag{5.177}$$

The above expression can then finally be written as

$$k(T, s) = \frac{k_B T}{h} \frac{\sigma_{symm} Q_{vib}^{\dagger} Q_{rot}^{\dagger} \exp(-\beta E_0)}{Q_{vib} Q_{rot} Q_{trans}} \tag{5.178}$$

where

$$Q_{vib}^{\dagger} = \sum_{n_1 n_2} \exp(-E_{n_1 n_2}(s)\beta) \tag{5.179}$$

Thus, the rate constant depends upon s. In transition state theory, the value of s is taken to be zero, that is, the rate is evaluated at the saddle point. However, in variational transition state theory, s is varied such that the value of $k(T, s)$ is minimized.

We have shown that we can come from collision theory to the transition state result by assuming (in accordance with classical transition state theory) that the reaction probability is unity for kinetic energies in the reaction path motion above the potential activation barrier. It is now also possible to demonstrate that the transition state result (5.178) can be obtained from the reaction path hamiltonian directly (see appendix of [189]) or ref. [138], that is, without any reference to the reactant hamiltonian. This derivation shows that the total TST reaction rate is independent of the Coriolis coupling coefficients B_{kF}, $B_{kk'}$, and $C_{kk'}$. These coefficients only affect the internal distribution on vibrational levels.

Thus, this line of argument justifies the use of eq. (5.170) for the dynamical treatment of the problem in which $P_{react}(E, J, K)$ is evaluated numerically using the RP hamiltonian.

5.4.5 Second quantization approach

In the RP method, the motion along the reaction path is projected out and treated separately. Aside from this motion, three degrees of freedom are used to describe the overall rotation of the system. Since the center-of-mass fixes three coordinates, we have $3N - 7$ degrees of freedom left for the vibrational motion. For polyatomic systems or reactions in clusters or solutions, the number of atoms N is large. It is, therefore, convenient to introduce a methodology which is capable of handling the dynamics even for large values of N. Such an approach is obtained if the potential is expanded just to second order in the normal mode coordinates Q_k. If this is done, it is consistent also to expand the denominator $\tilde{N}^{-1}$ in eq. (5.160) to second order. The second quantization approach is now introduced by replacing the normal mode coordinates Q_k and momenta p_k by creation/annihilation operators, such that

$$Q_k = \sqrt{\frac{\hbar}{2\omega_k}}(a_k + a_k^+) \tag{5.180}$$

$$p_k = i\sqrt{\hbar\omega_k/2}(a_k^+ - a_k) \tag{5.181}$$

The RP hamiltonian can then be expressed as

$$H = H_{rot}^0 + H_{vib}^0 + \frac{p_s^2}{2} + V_0(s) + \sum_k f_k(a_k + a_k^+) + \sum_{kk'} f_{kk'} a_k a_{k'}^+ \tag{5.182}$$

where

$$H_{vib}^0 = \sum_k \hbar\omega_k(s)(a_k^+ a_k + \tfrac{1}{2}) \tag{5.183}$$

The functions f_k and f_{kl} are coefficients in front of the linear and quadratic operators respectively. We now assume a product-type wave function of the type

$$\Psi(s, \{Q_k\}, t) = \Phi(s, t)\psi(\{Q_k\}, t) \tag{5.184}$$

where we have introduced a quantum-classical approach and treated the overall rotational motion by classical mechanics and quantized the motion along the reaction path and also the perpendicular vibrations. Inserting the product-type wave function in the time-dependent Schrödinger equation, we get the following two equations

$$i\hbar\frac{\partial\Phi}{\partial t} = \langle\psi|H|\psi\rangle\Phi - i\hbar\left\langle\psi\left|\frac{\partial\psi}{\partial t}\right.\right\rangle\Phi \qquad (5.185)$$

$$i\hbar\frac{\partial\psi}{\partial t} = \langle\Phi|H|\Phi\rangle\psi - i\hbar\left\langle\Phi\left|\frac{\partial\Phi}{\partial t}\right.\right\rangle\psi \qquad (5.186)$$

where the bracket in the first equation indicates integration over the coordinates Q_k, and in the second over s. We notice that the last term in both equations depend only upon time, that is, they may be transformed out of the equations by a simple phase change. Since we are not interested in an overall phase factor, we will therefore neglect these terms.

In eq. (5.186) we shall replace s by the expectation value $\langle s\rangle = \langle\Phi|s|\Phi\rangle$. We notice in passing that the classical limit can be approached by taking Φ as a Gaussian wave packet centered around a trajectory $s(t)$. A systematic approach involves an expansion of the wave function in a Gauss-Hermite basis set as discussed in chapter 2. We mention, below, other solution schemes. Equation (5.186) is solvable in operator space (see chapter 4 and ref. [147]) if the hamiltonian contains operators up to second order, that is, operators of the type a_k^+, a_k and $a_k^+ a_l$. That is, the wave function can be obtained as

$$\psi(t) = U(t, t_0)\psi(t_0) \qquad (5.187)$$

where the evolution operator $U(t, t_0)$ can be expressed in terms of the operators a_k, a_k^+ and $a_k a_l^+$. What is also important is that the expectation values needed for solving eq. (5.185) can also readily be evaluated. This evaluation gives the following expression for the effective hamiltonian governing the reaction path motion [179, 204, 188, 189]

$$H_{eff} = \frac{1}{2}p_s^2 + V_0(s) + H_{rot} + \sum_k \hbar\tilde{\omega}_k\left(\frac{1}{2} + n_k^0 + \rho_k(t)\right)$$
$$+ i\sum_k\left(F_k^- P_k^+ - F_k^+ P_k^-\right) + \sum_{kl, k>l}\left\{\left[F_{kl}^+\left(n_k^0 Q_{kl} + n_l^0 Q_{lk}^* + \alpha_l^+\alpha_k^-\right)\right]\right.$$
$$\left. + \left[F_{kl}^-\left(n_k^0 Q_{kl}^* + n_l^0 Q_{lk} + \alpha_l^-\alpha_k^+\right)\right]\right\} \qquad (5.188)$$

The first term in eq. (5.188) is the kinetic energy for the translational motion along s. $V_0(s)$ is the potential as a function of s and H_{rot} is the Hamiltonian for the rotational motion, which to lowest order is defined as

$$H_{rot} = \frac{1}{2}\sum_\alpha P_\alpha^2/I_{\alpha\alpha}^e(s); (\alpha = x, y, z) \qquad (5.189)$$

where the $I_{\alpha\alpha}^e(s)$s are the principal axes moments of inertia of the complex, evaluated along the reaction path. The momenta P_x, P_y, and P_z may be expressed in terms of Euler angles (see section 5.1) describing the overall rotation of the complex and their conjugate momenta. We notice that coupling between the overall rotational motion and the perpendicular vibrations (Coriolis coupling) is included in the linear "forces" f_k (see also below). The fourth term in eq. (5.188) represents the internal energy of the molecular system in the perpendicular vibrational modes, including the zero-point vibrational energy. The modulated vibrational frequency $\tilde{\omega}_k$ is defined as

$$\tilde{\omega}(s) = \omega_k(s) + \tfrac{1}{2}p_s^2\tilde{C}_{kk}(s)/\omega_k(s) \tag{5.190}$$

where the frequency modulation terms $\tilde{C}_{kk}$ will be defined below. The $\omega_k(s)$ are diabatic vibrational frequencies obtained by solving the restricted eigenvalue problem along the reaction path.

The parameter ρ_k is a measure of the excitation of each mode, and is given by

$$\rho_k = \sum_j n_j^0 Q_{jk}Q_{jk}^* + P_k^+ P_k^- \tag{5.191}$$

where the functions Q_{jk} are determined from the solution of eq. (5.208) below. n_k^0 is the initial normal mode vibrational quantum number in mode k. P_k^+ is defined as

$$P_k^+ = \alpha_k^+ + \sum_{j=1}^{M} Q_{jk}\alpha_j^+ \tag{5.192}$$

where

$$\alpha_k^+ = -\frac{1}{\hbar}\int dt\left(F_k^+ + \sum_{j}^{M} F_j^+ Q_{jk}^*\right) \tag{5.193}$$

P_k and α_k are complex quantities such that $P_k^- = (P_k^+)^*$ and $\alpha_k^- = (\alpha_k^+)^*$. F_k is defined by

$$F_k^\pm = f_k \exp\left(\pm i\theta_k\right) \tag{5.194}$$

where the phase factor θ_k is defined in terms of the modulated vibrational frequency $\tilde{\omega}_k$ as

$$\theta_k = \int dt\, \tilde{\omega}_k(s(t)) \tag{5.195}$$

The quantity f_k in eq. (5.194) is the generalized force for mode k linear in normal mode coordinates

$$f_k = -b_k(s)\left(p_s^2 B_{kF}(s) + \sum_{\alpha\beta}\frac{P_\alpha P_\beta a_k^{\alpha\beta}}{I_{\alpha\alpha}^e I_{\beta\beta}^e}\right) \tag{5.196}$$

where

$$b_k(s) = \sqrt{\frac{\hbar}{2\omega_k(s)}} \tag{5.197}$$

and

$$\alpha_k^{\alpha\beta} = \sum_i \sqrt{m_i} \left[a_{i\beta} l_{ik}^{(\alpha)} + a_{i\alpha} l_{ik}^{(\beta)} \right] \tag{5.198}$$

In the above equation, $\mathbf{l}_{ik} = (l_{ik}^x, l_{ik}^y, l_{ik}^z)$ are the orthogonal displacement vectors for atom i due to vibration of the kth normal mode. Similarly, $\mathbf{a}_i(s) = (a_{ix}, a_{iy}, a_{iz})$ is the position vector of the atom i along the reaction path. Numerical procedures to compute $\mathbf{l}_{ik}$ and $\mathbf{a}_i(s)$ along the reaction path are discussed in detail in the appendix of ref. [189] (see also appendix D). Once $\mathbf{a}_i(s)$ is calculated, the moments of inertia may be evaluated easily as

$$I_{\alpha\alpha}^e = \sum_i m_i(a_{i\beta}^2 + a_{i\gamma}^2) \, ; (\alpha \neq \beta \neq \gamma) \tag{5.199}$$

Thus, the fifth term in eq. (5.188) arises from the expectation value of the linear term of the RP hamiltonian function expanded up to second order in the oscillator coordinates. The term is referred to as the V-T (vibration-translation) coupling term (VT$_{eff}$ on fig. 5.12). The V-T term mainly depends on the linear force f_k, the first term of which contains p_s^2, the coupling term B_{kF} which couples the reaction path to the perpendicular vibrational mode k, and a matrix

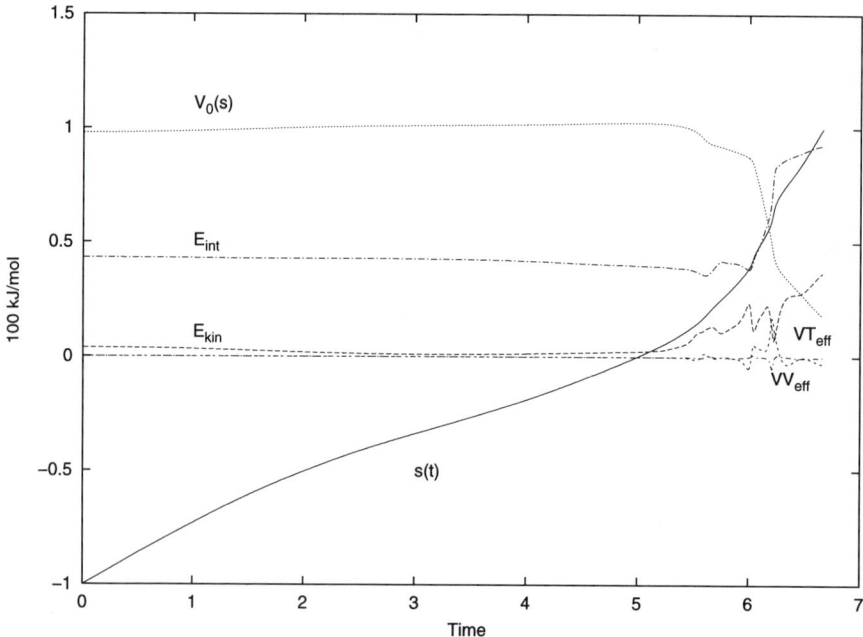

Fig. 5.12:   The various energy terms entering the reaction path model for chemical reaction: The kinetic energy $E_{kin} = p_s^2/2$, potential energy $V_0(s)$, vibrational energy $E_{int} = \sum_k \hbar\omega_k(s)(\rho_k + \frac{1}{2})$, the effective potential from "VT" forces VT$_{eff}$ and "VV" forces VV$_{eff}$ (see eq. (5.188)). H_{rot} is zero for the initial condition $j = J = 0$, chosen for the $H_2 + CN$ system. The initial kinetic energy is 17.5 kJ/mol.

element $b_k(s)$ of the creation operator. B_{kF} is a measure of the "curvature" of the reaction path and determines the amount of coupling to a given vibrational degree of freedom.

The second term in f_k is a Coriolis-type term which does not induce vibrational transitions but which tries to distort (stretch or bend) the molecule during the rotation of the complex and is referred to as the "centrifugal stretch term." The last term $\{\cdots\}$ in eq. (5.188) arises from the expectation value of the quadratic term of the hamiltonian and is the V-V (vibration-vibration) coupling term (VV_{eff} in fig. 5.12). The quadratic forces are given by the expression

$$F_{kl}^+ = f_{kl} \exp\left[i(\theta_k - \theta_l)\right] \tag{5.200}$$

where

$$f_{kl} = p_s\left[b_k b_l\left(\frac{1}{2}p_s\tilde{C}_{kl} + \sum_{k'} B_{k'l} \sum_{\alpha=x,y,z} P_\alpha \zeta_{kk'}^\alpha / I_{\alpha\alpha}^e(s)\right) - i\hbar\omega_+ B_{kl}\right] \tag{5.201}$$

Here, ω_+ and $\tilde{C}_{kl}$ are defined as

$$\omega_+ = (\omega_k + \omega_l)/\sqrt{\omega_k\omega_l} \tag{5.202}$$

and

$$\tilde{C}_{kl} = 4B_{kF}B_{lF} - C_{kl} + \sum_{k'} B_{k'k}B_{k'l} \tag{5.203}$$

The so-called direction cosines $\zeta_{kk'}^\alpha$ are defined in terms of the orthogonal displacement vectors $\mathbf{l}_{ik}$

$$\zeta_{kk'}^\alpha = \sum_i (l_{ik}^\beta l_{ik'}^\gamma - l_{ik}^\gamma l_{ik'}^\beta) \tag{5.204}$$

The different coupling elements B_{kF}, $B_{kk'}$ and $C_{kk'}$ are defined as follows

$$B_{kF} = \sum_i \frac{d}{ds}\mathbf{l}_{ik} \cdot \mathbf{l}_{iF} \tag{5.205}$$

$$B_{kk'} = \sum_i \frac{d}{ds}\mathbf{l}_{ik} \cdot \mathbf{l}_{ik'} \tag{5.206}$$

$$C_{kk'} = \sum_i \frac{d}{ds}\mathbf{l}_{ik} \cdot \frac{d}{ds}\mathbf{l}_{ik'} \tag{5.207}$$

where the subscript F indicates reaction path and k denotes one of the $3N-7(6)$ vibrational degrees of freedom. Numerical procedures to compute the orthogonal displacement vectors $\mathbf{l}_{ik}$ and $\mathbf{l}_{iF}$ follow the scheme given in the appendix D.

The potential, harmonic vibrational frequencies, position vectors of different atoms, and their orthogonal displacement vectors, are computed along the reaction path (s) at 2–400 points and are spline fitted in a convenient form for use in reaction path dynamics. Once the displacement vectors are known, the coupling elements can be evaluated at the appropriate s values during the dynamics. As mentioned above, the quadratic coupling terms (F_{kl}) are responsible for energy transfer to perpendicular vibrational modes. Thus, in order to obtain the amplitudes of vibrational transitions, one has to solve the system of coupled equations [147]

$$i\hbar\dot{\mathbf{R}} = \mathbf{AR} \tag{5.208}$$

where $\mathbf{R} = \mathbf{Q} + \mathbf{I}$ with $\mathbf{R}(-\infty) = \mathbf{I}$ and $\mathbf{I}$ is a unit matrix. The elements A_{kl} are obtained as $A_{kl} = F_{kl}^+$ for $(k > l)$ and $A_{lk} = F_{kl}^- = (F_{kl}^+)^*$ for $(l < k)$. The V-V transition amplitudes are extracted from the $\mathbf{Q}$ matrix by solving the inversion problem discussed in appendix C (see also [203]). The above system of coupled equations, together with the equations for θ_k, α_k, and β_k where the phase factor β_k is given by [204]

$$\beta_k = 2i \int dt\, (\alpha_k^+ \dot{\alpha}_k^- - \alpha_k^- \dot{\alpha}_k^+) \tag{5.209}$$

are integrated along with the equations of motion for the classical variables $(s, p_s,$ and the Euler angles and corresponding momenta). The amplitude for energy transfer from a set of initial vibrational states $\{n_i\}$ to final vibrational states $\{n_f\}$ is given by

$$\langle\{n_f\}|U|\{n_i\}\rangle = \sum_{\{k\}}\langle\{n_f\}|U_{VV}|\{n_k\}\rangle\langle\{n_k\}|U_{VT}|\{n_i\}\rangle \tag{5.210}$$

where the evolution operator U has been decomposed into a V-T and a V-V part. The matrix elements can be evaluated analytically and are given by eq. (4.40) in chapter 4 and eq. (A.21) in appendix B.

5.4.6 Initialization of the reaction path dynamics

The reaction path dynamics are initialized at a value of $s = s_0$ where the various coupling terms $B_{kF}, B_{kk'}$, and $C_{kk'}$ vanish. A rotational adiabatic assumption [204] is made at s_0, where the hamiltonian $H_{rot} = H_{rot}(s_0)$ is set equal to the rotational energy of the reactant molecules (here, taken to be AB and CD) and the orbital energy:

$$H_{rot}(s_0) = \frac{\hbar^2 l(l+1)}{2\mu R^2(s_0)} + \sum_i \frac{\hbar^2}{2I_i} j_i(j_i+1) \tag{5.211}$$

where $R(s_0)$ is the center-of-mass distance at $s = s_0$, μ the reduced mass for the relative motion, and I_k the moments of inertia of the two diatomic molecules.

The initial kinetic energy is defined as

$$E_{kin}(s_0) = E_{kin} - \frac{\hbar^2 l(l+1)}{2\mu R^2(s_0)} - V(s_0) - \Delta E_{vib}(s_0) \tag{5.212}$$

where $\Delta E_{vib}(s_0)$ is the difference in vibrational energy at $s = -\infty$ and s_0:

$$\Delta E_{vib}(s_0) = \hbar \sum_k \left[\omega_k(s_0)(n_k^0 + 1/2) - \omega_k(-\infty)(n_k^0 + 1/2)\right] \tag{5.213}$$

Since s_0 is chosen such that the coupling is negligible in the region $[-\infty, s_0]$, the above vibrational adiabatic assumption is not a serious restriction on the dynamics. The equations for the classical degrees of freedom and the coupled equations for energy transfer to the vibrational modes are integrated from s_0 to s_{max} where s_{max} is chosen to be well in the product channel for all couplings to cease. The reaction probability P^R is considered to be unity if the system smoothly transforms to products and if the final s value is $\geq s_{max}$. However, if there is a turning point for the relative translational motion before reaching the saddle point, it is necessary to evaluate a tunneling probability. Thus, for such trajectories, the reaction probability can, for instance, be approximated by the tunneling probability obtained using the WKB method [138]

$$P_{tunn} = \exp\left[-2I(s_1, s_2)\right] \tag{5.214}$$

where

$$I(s_1, s_2) = \frac{1}{\hbar} \int_{s_1}^{s_2} ds \left(\frac{2[H'_{eff}(s) - E]}{1 + 2\epsilon(s, t)}\right)^{1/2} \tag{5.215}$$

$$H'_{eff} = V_0(s) + H_{rot} + \hbar \sum_k \omega_k \left(\frac{1}{2} + \rho_k(t)\right) \tag{5.216}$$

and $\epsilon(s, t)$ contains coefficients of terms containing p_s^2 in eq. (5.188)

$$\epsilon(s, t) = \frac{\hbar}{2} \sum_k \left(\rho_k + \frac{1}{2}\right) \tilde{C}_{kk}/\omega_k(s) + \mathrm{i} \sum_k (R_k^- P_k^+ - R_k^+ P_k^-)$$

$$+ \frac{1}{2} \sum_{kl} \tilde{C}_{kl} b_k b_l \left\{\left(n_k^0 Q_{kl} + n_l^0 Q_{lk}^* + \alpha_l^+ \alpha_k^-\right)\right.$$

$$\left. + \left(n_k^0 Q_{kl}^* + n_l^0 Q_{lk} + \alpha_l^- \alpha_k^+\right)\right\} \tag{5.217}$$

where

$$R_k^\pm = -B_{kF}(s) b_k(s) \exp\left(\pm i\theta_k\right) \tag{5.218}$$

The tunneling integral is evaluated at the classical turning point s_1, where $p_s \approx 0$. At s_1 E is set equal to H_{eff} and the trajectory is continued in the complex momentum space with $p_s \to ip_s$ until $E = H_{eff}(s_2)$. The dynamical variables except s are frozen between s_1 and s_2. At s_2, the integration is continued in the real space with $p_s = 0$ and $s = s_2$.

Instead of using this simplified estimate of the tunneling probability and a classical mechanical treatment of the motion along the reaction path, we may treat this motion as quantum mechanical and initialize the wave function $\Phi(s, t)$ at s_0 as a Gaussian wave packet, and propagate the wave function in the effective potential arising from the coupling to the other modes, that is, by solving the TDSE

$$i\hbar \frac{\partial \Phi(s, t)}{\partial t} = \left[-\frac{1}{2}\frac{\partial^2}{\partial s^2} + V_{eff}(s, t)\right]\Psi(s, t) \tag{5.219}$$

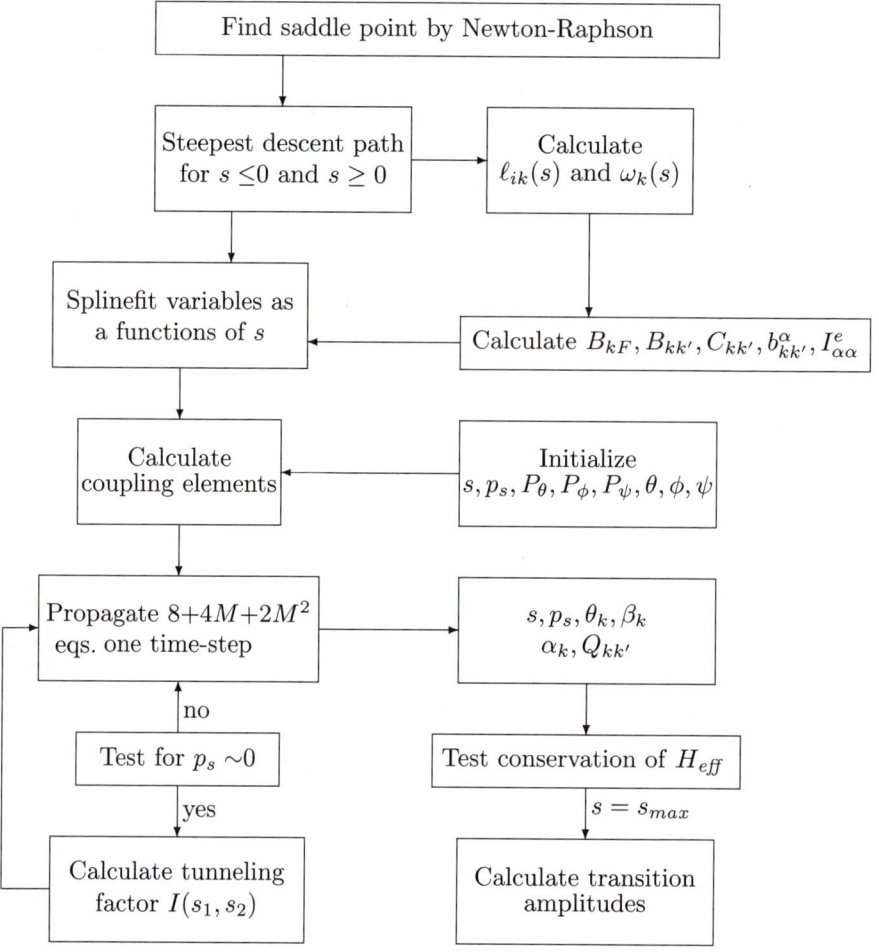

Fig. 5.13: Flow diagram for a quantum-classical reaction path approach to chemical reactions.

where V_{eff} is given by the last part of eq. (5.188). For further details, see ref. [189]. The operations involved in the reaction path method are shown in the flow-diagram (fig. 5.13).

5.4.7 Cross sections and rate-constants

The total reaction cross section is for an AB+CD reaction evaluated as [204]

$$\sigma(E_{kin}) = \frac{\pi\hbar^2}{2\mu E_{kin}(2j_1 + 1)(2j_2 + 1)} \sum_J (2J+1) \sum_{K=-J}^{J} P_{react}(E_{kin}, J, K)$$

(5.220)

where j_1 and j_2 are the initial rotational angular momenta of the two diatomics and μ the reduced mass.

The initialization of the dynamics is performed as discussed above. If product state-resolved reaction probabilities are desired, one has to compute transition amplitudes to different product states. This involves evaluating matrix elements of the V-V and V-T evolution operators at the end of the propagation. In this case, P^R should be replaced by P^R multiplied by the square of the transition amplitude given by eq. (5.210). If the initial quantum numbers $\{n_i^0\}$ are all zero, then one only needs to solve the V-T problem, as only elastic V-V transition is possible (V-V transitions include only those transitions where the vibrational quanta are conserved). Furthermore, in this case, the individual V-T transition probabilities reduce to the familiar Poisson distribution, as can easily be seen from eq. (4.39).

Thermal rate constants are obtained by Boltzmann averaging the reaction probability over the initial rotational and translational energy distribution of the reactant molecules

$$k(T) = \sigma_{symm}\sqrt{\frac{8k_BT}{\pi\mu}}(T_0/T)^3 \int_0^\infty dx\, \exp(-x)\langle\sigma(x)\rangle \qquad (5.221)$$

where $x = E_{cl}/k_BT$ and where σ_{symm} is a symmetry factor which counts the number of equivalent RP's. The average cross section is defined as

$$\langle\sigma(x)\rangle = \frac{\pi\hbar^6}{8m_1m_2(r_1^e r_2^e)^2\mu(k_BT_0)^3}J_{max}\frac{1}{N_t}\sum_{i=1}^{N_t}2J(2J+1)P_{react}(E_{cl},J) \quad (5.222)$$

where N_t is the number of collisions, T_0 a reference temperature, m_i is the reduced mass, r_i^e ($i = 1, 2$) are the equilibrium bond lengths of the two molecules, J is the total angular momentum, and J_{max} is the maximum angular momentum of the complex at s_0

$$J_{max} = \frac{\sqrt{2E_{cl}I_{xx}^e}}{\hbar} \qquad (5.223)$$

E_{cl} is the total classical energy which is the sum of rotational and kinetic energies. The moments of inertia are $I_{\alpha\alpha}$ ($\alpha = x, y, z$). The sum in (5.222) runs over randomly selected initial conditions. The total angular momentum is $\mathbf{J} = \mathbf{l}+\mathbf{j}_1+\mathbf{j}_2$ where the orbital angular momentum l is chosen randomly between zero and l_{max} where $l_{max} = \sqrt{(2\mu R(s_0)^2 E_{cl})}/\hbar$. The rotational angular momentum j is chosen randomly between 0 and j_{max}^p where $j_{max}^p = \sqrt{2I_p(E_{cl} - E_l)}/\hbar$ and $I_p(p = 1, 2)$ the moments of inertia of the reactant molecules. The orbital energy E_l is given by $E_l = \hbar^2 l(l + 1)/(2\mu R(s_0)^2)$. The projection quantum number K (see eq. (5.167)) is chosen randomly between $-J$ to $+J$ and the total rotational energy is obtained as given by eq. (5.211) The initial kinetic energy is then defined according to eq. (5.212). Trajectories with $J > J_{max}$ are rejected from the sampling space during the initialization.

In fig. 5.14, we have shown the reaction path and transition state rates as a function of temperature. Table 5.6 shows state-resolved and total reaction

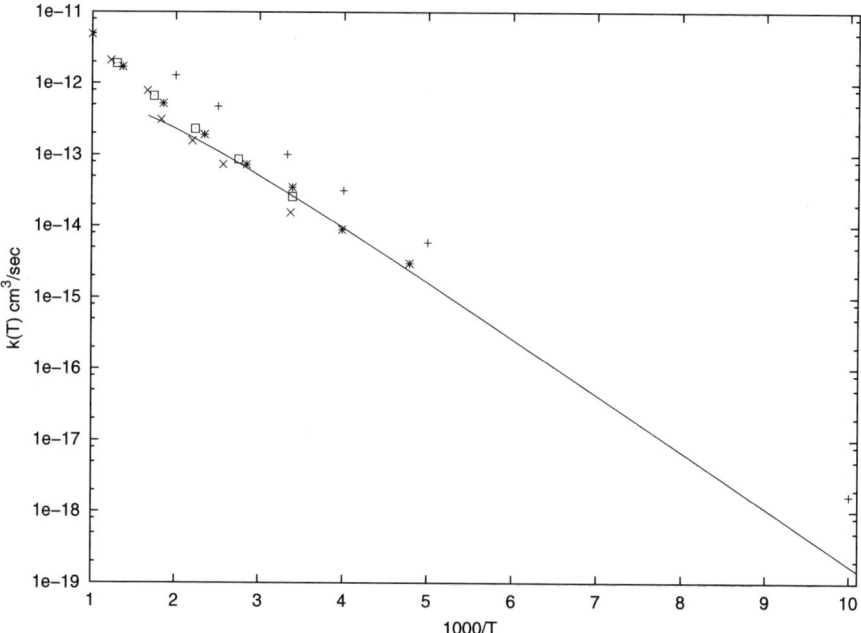

Fig. 5.14: Rate-constant for the reaction $H_2 + CN \rightarrow HCN + H$ obtained using the reaction path theory (solid line), and transition state theory (+). Experimental data from refs. [194] (x), [195] (asterisks), [196] (boxes) are also shown.

rates for $H_2 + CN$. We notice that the tunneling correction included in the TST theory is that due to Wigner $(1 + \hbar^2|\omega^*|^2/24(k_BT)^2)$, where ω^* is the imaginary frequency. The reaction path tunneling factor is estimated using the expression (5.214), which is more realistic since it does contain some curvature corrections. Hence, the agreement with experimental data is better at low temperatures.

At high temperatures, we have noticed that the RP hamiltonian has a tendency to underestimate the reaction rate. This is due to an overestimate of the above barrier reflection due to the harmonic approximation made in the normal mode expansion of the potential. Thus, the potential appears to be too narrow in the perpendicular degrees of freedom, and, hence, the increased

Table 5.6: State-resolved and total reaction rate-constants (in cm^3/sec) for the reaction $H_2+CN \rightarrow HCN+H$. The initial vibrational states of H_2 and CN are zero, and the final state (n_3, n_4, n_5, n_6) where the first two quantum numbers refer to the bending modes, n_5 to the symmetric and n_6 to the asymmetric stretch mode of HCN.

Temperature	0000	0010	0020	0001	0002	0011	Total
200 K	3.03(−16)	2.17(−16)	1.05(−16)	6.75(−17)	1.15(−17)	6.43(−17)	1.62(−15)
300 K	5.06(−15)	3.14(−15)	1.33(−15)	1.41(−15)	2.72(−16)	1.11(−15)	2.95(−14)
500 K	3.80(−14)	2.03(−14)	7.53(−15)	1.33(−14)	2.88(−15)	8.22(−15)	2.31(−13)

Experimental value at 300 K is $2.7 \pm 1.6(−14)\, cm^3$/sec [205]

reflection. This problem can be overcome by including anharmonic terms, or, alternatively, by switching to less approximate dynamical theories, such as the reaction surface or volume approach.

5.5   The reaction volume approach

In the reaction path method, one degree of freedom (s) is projected out and treated "exactly." It is now in principle straightforward to extend this methodology to more dimensions, that is, to obtain a reaction surface approach in which two degrees of freedom are treated correctly (see refs. [202] and [204]). However, here we wish to review the so-called reaction volume approach, in which three degrees of freedom are projected out. The reason is that most reactions are three center reactions, and, hence, the geometry of the three centers can be expressed in terms of three variables.

Thus, the hamiltonian we wish to derive allows for a free, unconstrained motion of three atoms. The motion of these atoms characterises the reaction. The remaining atoms follow paths in which they are rigidly bound to the moving atoms in some reference configuration—typically, defined by the equilibrium geometry. These atoms are, however, allowed to vibrate around their reference positions. The effect of these degrees of freedom on the motion of the three atoms is of interest for reactions in liquids, on surfaces, or in polyatomic systems.

Aside from the motion of the reference frame and a vibrational displacement from it, an overall rotation of the complete system is allowed for.

The three atoms in question form a triangle—the shape and size of which can be defined by three hyper-variables: the hyper-radius ρ and the two angles θ, ϕ.

Figure 5.15 shows a reaction center defined by the three atoms at which a bond is broken and one is formed under the reaction process. The three atom-

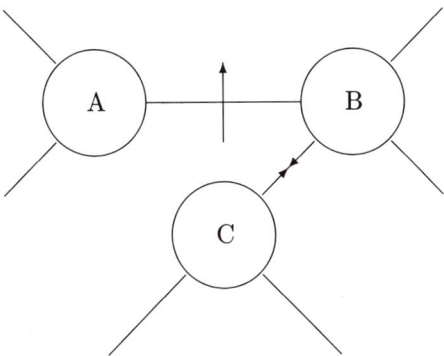

Fig. 5.15:   Three center reaction in a polyatomic system. A bond is broken between atoms A and B and formed between B and C.

atom distances of the center can be expressed in terms of hyperspherical variables (see also chapter 3).

$$R_{BC}^2 = \rho^2 \frac{d_1^2}{2}(1 + \sin\theta \cos\phi) \tag{5.224}$$

$$R_{AC}^2 = \rho^2 \frac{d_2^2}{2}(1 + \sin\theta \cos(\phi - \zeta_2)) \tag{5.225}$$

$$R_{AB}^2 = \rho^2 \frac{d_3^2}{2}(1 + \sin\theta \cos(\phi + \zeta_3)) \tag{5.226}$$

where $d_k^2 = (m_k/\mu)(1 - m_k/M)$, $M = m_1 + m_2 + m_3$, $\mu = \sqrt{m_1 m_2 m_3/M}$, $\zeta_2 = 2\tan^{-1}(m_3/\mu)$ and $\zeta_3 = 2\tan^{-1}(m_2/\mu)$. Thus ρ is a common translational coordinate.

It is not absolutely necessary to use hyper-spherical variables, but it makes obvious an interlink to the quantum wave packet approach, which was formulated in these coordinates in chapter 3 (see also [207]).

The reference vectors $\mathbf{a}_i(\rho, \theta, \phi)$ $i = 1, \ldots, N$ will depend parametrically on the variables ρ, θ, and ϕ. $\mathbf{a}_i = (a_{ix}, a_{iy}, a_{iz})$ are position vectors of atom i, measured from the center of mass of the system.

The usual Eckart condition is

$$\sum_{i=1}^{N} m_i(\mathbf{a}_i \times \mathbf{v}_i) = \mathbf{0} \tag{5.227}$$

where the summation runs over all atoms and where $\mathbf{v}_i = (d/dt)\rho_i$. (Note that ρ_i denotes a displacement of atom i from the reference frame and ρ the hyper-radius, associated with the size of the A-B-C triangle (see fig. 5.15).)

The position vector in a space fixed coordinate system is

$$\mathbf{r}_{0i} = \mathbf{R}_0 + \mathbf{a}_i(\rho, \theta, \phi) + \rho_i \tag{5.228}$$

where $\mathbf{R}_0$ is the position of the center-of-mass. Thus the center-of-mass condition is

$$\sum_{i=1}^{N} m_i \mathbf{a}_i = \mathbf{0} \tag{5.229}$$

The hyper-spherical variables are chosen to be those defined by Johnson [97]. Thus the relative position vectors in a local coordinate system, with z-axis perpendicular to the plane formed by atoms 1, 2, and 3 are given by $\mathbf{r} = \mathbf{a}_2 - \mathbf{a}_1$ and $\mathbf{R} = \mathbf{a}_3 - \epsilon \mathbf{a}_2 - \epsilon' \mathbf{a}_1$ where $\epsilon = m_2/(m_1 + m_2)$, $\epsilon' = 1 - \epsilon$, $\mathbf{r} = (r_x, r_y, 0)$, $\mathbf{R} = (R_x, R_y, 0)$ and

$$r_x = -\frac{\rho d_3}{\sqrt{2}}\left(\cos\frac{\theta}{2} + \sin\frac{\theta}{2}\right)\cos\frac{\phi}{2} \tag{5.230}$$

$$r_y = \frac{\rho d_3}{\sqrt{2}}\left(\cos\frac{\theta}{2} - \sin\frac{\theta}{2}\right)\sin\frac{\phi}{2} \tag{5.231}$$

$$R_x = \frac{\rho}{d_3\sqrt{2}}\left(\cos\frac{\theta}{2} + \sin\frac{\theta}{2}\right)\sin\frac{\phi}{2} \tag{5.232}$$

$$R_y = \frac{\rho}{d_3\sqrt{2}}\left(\cos\frac{\theta}{2} - \sin\frac{\theta}{2}\right)\cos\frac{\phi}{2} \tag{5.233}$$

where

$$d_3^2 = \frac{m_3}{\tilde{\mu}}(1 - m_3/M) \tag{5.234}$$

$$M = m_1 + m_2 + m_3 \tag{5.235}$$

$$\tilde{\mu} = \sqrt{m_1 m_2 m_3/M} \tag{5.236}$$

The position coordinates of atom j can now be written as:

$$\mathbf{a}_j = \frac{M_{j-1}}{M_j}\mathbf{a}_j^0 - \sum_{i=j+1}^{N}\frac{m_i}{M_i}\mathbf{a}_i^0 \tag{5.237}$$

where $M_j = \sum_{i=1}^{j} m_i$ and where $\mathbf{a}_i^0$ is the position vector of atom i measured from the mass midpoint of the system consisting of $i - 1$ atoms. We notice that the position vectors satisfy the center-of-mass condition (5.229). Thus, for atoms 1, 2, and 3, we obtain more specifically:

$$\mathbf{a}_1 = -\frac{m_2}{M_2}\mathbf{r} - \frac{m_3}{M_3}\mathbf{R} - \sum_{i=4}^{N}\frac{m_i}{M_i}\mathbf{a}_i^0 \tag{5.238}$$

$$\mathbf{a}_2 = \frac{m_1}{M_2}\mathbf{r} - \frac{m_3}{M_3}\mathbf{R} - \sum_{i=4}^{N}\frac{m_i}{M_i}\mathbf{a}_i^0 \tag{5.239}$$

$$\mathbf{a}_3 = \frac{M_2}{M_3}\mathbf{R} - \sum_{i=4}^{N}\frac{m_i}{M_i}\mathbf{a}_i^0 \tag{5.240}$$

The potential is expanded in the displacement coordinates as:

$$V(\mathbf{r}_1,\ldots,\mathbf{r}_N) = V(\mathbf{a}_1,\ldots,\mathbf{a}_N) + \sum_{i=1}^{N}\frac{\partial V}{\partial \mathbf{r}_i}\Big|_{\mathbf{a}_i}\cdot\rho_i$$

$$+ \frac{1}{2}\sum_{ij}\rho_j^T \cdot \frac{\partial^2 V}{\partial \mathbf{r}_i \partial \mathbf{r}_j}\Big|_{\mathbf{r}_i=\mathbf{a}_i}\cdot\rho_i + \ldots \tag{5.241}$$

where the expansion is truncated at the third term. The reference vectors $\mathbf{a}_i$ are, as mentioned, functions of the hyper-spherical coordinates ρ, θ and ϕ. The displacement vectors of the atomic positions due to changes in the "parameters" θ, ϕ, and ρ are given by $\Delta\mathbf{a}_i$ and displacements away from their positions by ρ_i. We introduce their "perpendicular" displacements such that

$$\sum_{i=1}^{N}m_i\Delta\mathbf{a}_i \cdot \rho_i = 0 \tag{5.242}$$

Since

$$\Delta\mathbf{a}_i = \frac{\partial \mathbf{a}_i}{\partial \rho}\Delta\rho + \frac{\partial \mathbf{a}_i}{\partial \theta}\Delta\theta + \frac{\partial \mathbf{a}_i}{\partial \phi}\Delta\phi \tag{5.243}$$

this requires for arbitrary variations in θ, ϕ and ρ that we introduce the following three constraints:

$$\frac{1}{\rho} \sum_i^N m_i \rho_i \cdot \frac{\partial \mathbf{a}_i}{\partial \theta} = 0 \tag{5.244}$$

$$\frac{1}{\rho} \sum_i^N m_i \rho_i \cdot \frac{\partial \mathbf{a}_i}{\partial \phi} = 0 \tag{5.245}$$

$$\sum_i^N m_i \rho_i \cdot \frac{\partial \mathbf{a}_i}{\partial \rho} = 0 \tag{5.246}$$

where ρ_i denote the perpendicular displacements from the "planar" motion. Together with the other 6 constraints we have altogether nine constraints, giving a total of $3N - 9$ vibrational degrees of freedom for the displacements away from the reference frame. In the so-called reaction path method, we use as the reference path the minimum energy path, which is defined by $\Delta \mathbf{a}_i \sim -\mathbf{g}_i$, that is, the displacements are in the negative gradient direction. Thus, the linear term in the expansion of the potential vanishes, due to eq. (5.242). However, since in the reaction volume theory, the reference frame is at least in principle arbitrary, the constraints one wishes to use in order to define the motion of the frame will not necessarily remove the linear expansion terms (see also discussion below).

The position coordinates are

$$\mathbf{r}_i = \mathbf{a}_i(\rho, \theta, \phi) + \rho_i \tag{5.247}$$

and we, furthermore, need to introduce the vectors

$$\ell_\rho(i) = \sqrt{m_i} \frac{\partial \mathbf{a}_i}{\partial \rho} \tag{5.248}$$

$$\ell_\theta(i) = \frac{2\sqrt{m_i}}{\rho} \frac{\partial \mathbf{a}_i}{\partial \theta} \tag{5.249}$$

$$\ell_\phi(i) = \frac{2\sqrt{m_i}}{\rho} \frac{\partial \mathbf{a}_i}{\partial \phi} \tag{5.250}$$

These three vectors are orthorgonal to the eigenvectors ℓ_{ik} by construction (see appendix D). In order to find the ρ, θ, and ϕ dependence of the reference vectors $\mathbf{a}_i$, we need some prescription. Thus, one could imagine that the atoms $(i \geq 4)$ are moved relative to the reference frame, with fixed bond angles or lengths. The atoms could also be moved in the direction of the negative gradient or such that the gradient is minimized. It should be noted that it is only the reference position which is "defined" in this manner. The actual motion involves a displacement away from this position. The size of this displacement is defined by the dynamical solution of the problem. However, since this displacement is only described approximately, it is important to have the reference position as realistical as possible, that is, it should reflect as much of the physics of the problem as possible.

Thus, it would be desirable, but not necessary, to minimize the linear term in the expansion of the potential around the reference position, that is, to minimize the absolute value of the gradient $|\mathbf{g}_i|$ and define the change in reference position of the atoms in this manner. Since the gradient depends upon the variables implicitly (through $\mathbf{a}_i(\rho, \theta, \phi)$), we have:

$$\frac{d|g_{i\gamma}(\{\mathbf{a}_j\})|}{d\rho} = \sum_{j\alpha} \frac{\partial |g_{i\gamma}|}{\partial a_{j\alpha}} \frac{da_{j\alpha}}{d\rho} + \sum_{k=1}^{3} \lambda_k \frac{dI_k}{d\rho} \tag{5.251}$$

with $i, j = 1, \ldots, N$ and $\alpha, \gamma = x, y, z$. We have also introduced the Lagrange multipliers λ_k connected to I_k the three constraints which relate the reference positions to the hyper-spherical variables, that is,

$$I_1 = r^2 - \frac{d_1^2}{2} \rho^2 (1 + \sin\theta \cos\phi) \tag{5.252}$$

$$I_2 = R^2 - \frac{1}{2d_1^2} \rho^2 (1 - \sin\theta \cos\phi) \tag{5.253}$$

$$I_3 = \mathbf{r} \cdot \mathbf{R} + \frac{1}{2} \rho^2 \sin\theta \sin\phi \tag{5.254}$$

where $\mathbf{r} = \mathbf{a}_2 - \mathbf{a}_1$ and $\mathbf{R} = \mathbf{a}_3 - (1/(m_1 + m_2))(m_1 \mathbf{a}_1 + m_2 \mathbf{a}_2)$. The extremum condition would be $d|\mathbf{g}_i|/d\rho = 0$. Thus we have $3N+3$ unknowns $da_{j\alpha}/d\rho$ and λ_k the Lagrange multipliers. These can be obtained from the $3N + 3$ equations (5.251) and (5.252)–(5.254). This yields the equations for the derivative of the reference vectors with respect to the variable ρ. A similar procedure can be used for the θ and ϕ derivatives. Another possibility is to remove nine variables by using the constraints. It has turned out that a minimization of the potential followed by a minimization of the gradient, is numerically preferable if a nonlinear least squares search for the minimum is attempted [209].

When the derivatives are known, the atom displacements are obtained from the equation

$$\Delta \mathbf{a}_j = \frac{\partial \mathbf{a}_j}{\partial \rho} \Delta \rho + \frac{\partial \mathbf{a}_j}{\partial \theta} \Delta \theta + \frac{\partial \mathbf{a}_j}{\partial \phi} \Delta \phi \tag{5.255}$$

Let us now turn to the eigenvectors, which enter the expression for the displacement in terms of normal mode coordinates Q_k, that is,

$$\rho_i = m_i^{-\frac{1}{2}} \sum_k \ell_{ik}(\theta, \phi, \rho) Q_k \tag{5.256}$$

where

$$\sum_i \ell_{ik} \cdot \ell_{ik'} = \delta_{kk'} \tag{5.257}$$

$k = 1, \ldots, 3N - 9$ and $i = 1, \ldots, N$.

These eigenvectors are obtained by diagonalizing the Hessian matrix subject to the nine constraints (see appendix D). The evaluation of the kinetic energy term is given in appendix E.

5.5.1   A simplified hamiltonian

In some cases we can simplify the expression (A.97) derived in appendix E. Thus, by using the fact that we often have $\ell_p \cdot \ell_q \sim \delta_{pq}$ where $(p, q = \theta, \phi, \rho)$, we obtain from eqs.(A.60)–(A.65) the following approximations:

$$A_{\rho\rho} = 1 + 2\sum_k Q_k B^\rho_{k\rho} \tag{5.258}$$

$$A_{\theta\theta} = \frac{\rho^2}{4} + \rho\sum_k Q_k B^\theta_{k\theta} \tag{5.259}$$

$$A_{\phi\phi} = \frac{\rho^2}{4} + \rho\sum_k Q_k B^\phi_{k\phi} \tag{5.260}$$

$$A_{\rho\theta} = \sum_k Q_k C^{\rho\theta}_k \tag{5.261}$$

$$A_{\rho\phi} = \sum_k Q_k C^{\rho\phi}_k \tag{5.262}$$

$$A_{\theta\phi} = \frac{\rho}{2}\sum_k Q_k C^{\theta\phi}_k \tag{5.263}$$

We may also introduce $A \sim \frac{1}{2}\rho^2(1 - \sin\theta)$, $B \sim \frac{1}{2}\rho^2(1 + \sin\theta)$ and $C \sim \rho^2$. We then get the following approximate hamiltonian:

$$2H = \sum_k (p_k^2 + a_k Q_k + \omega_k^2 Q_k^2) + V(\rho, \theta, \phi)$$

$$+ p_\rho^2\left(1 - 2\sum_k Q_k B^\rho_{k\rho}\right) + \frac{4p_\theta^2}{\rho^2}\left(1 - \frac{4}{\rho}\sum_k Q_k B^\theta_{k\theta}\right)$$

$$+ \frac{4p_\phi^2}{\rho^2\sin^2\theta}\left(1 - \frac{4}{\rho}\sum_k Q_k B^\phi_{k\phi}\right) - \frac{8p_\theta p_\rho}{\rho^2}\sum_k Q_k C^{\rho\theta}_k$$

$$- \frac{8p_\rho p_\phi}{\rho^2\sin^2\theta}\sum_k Q_k C^{\rho\phi}_k - \frac{16p_\theta p_\phi}{\rho^3\sin^2\theta}\sum_k Q_k C^{\theta\phi}_k$$

$$+ \frac{4p_\phi P_z \cos\theta}{\rho^2\sin^2\theta}\left(1 - \frac{4}{\rho\sin^2\theta}\sum_k Q_k B^\phi_{k\phi}\right) + \frac{P_x^2}{A} + \frac{P_y^2}{B}$$

$$+ \frac{P_z^2}{\rho^2\sin^2\theta}\left(1 - \frac{4\cos^2\theta}{\rho\sin^2\theta}\sum_k Q_k B^\phi_{k\phi}\right) \tag{5.264}$$

In the case of no coupling of the reaction center to the surroundings, that is, for the $N = 3$ system, we have:

$$2H = p_\rho^2 + V(\rho, \theta, \phi) + \frac{4}{\rho^2}\left(p_\theta^2 + \frac{p_\phi^2}{\sin^2\theta}\right)$$

$$+ \frac{4p_\phi P_z \cos\theta}{\rho^2\sin^2\theta} + \frac{2P_x^2}{\rho^2(1 - \sin\theta)} + \frac{2P_y^2}{\rho^2(1 + \sin\theta)} + \frac{P_z^2}{\rho^2\sin^2\theta} \tag{5.265}$$

which is the correct three-body hamiltonian [97]. We note that the hyper-spherical mass $\tilde{\mu}$ has been incorporated in the hyper-radius, that is, our hyper-radius is $\sqrt{\tilde{\mu}}\rho$.

The reaction volume hamiltonian can be used for large systems in which the reaction is a three center reaction. For such a system, it will be advantageous to treat the $3N - 9$ perpendicular vibrations within the second quantization approach, the overall rotational motion classically, and the three hyper-spherical coordinates using, for example, grid methods.

By minimizing the gradient we define a reference volume, in which all "reaction following" motions, as, for example, umbrella and other concerted motion, is taken care of through the reference position $\mathbf{a}_i(\theta, \phi, \rho)$. Small amplitude motions away from this reference position may, therefore, be treated approximately, for instance in the harmonic approximation. This also limits the requirements as far as the potential energy surface is concerned. We need the full surface in three dimensions, and in the remaining degrees of freedom, only gradients and hessians.

5.6 Summary

The reaction path method has been used to calculate state resolved reaction rates and total rates for a number of systems as, for instance: $H_2+OH \rightarrow H_2O+H$ [175], $O+O_3 \rightarrow 2O_2$ [176] and $Cl^-+CH_3Cl \rightarrow ClCH_3+Cl^-$ [177]. The method has been used to analyse the reaction path and the coupling elements B_{kF} for systems such as $CH_3+H_2 \rightarrow CH_4+H$, $C_2H_4+ FH \rightarrow CH_3CFH_2$, $C_4H_6+C_2H_4 \rightarrow C_6H_{10}$ that is, rather complex systems (for a more complete list see ref. [178]). It has been noted that in the harmonic approximation, that is, the SQ formulation, the reaction rates are a bit underestimated at high temperatures. This is attributed to the widening of the transition state region caused by anharmonic terms, that is, in the harmonic approximation the valleys in the perpendicular vibrational degrees of freedom are too narrow—hence, the reaction rate is lowered. However, at lower energies, good agreement with other dynamical methods is obtained. The reaction path theory offers, as mentioned, the possibility of obtaining vibrational state-resolved rate constants (see, for example, [208]). This information is important for modeling, for instance, atmospheric reactions or other kinetic systems where non-equilibrium vibrational distributions play a role (lasers, plasmas, and so on).

With the RP method, we may treat reactions involving many atoms, and, from a computational point of view, the method is the most economic way of doing so. The reason is that the information needed from electronic structure calculations is just gradients and hessians along the path. Thus, we may avoid the construction of the full potential energy surface. The method offers dynamical corrections to the transition state theory, that is, the theory includes tunneling and above barrier reflection.

However, the restrictions on the dynamics posed by the inclusion of only quadratic terms in the expansion of the perpendicular vibrational motions, the fact that the reaction path itself may be poorly defined in certain regions, as well as bifurcations of the path, are restrictions which should not be overlooked. In situations, where these problems are present, it is possible to switch to the reaction surface or the reaction volume methods. These methods are more demanding from the point of view of the potential energy surface, and also as far as the dynamical requirements are concerned. But they nevertheless reduce the problem considerably by demanding only the full information in a 2 or 3D space, and the harmonic displacement of the spectator atoms connected to the reaction center. The reaction volume approach has so far been used to calculate reaction cross sections by solving the dynamics using classical mechanics [208], whereas only little work has been performed as far as a quantum-classical solution is concerned. But preliminary calculations in which the three hyperspherical coordinates ρ, θ, and ϕ are treated quantally using grid methods [209] have appeared.

5.7 Chemical processes in clusters and solution

Chemical processes, that is, energy transfer and reactions in solution or in clusters (microsolution) is an area where the quantum-classical theory is, by mere construction, almost a must. The bulk of these systems is the cluster or the solvent itself, and since these atoms or molecules do not perform small amplitude motions, they will have a significant influence on relaxation or reaction processes of the solute. Hence, it is important to be able to study the influence of solvent motion on the dynamics of the small molecular system, the solute. The most realistical of the possible avenues is to use classical mechanics for the solvent motion. Other, but less realistic, possibilities are to use external model fields, and stochastic or statistical methods in order to include the effect of the environment. To the latter category belong, for instance, the mean field effective potential developed in section 5.2 on surface scattering, Redfield theories [210], path-integral approaches (in the harmonic approximation), locally propagating gaussians [211], TDSCF methods, and others.

The external or environmental influence on, say, a chemical reaction may be of a random or statistical nature but it may also be highly directional (stereospecific) as it is believed to be in some biochemical reactions. In such cases, an external field based on statistical concepts has little to do with reality. The fact which makes things somewhat easier is that we normally are interested in quantum resolution, that is, state resolution in the small system and not in the solvent or cluster. The influence of the latter is a structural and/or dynamical influence which can be modeled by classical dynamics. Thus the classical path or the quantum-classical approach with a classical treatment of the solvent molecules appears, at least at first sight, to be an ideal choice.

Consequently, methods such as the $V_q R_c T_c$ method have been used to study vibrational relaxation of CO in an Ar-cluster [212]. These calculations are carried out by first constructing the shape of the Ar-cluster according to the known structure, and then removing an Ar atom at the center of the cluster. Cluster sizes of 6, 27, and 73 atoms were considered in [212]. The CO molecule is placed at the center with a random orientation. Then the equilibrium structure is found by minimizing the potential using the Newton-Raphson technique to move the Ar-atoms with the CO in a fixed position. Next, the Ar-atoms are given kinetic energy according to the cluster temperature and the system is allowed to follow the equations of motion, quantum for the vibrational motion of the CO molecule and classical for the Ar-atom motion as well as the rotational and translational motions of the molecule. It was found that the vibrational transitions were enhanced by increasing the cluster size and decreasing the temperature. The numerical simulations revealed that the CO molecule interacted with the cluster as a whole.

A method which is similar in spirit, the MDQT method, has been used for proton transfer reactions in solution [119]; and for $S_N 2$ reactions in solution, the so-called onion model has been used [215] (see fig. 5.16). Here, the solvent molecules in a sphere are equilibrated according to a certain temperature using a standard molecular dynamics program. A cavity is created at the center of the sphere, in which the reacting system is placed. The solvent molecules are again equilibrated, keeping the reactants at fixed positions. As an outer sphere, we have placed a dielectric phonon heat bath. In this heat bath, the atoms are allowed to have harmonic oscillations and the forces exerted on the molecules in the inner spheres are included when solving the dynamics for the reactant and non-rigid part of the solvent.

Vibrational relaxation in liquids has often been treated within a time-dependent perturbation theory through Fermi's Golden Rule expression. Thus in this approximation the relaxation time for the system from state i to state j is given as [213]

$$\tau_{ij}^{-1} = \frac{2\pi}{\hbar} \sum_{nm} P_n(T) |V_{in;jm}|^2 \delta(E_i - E_j + E_n - E_m) \qquad (5.266)$$

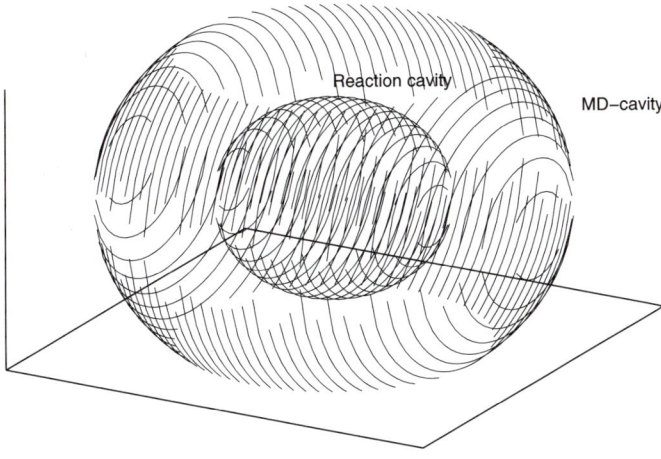

Fig. 5.16: Schematic approach for a molecular dynamics solution of chemical reactions in solutions or in clusters.

where $P_n(T)$ is the probability for state n of the bath and $V_{in;jm}$ the matrix element for a transition from state (in) of the system (i) and bath (n) to state (jm). The expression can be rewritten by using the fact that the delta function can be expressed as an integral over time, that is,

$$\delta(\epsilon_n - \epsilon_m) = \frac{1}{2\pi\hbar} \int_{-\infty}^{\infty} dt \exp(it(\epsilon_n - \epsilon_m)/\hbar) \qquad (5.267)$$

so that [213]

$$\tau_{ij}^{-1} = \frac{1}{\hbar^2} \int_{-\infty}^{\infty} dt \exp(i\omega_{ij}t)\langle V_{ij}(t)V_{ji}(0)\rangle \qquad (5.268)$$

where

$$V_{ij}(t) = \exp(iH_b t/\hbar)V_{ij}\exp(-iH_b t/\hbar) \qquad (5.269)$$

H_b is the bath hamiltonian, and the brackets in (5.268) indicate an average (quantum trace) over bath modes. It is now desirable to treat the bath modes classically while quantizing the system. If in the above equation we replace $V_{ij}(t)$, by $V_{ij}^{clas}(t)$, we will get $\tau_{ij} = \tau_{ji}$ rather than the correct detailed balance relation

$$\tau_{ij}^{-1} = \exp(\hbar\omega_{ij}\beta)\tau_{ji}^{-1} \qquad (5.270)$$

where $\beta = 1/k_B T$.

It was, therefore, suggested [213, 216] that one should make the substitution in a symmetrized expression for τ

$$\tau_{ij}^{-1} = \frac{2}{\hbar^2(1 + \exp(-\beta\hbar\omega_{ij}))} \int_{-\infty}^{\infty} dt \exp(i\omega_{ij}t)\frac{1}{2}\langle [V_{ij}(t), V_{ji}(0)]_+\rangle \qquad (5.271)$$

where $[,]_+$ denotes the anticommutator. Thus the quantum-classical expression with the bath modes treated classically is

$$\tau_{ij}^{-1} = \frac{2}{\hbar^2(1 + \exp(-\beta\hbar\omega_{ij}))} \int_{-\infty}^{\infty} dt \exp(i\omega_{ij}t)\langle V_{ij}^{clas} V_{ji}^{clas}(0)\rangle \qquad (5.272)$$

This expression satisfies the detailed balance principle in the limit $|\beta\hbar\omega_{ij}| \gg 1$, that is, in the limit where the average bath energy is small compared to the quantum jump, and where a large difference between up and downwards transition rates is expected.

In the above expression, we could consider the factor $(1+\exp(-\beta\hbar\omega_{ij}))^{-1} = \cosh(\beta\hbar\omega/2)/\sinh(\beta\hbar\omega/2)$ as a correction factor to be introduced when treating the bath classically. An almost similar factor was found by Berne et al. [214] when considering a system with hamiltonian

$$H = H_q + H_b + H_{qb} \qquad (5.273)$$

where

$$H_q = \frac{p^2}{2\mu} + \frac{1}{2}\mu\omega^2 q^2 \qquad (5.274)$$

that is, a harmonic oscillator coupled to a bath of oscillators

$$H_b = \sum_k \left(\frac{p_k^2}{2m_k} + \frac{1}{2} m_k \omega_k^2 q_k^2 \right) \tag{5.275}$$

by a linear coupling term

$$H_{qb} = -q \sum_k c_k q_k \tag{5.276}$$

In this case [214], the quantum correlation function

$$C_{qq}^q(\omega) = \int_{-\infty}^{\infty} dt \exp(-i\omega t) \frac{1}{2} \langle [\hat{q}(t), q(0)]_+ \rangle \tag{5.277}$$

could be written as a correction factor times the "classical" correlation function, that is,

$$C_{qq}^q(\omega) = x \coth(x) C_{qq}^{cl}(\omega) \tag{5.278}$$

where $x = \beta\hbar\omega/2$ and where the classical correlation function

$$C_{qq}^{cl}(\omega) = \int_{-\infty}^{\infty} dt \exp(-i\omega t) \langle q(t) q(0) \rangle_{cl} \tag{5.279}$$

can be obtained by running trajectories.

In both cases, a detailed balance correction factor appears. The latter factor $x \coth x$ is valid if a harmonic bath system coupled linearly to the oscillator is considered [216]. The derivation furthermore rests on the application of the Golden Rule expression for the relaxation process.

It is important to realize that the reason that these factors appear is not that the bath has been treated by classical mechanics. Classical mechanics is usually perfect for the bath modes (with the exception of low mass systems at low temperatures). The reason is rather that the quantum-classical correlation (as defined previously) has not been incorporated correctly.

In chapter 3, we considered a detailed balance correction to the collision between an atom and an oscillator. It was shown that detailed balance was approximately restored by the introduction of a mean velocity connected to a given transition in the quantum system. Consider for the above case the bath mode k with energy E_0. The solution of the classical equations of motion for such a mode would give

$$q_k(t) = \frac{1}{\omega_k} \sqrt{E_0/m_k} \cos(\omega_k t + \phi_k) \tag{5.280}$$

where ϕ_k is an arbitrary phase factor. The amplitude for de-excitation of the oscillator due to coupling to this bath mode is

$$\alpha \sim \int dt c_k q_k(t) \exp(i\omega t) \tag{5.281}$$

For a one-quantum $\hbar\omega$ de-excitation we have that the mean velocity of the bath mode is $v_{av} = \frac{1}{2}(v_0 + v_0\sqrt{1 + \hbar\omega/E_0})$ where $v_0 = \sqrt{2E_0/\mu}$. If we introduce a Boltzmann averaged amplitude we obtain a correction factor to the rate-constant for deactivation of $2x/\pi$ for large x values and unity for $x = 0$. Since the correction ($x\coth x$) discussed previously gives factors x and unity respectively, we see that introduction of the mean velocity or even better the variationally determined velocity (see chapter 3) could be interesting to use for relaxation of a quantum oscillator in a classical bath. The advantage of this approach is that it is independent of whether the system is a linearly coupled harmonic bath or not.

As mentioned previously, the detailed balance correction increases the de-excitation and decreases the excitation probability. The present analyses indicates that it is not necessary to treat the bath modes quantally if the proper quantum-classical correlation is introduced.

Another problem which has been discussed in the literature [213] is whether a binary collision model can be applied in solution. Arguments against it have been given by for example Zwanzig [217]. We would support these arguments by some calculations on the vibrational relaxation of nitrogen at liquid nitrogen temperatures. These calculations [218] were carried out using gas-phase dynamics and they showed that about the same rate-constant for relaxation as in the liquid phase [219] was obtained. But only so if trajectories which in the bimolecular collision quenched the contribution from the quadrupole coupling terms were integrated. These trajectories had to be started and ended at a center of mass distance of about 30–50 Å. Since such collisions do not occur in solution, the quadrupole coupling must be quenched in a different manner, that is, by coupling to many bath molecules in such a fashion that the angular dependent quadrupole term tends to average out. This analyses taken together with the result of the above-mentioned CO-$(Ar)_N$ cluster calculations indicate that vibrational relaxation in solution is influenced by couplings to at least the first and probably even several solvation shells.

For systems of meso- and macroscopic size, it is advantageous to consider the density of systems and introduce ensemble distributions, that is, equilibrium concepts and, hence, to introduce the distribution function. If quantum effects are important, the quantum

$$Q_{qm} = \mathrm{Tr}\{\exp(-\beta\hat{H})\} \tag{5.282}$$

rather than the classical distribution function is relevant. "Tr" means the quantum mechanical trace and $\beta = 1/k_B T$.

One of the best known phase space distribution functions is the one due to Wigner [81, 223]. For a system with one degree of freedom, we have

$$\Gamma(q, p, t) = \frac{1}{\pi\hbar}\int_{-\infty}^{\infty} dy\, \Psi^*(q + y, t)\Psi(q - y, t)\exp(2ipy/\hbar) \tag{5.283}$$

where $\Psi(q,t)$ is the wave function. The Wigner function has the property that is to order $\hbar^2$ obeys the classical equations of motion, that is,

$$\frac{\partial \Gamma(q,p,t)}{\partial t} = -\frac{\partial H}{\partial p}\frac{\partial \Gamma}{\partial q} + \frac{\partial H}{\partial q}\frac{\partial \Gamma}{\partial p} + \text{terms of order } \hbar^2 \tag{5.284}$$

where H is the classical hamiltonian $H(p,q) = (p^2/2m) + V(q)$. The drawback of this function is that it has negative regions in phase-space, that is, it is not positive definite. In practical applications this has led to slow convergence of the weak transition probabilities (see, for instance, [222]). Hence, alternative routes have formulated the quantum statistical mechanics through the path integral approach [3].

We can introduce the average probability of observing a particle at x as a Boltzmann weighted sum over the states n

$$P(x) = \frac{1}{Q}\sum_n \phi_n^*(x)\phi_n(x)\exp(-\beta E_n) \tag{5.285}$$

where the partition function is $Q = \sum_n \exp(-\beta E_n)$. More generally, we will introduce the density matrix

$$\rho(x,x') = \sum_n \phi_n(x)\phi_n^*(x')\exp(-\beta E_n) \tag{5.286}$$

such that $P(x) = \rho(x,x)/Q$ and $Q = Tr[\rho]$ is the quantum mechanical trace of the density matrix.

We now notice that the density matrix resembles the quantum mechanical propagator, which when resolved in eigenstates ϕ_n of the system is

$$K(x',t';x,t) = \sum_n \phi_n(x)\phi_n^*(x')\exp(-iE_n(t'-t)/\hbar) \tag{5.287}$$

Thus, by letting $t = 0$ and $t' = -i\beta/\hbar$ we can obtain the expression for the density matrix.

In chapter 2 we discussed the possibility of evaluating the propagator by path integral technique. Thus a similar technique can also be used for the evaluation of the density matrix [3], that is, we have

$$\rho(x_2, x_1) = \int \mathcal{D}x(t)\exp\left[-\frac{1}{\hbar}\int_0^{\beta\hbar}(m\dot{x}^2/2 + V(x))dt\right] \tag{5.288}$$

for a particle with mass m in a potential $V(x)$. This integral is much easier to evaluate than the real time path integrals, due to the exponential decay for paths away from the maximum contribution. Since we know that path integrals with up to second-order terms can be evaluated analytically, we could truncate the potential after the second-order term, such that

$$V(x) = V(\bar{x}) + V'(\bar{x})(x - \bar{x}) + \tfrac{1}{2}V''(\bar{x})(x - \bar{x})^2 \tag{5.289}$$

where it has been suggested to use an average distance [3] as the reference

$$\bar{x} = \frac{1}{\beta\hbar} \int_0^{\beta\hbar} x(t)dt \qquad (5.290)$$

Using eq. (2.12) with P time slices, so that $\Delta t = t/P = -i\hbar\beta/P$, we obtain

$$\rho(x_2; x_1) = \lim_{P\to\infty} (mP/2\pi\hbar^2\beta)^{P/2}$$

$$\times \int dy_1 \dots \int dy_P \exp\left[-\beta\left(\sum_{i=2}^{P} \frac{mP}{2\hbar^2\beta^2} \times ((y_i - y_{i-1})^2 + (y_1 - x_1)^2\right.\right.$$

$$\left.\left. + (x_2 - y_P)^2) + \sum_{i=1}^{P} V(y_i)/P\right)\right] \qquad (5.291)$$

If the potential is expanded to second order around the average distance

$$\bar{x} = \frac{1}{P} \sum_{i=1}^{P} y_i \qquad (5.292)$$

we can evaluate the expression for $\rho(x_2; x_1)$ and hence also the partition function

$$Q(T) = \int d\bar{x}\rho(x_1; x_1) \qquad (5.293)$$

For further details, see Feynman and Hibbs [3]. The above multi-dimensional phase-space integral (5.291) can more generally be evaluated using Monte Carlo methods or by standard molecular dynamics calculations using the hamiltonian

$$H = \sum_{i=1}^{P} \left[\frac{p_i^2}{2m_f} + \frac{1}{2}m\omega_P^2(y_i - y_{i+1})^2 + \frac{1}{P}V(y_i)\right] \qquad (5.294)$$

where m_f is a fictitious mass and the frequency $\omega_P = \sqrt{P}/(m\beta)$. This so-called path-integral molecular dynamics (PIMD) method has been reviewed in [44]. An approximate but numerically easier method is to use so-called centroid molecular dynamics.

5.7.1 Centroid molecular dynamics

The problems mentioned above concerning the proper description of the quantum-classical correlation has increased the interest in treating the dynamics of the solvent using quantum mechanical methods. The path-integral evaluation of quantum distribution functions has been used by Voth and coworkers to formulate a so-called centroid molecular dynamics (CMD) approach [220, 221]. CMD is an approximate quantum dynamical method for studying nuclear quantum effects in condensed phase. The centroid position is defined by eq. (5.290).

In classical statistical mechanics, the partition function for a one-dimensional system can be expressed as

$$Q_{cl}(T) = \frac{1}{2\pi\hbar} \int dx \int dp \exp(-H(x,p)\beta) \qquad (5.295)$$

where $\beta = 1/k_B T$. In centroid MD the quantum mechanical analog is expressed in a similar manner

$$Q_{qm}(T) = \frac{1}{2\pi\hbar} \int dx_c \int dp_c \rho_c(x_c, p_c) \qquad (5.296)$$

where the centroid density function is given as

$$\rho_c(x_c, p_c) = \exp(-\beta p_c^2/2m)\rho_c(x_c) \qquad (5.297)$$

Thus, we can define a centroid hamiltonian as

$$H_c = \frac{p_c^2}{2m} + V(x_c) \qquad (5.298)$$

where $V(x_c) = -k_B T \ln(\rho_c(x_c))$. The centroid variables follow classical mechanics, however, in an effective potential $V(x_c)$ which involves the determination of the density function $\rho_c(x_c)$. The centroid density function is obtained as a path integral, such that

$$\rho(x_c) = \int \mathcal{D}x(t)\delta(x_0 - x_c) \exp\left(-\beta \int_0^{\beta\hbar} dt V(x)\right) \qquad (5.299)$$

or by introducing $\int_0^{\beta\hbar} dt V(x) = \sum_{i=1}^P V(y_i)/P$ we have

$$\rho(x_c) = \prod_{i=1}^P \int dy_i \exp(-\beta V(y_i)/P)\delta(x_0 - x_c) \qquad (5.300)$$

Time correlation functions are obtained by running trajectories on the effective potential. Thus, the quantum correlation function

$$\langle A(0)B(t)\rangle = \frac{1}{Q}\frac{1}{\beta} \int_0^\beta ds \mathrm{Tr}(\exp(-\beta H)\exp(sH)A\exp(-sH)B(t)) \qquad (5.301)$$

can be approximated by

$$\langle A(0)B(t)\rangle_c = \frac{1}{2\pi\hbar} \int dx_c \int dp_c \rho(x_c, p_c)B_c(t)A_c(0) \qquad (5.302)$$

where the classical expressions for B and A are evaluated at the centroid variables.

Although the centroid dynamics is in this approximation classical some quantum delocalization is still kept.

The centroid dynamics have recently been used to simulate proton transport in water [225]. Although the centroid molecular dynamics and also path integral approaches coupled to DFT (density functional) calculations of the interaction potential are feasible [226], they are computationally heavy and, hence, proton transport reactions have also been simulated using more approximate models of the mean field type [227]. Molecular mechanics methods, in which packages of force-field data and other semi-empirical or empirical data are used to construct the potential energy surface and classical mechanics used for the nuclear motion, will still be the method of choice for treating a large body of the problems involving large molecules in solution.

Chapter 6
Conclusion

In this book, we have discussed the problems concerning mixing of classical and quantum mechanics, and we have given several possible solutions to the problem and a number of suggestions for the setup of working computational schemes. In the present chapter, we give some recommendations as to which methods one should use for a given type of system and problem (see tables 6.1 and 6.2). As can be seen from the tables and what is apparent from the discussion in the previous chapters, the quantum-classical method has been and is used for solving many different molecular dynamics problems. Recommendations, as far as molecule surface or processes in solution are concerned, have not been incorporated, the reason being that the methods here are still to some extent under development.

We have seen that the quantum-classical approach can be derived in two different fashions. In one method the classical limit $\hbar \to 0$ is taken in some degrees of freedom. In the other approach the quantum mechanical equations are parameterized in such a fashion that classical equations of motions are either pulled out of or injected into the quantum mechanical. Thus the first method involves and introduces the classical picture in certain particular degrees of freedom—in the second method the classical picture is in principle not introduced—it is just a reformulation of quantum mechanics. This reformulation has the exact dynamics as the limit. However, if exact calculations are to be performed, the reformulation may not be advantageous from a computational point of view, and, hence, standard methods are often more conveniently applied. We prefer the second approach for introducing the quantum-classical scheme because, as mentioned, it automatically has the exact formulation as the limit. The approach is most conveniently implemented through the trajectory driven DVR, or the so-called TDGH-DVR method, which gives the systematic way of approaching the quantum mechanical limit from the classical one. Thus, the method interpolates continuously between the classical and the quantum limit—a property it shares with, for instance, the FMS method and the Bohm formulation.

The TDGH-DVR method is convenient to use for large systems if many degrees of freedom can be described using just a single grid point (the classical limit). By working with parameters such as momentum and width parameters, by appropriate choices of these parameters we can approach the stationary grid method for some degrees of freedom (DOF), while using a moving grid for others

Table 6.1: Overview of approximate quantum-classical methods and computational schemes for treating inelastic molecular dynamics problems.

System	Quantity	Recommendation
$A + BC$	Vibrational/rotational transitions	$V_q R_q T_c$ method
$A + BC$	Vibrational transitions	$V_q R_c T_c$ method
	small to intermediate vibrational level	
	Highly excited states (near dissociation)	2D-grid + trajectory
$A + BC$	Vibrational/rotational transitions	Classical trajectories+
	Highly excited states and heavy	binning or 2D-grid+
	molecules	trajectory
$AB + CD$	Vib/rot + vib-transitions	$V_q R_q T_c + V_q R_c T_c$ method
	One light and one heavy molecule	
$AB + CD$	Vibrational transitions	$V_q R_c T_c$ method
	Two heavy molecules	
$A + BCD$	Vibrational excitation	Vibrational state
	Light polyatom	expansion + SQ method
A + polyatom	Vibrational excitation	SQ-treatment of
	heavy molecules	vibrations + classical
		mechanics for $R + T$

V = Vibration, R = Rotation, T = Translation, q = quantum, c = classical, SQ = second quantization.

and a classical trajectory description in the third set of DOFs. The method gives the answer to the question: *given that the molecular dynamics is described by classical mechanics, what is then the systematic way of improving the description so as to approach the quantum limit?*

In principle, methods which are based on taking the $\hbar \to 0$ limit are inappropriate as a starting point for a quantum-classical theory. The reason is, of course, the uncertainty principle $\Delta E \Delta t > \hbar$, with which a given accepted error in ΔE restricts $\hbar \to 0$ methods to be valid in a short time period only. In some methods, the $\hbar \to 0$ limit is not taken (as, for instance, the FMS or MEM)

Table 6.2: Overview of quantum-classical methods for reactive and non-adiabatic processes.

Single surface $A + BC$	State-to-all	1D-grid + trajectory
Single surface $AB + CD$	State-to-all	2D-grid + trajectory
Multi-surface $A + BC$	Vib-resolved	3D-hyperspherical
		+ classical rotation
Multi-surface	Total rates + cross	FMS, TSH or multi-surface
	sections	TDGH-DVR methods
Single surface, three	Partially state-resolved	Reaction volume approach
center reactions		or TDGH-DVR method

methods, but classical mechanics is injected in the quantum mechanical equations. In these theories it is difficult, however, to invoke full classical concepts in some DOFs.

Methods which do divide the DOFs in two sets—a classical set and a quantum set—have the problem that the correlation between them has to be incorporated. Only if the coupling between the two sets of DOFs is weak can we ignore what we have called the quantum-classical correlation. In the quantum dressed classical mechanics method, on the other hand, the DOFs are not separated, but classical mechanics is pulled out of the quantum equations of motion and solved in the full space! What is solved in quantum space is, therefore, a matter of choice connected to the situation under investigation. The method gives the recipe for the working quantum mechanical equations of motion in this situation. We can put the grid points in any subset of coordinates and even in coordinates which do not coincide with the classical ones. The classical equations of motion are most conveniently solved in cartesian coordinates. For the quantum coordinates, we will often choose one or several chemical bonds—so as to be able to describe bond breaking, for instance. Since all we need from the classical equations of motion in order to solve the quantum equations are the classical forces in the quantum variables, we can use chain rules to obtain them from the classical forces in cartesian space.

This method is, therefore, the easiest and most obvious way of mixing quantum and classical mechanics. From a puristic point of view, it is also the only possibility, but this does not mean that it will be the method of choice for solving a given problem. Note, also, that the inclusion of quantum-classical correlation effects might imply using more than a single grid point in the degrees of freedom which are more strongly coupled to the quantum DOFs.

For low dimensional systems, effective "exact" quantum methods do exist, and for larger systems, methods which are not necessarily derived from first principles but constructed to solve a specific problem or class of problems may be more appropriate to use. To this latter category belong trajectory surface hopping methods, mean field methods, the classical path method, and so on.

For processes in solution and on solids, the quantum-classical methods will still need further development. At least there is more need for testing of the available methods against experimental data, or reliable less approximate calculations. The situation at present, that is, with present computational facilities, is that the methods which work for gas-phase problems may still be too expensive to use for performing routine calculations in systems where the small quantum system is coupled to a large "heat bath." Other problems are that the simple detailed balance correction schemes such as the symmetrized Ehrenfest approach—although, in principle, applicable—are more complex to work with in molecule-surface or molecule-cluster or solution cases.

Appendices

Appendix A: Switch to a frozen Gaussian expansion

In certain regions, it may be advantageous to switch to a frozen, that is, a fixed-width Gauss-Hermite expansion defined by $\mathrm{Re}A = 0$, $p(t) = 0$ and $\mathrm{Im}A$ constant, for instance, when the trajectory hits a turning point $p(t_0) = 0$. Here, we wish to re-expand the wave function in the following manner

$$\pi^{1/4} \sum_n c_n(t_0) \exp\left(\frac{i}{\hbar}(\gamma + \mathrm{Re}A(t_0)(x - x_0)^2)\right) \phi_n(\xi(t_0))$$

$$= \pi^{1/4} \sum_m a_m(t_0) \exp\left(\frac{i}{\hbar}\gamma\right) \phi_m(\xi(t_0)) \tag{A.1}$$

where we have used the fact that $p(t_0) = 0$. Thus the new expansion coefficients can be obtained as

$$a_m(t_0) = \sum_n c_n(t_0) \int_{-\infty}^{\infty} d\xi \exp\left(\frac{i}{\hbar}\mathrm{Re}A(t_0)(x - x_0)^2\right)$$

$$\times \frac{1}{\sqrt{n!m!\pi 2^{m+n}}} \exp(-\xi^2) H_m(\xi) H_n(\xi) \tag{A.2}$$

where $x_0 = x(t_0)$. The integral can be evaluated to give (see [71])

$$a_m(t_0) = \sum_n c_n(t_0) C_{mn} \tag{A.3}$$

where $C_{mn} = 0$ if $m + n$ is odd and

$$C_{mn} = \sqrt{\frac{n!m!}{a}} \sum_{k=0}^{\min(m,n)} \frac{(m + n - 2k - 1)!!}{k!(n - k)!(m - k)!}(1/a - 1)^{(m+n)/2-k} \tag{A.4}$$

for $m + n$ even. We notice that for $a \to 1$, that is, for $\mathrm{Re}A(t_0) \to 0$, we have $C_{mn} \to \delta_{nm}$. The quantity a is defined as $a = 1 - \frac{i}{2}\frac{\mathrm{Re}A(t_0)}{\mathrm{Im}A(t_0)}$. In the fixed-width basis set, we solve the following set of equations for the expansion coefficients

$$i\hbar\dot{a}_k(t) = \sum_n a_n(t)\left[\delta_{kn}\left(n + \frac{1}{2}\right)\frac{2\hbar\mathrm{Im}A}{m} + V_{kn} - \frac{2\mathrm{Im}A^2}{m}M_{kn}^{(2)}\right] \tag{A.5}$$

where V_{kn} is a matrix element $\langle\Phi_k|V|\Phi_n\rangle$.

From the force constant $k(t_0)$, and the fact that $\mathrm{Im}A(t_0)$ is a constant, we deduce that the Gauss-Hermite basis set Φ_n with $n \in [0 : n_{max}]$ covers an expansion range of approximately $[x_0; x_f]$, where x_f is defined by

$$x_f = x_0 + \sqrt{\hbar(n_{max} + 1/2)/\mathrm{Im}A} \tag{A.6}$$

and n_{max} is the highest basis function included in the expansion. Thus, it is advantageous to have $\mathrm{Im}A$ as small as possible when the switch to the fixed-width Gauss-Hermite basis is made. However, if $\mathrm{Im}A$ is too small, the basis set has no support in the region it is supposed to cover.

The basis set can be used to propagate the wave function as long as it is not reflected by the boundaries x_f and x_0. Thus, the procedure is the following: At the turning point for the motion in $x(t)$ we switch to a basis set with fixed-width $\mathrm{Im}A(t_0)$, which is used until the wave function has been reflected from the potential or sometimes bifurcated in two parts in the case of tunneling through a barrier. Then it is again advantageous to re-expand in the time-dependent Gauss-Hermite basis set. This is most conveniently done by performing a Fourier transform of the now localized wave packet. The Fourier transform yields the coefficients in the expansion

$$\pi^{1/4} \frac{1}{\sqrt{2\pi}} \sum_k b_k \exp(\pm ikx) \tag{A.7}$$

where the sign depends on which part of the wave function we consider (see fig. A.1). We now want to represent this wave function as

$$\pi^{1/4} \sum_m c_m \exp\left(\frac{i}{\hbar}(\gamma_a + p_a(x - x_a))\phi_m(\xi)\right) \tag{A.8}$$

where $x_a = x(t_a)$ and $\mathrm{Re}A(t_a)$ has been set to zero.

If the wave function has bifurcated, we have two parts centered around x_a and x_b with opposite signs of p_a and p_b. Thus, the Fourier coefficients can also be expressed as

$$b_k = \pi^{-1/4}\exp(\mp ikx_a + \mathrm{Im}\gamma_a/\hbar)\int d\xi \, \exp\left[\frac{i}{\hbar}(p_a/\hbar \mp k)(x - x_a)\right]\phi_m(\xi) \tag{A.9}$$

where $-k$ goes with p_a and $+k$ with p_b (see fig. A.1). Introducing $x - x_a = \sqrt{\hbar/2\mathrm{Im}A_a}\xi$, we get [71]

$$b_k = \exp[\mp ikx_a - \hbar(p_a/\hbar \mp k)^2/(4\mathrm{Im}A_a)](2\pi\hbar/\mathrm{Im}A_a)^{1/4}$$
$$\times \sum_n c_n \frac{1}{\sqrt{n!}}(i\sqrt{\hbar/2\mathrm{Im}A_a}(p_a/\hbar \mp k))^n \tag{A.10}$$

where $c_n \sim \delta_{n0}$ is expected for a localized nearly Gaussian wave packet. From the known coefficients b_k, we can now deduce the unknown parameters c_n, x_a, p_a, and A_a. If the wave packet has bifurcated, a similar expression is obtained for the b part of it. Thus, the value of x_a is estimated from the maximum in the

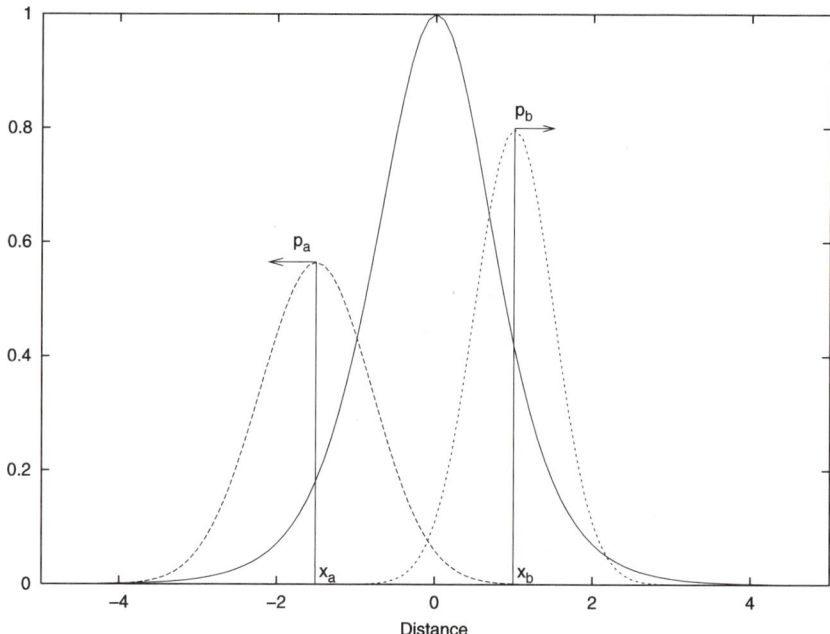

Fig. A.1: Splitting of a wave packet by a potential barrier. Each wave packet can be represented by a local Gauss-Hermite basis and propagated separately to reactant and product side, respectively.

coordinate space distribution, p_a from the maximum in the momentum space distribution and $\mathrm{Im}A_a$ from the half-width of the two distributions, that is,

$$\mathrm{Im}A_a = 0.5(p_a - p_{1/2})/(\hbar \ln 2) \tag{A.11}$$

$$\mathrm{Im}A_a = 0.5\hbar \ln 2/(x_a - x_{1/2})^2 \tag{A.12}$$

where $p_{1/2}$ and $x_{1/2}$ denote the values of the momentum and coordinate for which the momentum and coordinate space distributions have reached half the maximum value.

The momentum distribution is most conveniently obtained by the Fourier transform of the wave function, that is, from the b_k coefficients of the series

$$\Psi_j = \frac{1}{\sqrt{N}} \sum_{k=-N/2}^{N/2-1} b_k \exp(-2i\pi kj/N) \tag{A.13}$$

where N is the number of grid points distributed in the range of the wave function $[x_b; x_c]$ and

$$x_j = x_b + (j-1)\Delta x \tag{A.14}$$

$j = 1, \ldots, N$, and $\Delta x = (x_c - x_b)/N$.

Once an estimate of the parameters x_a, p_a, and $\text{Im}A_a$ has been obtained, we can obtain the expansion coefficients in eq. (A.8) as

$$c_m = \frac{1}{\sqrt{N}} \frac{i^m}{\sqrt{m! 2^m}} (2\hbar/\text{Im}A_a)^{1/4}$$

$$\times \sum_{k=-N/2}^{N/2-1} b_k \exp\left(-\frac{1}{2}z_k^2 + \frac{2i\pi}{L}k(x_b - \Delta x) - iKx_a\right) H_m(z_k) \qquad (A.15)$$

where $L = x_c - x_b$, $K = 2\pi k/L$ and

$$z_k = -\sqrt{\hbar/2\text{Im}A_a}(K + p_a/\hbar) \qquad (A.16)$$

Again, since the basis set is complete, we only need an estimate of the three parameters x_a, p_a and $\text{Im}A_a$.

We notice, also, that the procedure described above can be used even if the DVR propagation method is used, by first switching from the DVR to the Gauss-Hermite representation by using the fact that the connection between the expansion coefficients $a_n(t)$ in the G-H basis and the $c_i(t)$ coefficients in the DVR representation is

$$a_n(t) = \sum_i c_i(t)\phi_n(z_i) \qquad (A.17)$$

Once the new G-H basis is defined locally, we may switch back to the DVR scheme and propagate two wave packets independently in the DVR representation.

Appendix B: Matrix elements over VV operators

In the second quantization formulation of molecular dynamics, as well as in energy transfer problems, in order to obtain state resolved transition amplitudes we have to evaluate summations of the types below:

For $M = 2$ we have

$$\langle n_1' n_2' | U_{VV} | n_1 n_2 \rangle = \sqrt{\frac{n_2! n_2'!}{n_1' ! n_1!}} (1 + f_{11})^{n_1'} (1 + f_{22})^{n_2'}$$

$$\times \alpha_{21}^{n_1 - n_1'} \sum_k \frac{(\alpha_{12}\alpha_{21})^k (n_1 + k)!}{k!(n_1 - n_1' + k)!(n_2 - k)!} \qquad (A.18)$$

and for $M = 3$

$$\langle n_1' n_2' n_3' | U_{VV} | n_1 n_2 n_3 \rangle$$

$$= \sqrt{\frac{n_3! n_2'! n_3'!}{n_1' ! n_1! n_2!}} \, \Pi(1 + f_{ii})^{n_i'}$$

$$\times \sum_{k_{ij}} \frac{(n_1 + k_{12} + k_{13})!(n_2 + k_{23})! \, f_{21}^{k_{21}} f_{12}^{k_{12}} f_{31}^{k_{31}} f_{13}^{k_{13}} f_{23}^{k_{23}} f_{32}^{k_{32}}}{(n_2' - k_{21})!(n_3' - k_{31} - k_{32})! \ k_{21}! k_{12}! k_{31}! k_{13}! k_{23}! k_{32}!} \qquad (A.19)$$

This sum is subject to 2 independent constraints and hence reduces to a 4 fold sum (see below). For $M = 4$ we have

$$\langle n_1' n_2' n_3' n_4' | U_{VV} | n_1 n_2 n_3 n_4 \rangle$$

$$= \sqrt{\frac{n_4! n_2'! n_3'! n_4'!}{n_1'! n_1! n_2! n_3!}} \, \Pi (1 + f_{ii})^{n_i'}$$

$$\times \sum_{k_{ij}} \frac{(n_1 + k_{12} + k_{13} + k_{14})!(n_2 + k_{23} + k_{24})!(n_3 + k_{34})!}{(n_2' - k_{21})!(n_3' - k_{31} - k_{32})!(n_4' - k_{41} - k_{42} - k_{43})!}$$

$$\times \frac{f_{12}^{k_{12}} f_{13}^{k_{13}} f_{14}^{k_{14}} f_{21}^{k_{21}} f_{23}^{k_{23}} f_{24}^{k_{24}} f_{31}^{k_{31}} f_{32}^{k_{32}} f_{34}^{k_{34}} f_{41}^{k_{41}} f_{42}^{k_{42}} f_{43}^{k_{43}}}{k_{12}! k_{13}! k_{14}! k_{21}! k_{23}! k_{24}! k_{31}! k_{32}! k_{34}! k_{41}! k_{42}! k_{43}!} \tag{A.20}$$

This sum is subject to 3 independent constraints and is, therefore, a nine-fold sum. In general, the expression for the transition amplitude from a state $|n_1, \ldots, n_M\rangle$ to $|n_1', \ldots, n_M'\rangle$ induced by the operator U_{VV} is given by [130]

$$\langle n_M', \ldots, n_1' | U_{VV} | n_1, \ldots, n_M \rangle$$

$$= \exp \left(\sum_{j=1}^{M} \alpha_{jj} n_j' \right) \sqrt{\frac{n_M! \prod_{j=2}^{M} n_j'!}{n_1'! \prod_{j=1}^{M-1} n_j!}} \times \sum_{k_{ji}=0, i \neq j}^{\infty} \prod_{i \neq j} \left(\frac{\alpha_{ji}^{k_{ji}}}{k_{ji}!} \right)$$

$$\times \frac{\prod_{p=1}^{M-1} \left(n_p + \sum_{q=p+1} k_{pq} \right)!}{\prod_{p=2}^{M} \left(n_p + \sum_{q=p+1}^{M} k_{pq} - \sum_{q \neq p} k_{qp} \right)!} \tag{A.21}$$

where the time-dependent functions $\alpha_{ji} = f_{ji}$ $(j \neq i)$ and $\exp(\alpha_{ii}) = 1 + f_{ii}$. These functions are related to the solution of the equation (4.33), see [130] and appendix C. The sums run over terms containing non-negative factorials and due to the constraints

$$n_i' = n_i + \sum_{j \neq i} k_{ij} - \sum_{j \neq i} k_{ji} \tag{A.22}$$

for $i = 1, \ldots, M$ the $M^2 - M$-fold sum is reduced to an $(M - 1)^2$-fold sum. Note that the M constraints are reduced to $M - 1$ due to the fact that the VV operators conserve the number of quanta and, hence,

$$\sum_i n_i = \sum_i n_i' \tag{A.23}$$

which is contained in the M equations above and is, therefore, automatically fulfilled. If it had not been for the constraints, we could have carried out the multiple sums analytically, the result being

$$\langle n_M', \ldots, n_1' | U_{VV} | n_1, \ldots, n_M \rangle = \prod_{i=1}^{M} (1 + f_{ii})^{n_i'} \sqrt{\frac{\prod n_i!}{\prod n_i'!}} \, \frac{\prod_{i=2}^{M} \left(1 + \sum_{k=1}^{i-1} f_{ik} \right)^{n_i'}}{\prod_{i=1}^{M-1} \left(1 - \sum_{k=i+1}^{M} f_{ik} \right)^{n_i+1}} \tag{A.24}$$

However, this expression is not very accurate for small values of M.

In the limit of small perturbations, f_{ji} is small and we can expect a first order expression to be valid. Thus, only terms of order f_{ji}^p $(p = 0, 1)$ are included in the sum. To second order, also $p = 2$ terms are included, and so on.

Appendix C: The inversion problem

In the SQ formulation of dynamical problems, the matrix representation of the so-called VV operators yields a set of equations

$$i\hbar\dot{\mathbf{R}}(t, t_0) = \mathbf{B}\mathbf{R}(t, t_0) \tag{A.25}$$

where $\mathbf{R}(t_0, t_0) = \mathbf{I}$, the unit matrix. In order to obtain the amplitude for a transition from an initial state vector $|n_1, \ldots, n_M\rangle$ to $|n_1', \ldots, n_M'\rangle$, we use the evolution operator U_{VV} and, hence, we need to evaluate the functions α_{ji}. We introduce the new functions

$$f_{ij} = \alpha_{ij} \ (i \neq j) \tag{A.26}$$

$$f_{ii} = \exp(\alpha_{ii}) - 1 \tag{A.27}$$

and obtain the functions f_{ij} in the following manner:

For $m < n$ we have:

$$f_{mn} = |\mathbf{A}|/|\mathbf{D}| \tag{A.28}$$

where $|\mathbf{A}|$ indicates the determinant of the matrix $\mathbf{A}$. The matrices $\mathbf{A}$ and $\mathbf{D}$ are of order $(n-1) \times (n-1)$, and the elements of the two matrices given as:

$$D_{kl} = Q_{kl} \ (k \neq l) \tag{A.29}$$

$$D_{kk} = 1 + Q_{kk} \tag{A.30}$$

and

$$A_{km} = Q_{kn} \tag{A.31}$$

$$A_{kl} = D_{kl} \ (l \neq m) \tag{A.32}$$

where k and l run from 1 to $n - 1$. For the element f_{nn} we have:

$$f_{nn} = Q_{nn} - \sum_{m=1}^{n-1} f_{mn} Q_{nm} \tag{A.33}$$

where the elements f_{mn} have just been determined. For f_{nm} where $m < n$ we obtain:

$$f_{nm} = \frac{Q_{nm} - \sum_{k=1}^{m-1} Q_{nk} f_{km}}{1 + f_{nn}} \tag{A.34}$$

Thus, the above equations enable us to determine all the elements f_{nm} from the solution to the differential equation (A.25) using that $\mathbf{Q} = \mathbf{R} - \mathbf{I}$. The expectation value of the hamiltonian only involves the functions Q_{ij}. Thus the above "inversion" problem need not to be solved in order to evaluate the effective hamiltonian H_{eff}. This has been utilized when calculating the effective hamiltonians arising from the coupling to phonon and electron excitation in solids.

Appendix D: Constrained eigenvalue problems

The eigenvalue problem which is solved along the reaction path is constrained such that for systems with a linear configuration along the reaction path we have six constraints, that is, the solution of the eigenvalue problem leaves us with $3N - 6$ non-zero vibrational frequencies $\omega_k(s)$. For non-linear configurations the number of constraints is seven, leaving us with $3N - 7$ "perpendicular" frequencies.

In order to include the constraints for the reaction path dynamics, we use the fact that the center of mass and the Eckart conditions are

$$\sum_i m_i \rho_i = 0 \tag{A.35}$$

$$\sum_i m_i \mathbf{a}_i \times \rho_i = 0 \tag{A.36}$$

In addition, we have the reaction path constraint. Introducing the normal mode coordinates Q_k, we express the displacement vector as

$$\rho_i = m_i^{-1/2} \sum_k \ell_{ik} Q_k \tag{A.37}$$

The solution of the eigenvalue problem defines the eigenvectors ℓ_{ik}, as well. The seven constraints now read

$$\sum_i \sqrt{m_i} \ell_{ik} = 0 \tag{A.38}$$

$$\sum_i \sqrt{m_i} \mathbf{a}_i \times \ell_{ik} = 0 \tag{A.39}$$

$$\sum_i \mathbf{g}_i \cdot \ell_{ik} = 0 \tag{A.40}$$

where $\mathbf{g}_i$ is the gradient

$$\mathbf{g}_i = m_i^{-1/2} \left. \frac{\partial V}{\partial \mathbf{x}_i} \right|_{\mathbf{x}_i = \mathbf{a}_i} \tag{A.41}$$

For the reaction volume approach, we have nine constraints. Thus introducing the expression (5.255) the constraints will be

$$\sum_i \sqrt{m_i} \ell_{ik} = 0 \tag{A.42}$$

$$\sum_i \sqrt{m_i} \mathbf{a}_i \times \ell_{ik} = 0 \tag{A.43}$$

$$\frac{1}{\rho} \sum_i \sqrt{m_i} \ell_{ik} \cdot \frac{\partial \mathbf{a}_i}{\partial \theta} = 0 \tag{A.44}$$

$$\frac{1}{\rho} \sum_i \sqrt{m_i} \ell_{ik} \cdot \frac{\partial \mathbf{a}_i}{\partial \phi} = 0 \tag{A.45}$$

$$\sum_i \sqrt{m_i} \ell_{ik} \cdot \frac{\partial \mathbf{a}_i}{\partial \rho} = 0 \tag{A.46}$$

With the nine constraints, we introduce the matrix $\mathbf{P}$, where

$$
\mathbf{P} = \begin{bmatrix}
\sqrt{m_1} & 0 & 0 & \sqrt{m_2} & 0 \\
0 & \sqrt{m_1} & 0 & 0 & \sqrt{m_2} \\
0 & 0 & \sqrt{m_1} & 0 & 0 \\
0 & -\sqrt{m_1}a_{1z} & \sqrt{m_1}a_{1y} & 0 & -\sqrt{m_2}a_{2z} \\
\sqrt{m_1}a_{1z} & 0 & -\sqrt{m_1}a_{1x} & \sqrt{m_2}a_{2z} & 0 \\
-\sqrt{m_1}a_{1y} & \sqrt{m_1}a_{1x} & 0 & -\sqrt{m_2}a_{2y} & \sqrt{m_2}a_{2x} \\
\ell_\rho^{1x} & \ell_\rho^{1y} & \ell_\rho^{1z} & \ell_\rho^{2x} & \ell_\rho^{2y} \\
\ell_\theta^{1x} & \ell_\theta^{1y} & \ell_\theta^{1z} & \ell_\theta^{2x} & \ell_\theta^{2y} \\
\ell_\phi^{1x} & \ell_\phi^{1y} & \ell_\phi^{1z} & \ell_\phi^{2x} & \ell_\phi^{2y}
\end{bmatrix}
$$

$$
\begin{bmatrix}
0 & \sqrt{m_3} & 0 & 0 \\
0 & 0 & \sqrt{m_3} & 0 \\
\sqrt{m_2} & 0 & 0 & \sqrt{m_3} \\
\sqrt{m_2}a_{2y} & 0 & -\sqrt{m_3}a_{3z} & \sqrt{m_3}a_{3y} \\
-\sqrt{m_2}a_{2x} & \sqrt{m_3}a_{3z} & 0 & -\sqrt{m_3}a_{3x} \\
0 & -\sqrt{m_3}a_{3y} & \sqrt{m_3}a_{3x} & 0 \\
\ell_\rho^{2z} & \ell_\rho^{3x} & \ell_\rho^{3y} & \ell_\rho^{3z} \\
\ell_\theta^{2z} & \ell_\theta^{3x} & \ell_\theta^{3y} & \ell_\theta^{3z} \\
\ell_\phi^{2z} & \ell_\phi^{3x} & \ell_\phi^{3y} & \ell_\phi^{3z}
\end{bmatrix} \tag{A.47}
$$

and the matrix $\mathbf{N}$

$$
\mathbf{N} = \begin{bmatrix}
-\sqrt{m_4} & 0 & 0 & \cdot & \cdot \\
0 & -\sqrt{m_4} & 0 & \cdot & \cdot \\
0 & 0 & -\sqrt{m_4} & \cdot & \cdot \\
0 & \sqrt{m_4}a_{4z} & -\sqrt{m_4}a_{4y} & \cdot & \cdot \\
-\sqrt{m_4}a_{4z} & 0 & \sqrt{m_4}a_{4y} & \cdot & \cdot \\
\sqrt{m_4}a_{4y} & -\sqrt{m_4}a_{4x} & 0 & \cdot & \cdot \\
-\ell_\rho^{4x} & -\ell_\rho^{4y} & -\ell_\rho^{4z} & \cdot & \cdot \\
-\ell_\theta^{4x} & -\ell_\theta^{4y} & -\ell_\theta^{4z} & \cdot & \cdot \\
-\ell_\phi^{4x} & -\ell_\phi^{4y} & -\ell_\phi^{4z} & \cdot & \cdot
\end{bmatrix} \tag{A.48}
$$

where the dots indicate the obvious extension to $N > 4$ problems, and where we have introduced the notation $\ell_\rho^{1x} = \sqrt{m_i}(\partial a_{1x}/\partial\rho)$ etc.

   If we have only seven constraints, the $\mathbf{P}$ matrix is defined as the upper left 7×7 part of the matrix (A.47), with the elements $g_{1x}, g_{1y}, \ldots, g_{3x}$ in the bottom row. The $\mathbf{N}$ matrix is changed accordingly to

$$\mathbf{N} = \begin{bmatrix} 0 & 0 & -\sqrt{m_4} & 0 & 0 \\ -\sqrt{m_3} & 0 & 0 & -\sqrt{m_4} & 0 \\ 0 & -\sqrt{m_3} & 0 & 0 & -\sqrt{m_4} \\ \sqrt{m_3}a_{3z} & -\sqrt{m_3}a_{3y} & 0 & \sqrt{m_4}a_{4z} & -\sqrt{m_4}a_{4y} \\ 0 & \sqrt{m_3}a_{3x} & -\sqrt{m_4}a_{4z} & 0 & \sqrt{m_4}a_{4y} \\ -\sqrt{m_3}a_{3x} & 0 & \sqrt{m_4}a_{4y} & -\sqrt{m_4}a_{4x} & 0 \\ -g_{3y} & -g_{3z} & -g_{4x} & -g_{4y} & -g_{4z} \end{bmatrix} \tag{A.49}$$

With just six constraints (the linear case), we remove one row in the $\mathbf{P}$, as well as in the $\mathbf{N}$, matrix (for instance row no. 6).

The eigenvectors ℓ_k are now divided into two components, ℓ_k^a containing the first nine (seven or six) elements, and ℓ_k^b containing $3N - 9$ (7 or 6) elements. The constraints may now be written as

$$\mathbf{P}\ell_k^a = \mathbf{N}\ell_k^b \tag{A.50}$$

Thus, we have

$$\ell_k^a = \mathbf{P}^{-1}\mathbf{N}\ell_k^b = \mathbf{A}\ell_k^b \tag{A.51}$$

Introducing now the matrix elements of the hessian matrix $\mathbf{W}$ as

$$W_{i\alpha;j\beta} = \frac{1}{\sqrt{m_i m_j}} \frac{\partial^2 V}{\partial r_{i\alpha}\partial r_{j\beta}} \tag{A.52}$$

where i and $j = 1, \ldots, N$ and α and β are x, y, z, we now obtain the components a and b of the eigenvectors from the columns of the two matrices $\mathbf{L}^a$ and $\mathbf{L}^b$ where [179]

$$\mathbf{L}^a = \mathbf{AGD}^{-1/2}\mathbf{H} \tag{A.53}$$

$$\mathbf{L}^b = \mathbf{GD}^{-1/2}\mathbf{H} \tag{A.54}$$

Here, $\mathbf{A}$ is defined above and $\mathbf{G}$ is defined as the matrix which diagonalizes $\mathbf{A}^T\mathbf{A} + \mathbf{I}$, that is,

$$\mathbf{G}^T(\mathbf{A}^T\mathbf{A} + \mathbf{I})\mathbf{G} = \mathbf{D} \text{ (diagonal matrix)} \tag{A.55}$$

where $\mathbf{I}$ is the unit matrix and $\mathbf{H}$ is a matrix which diagonalizes

$$\mathbf{D}^{-1/2}\mathbf{G}^T\mathbf{BGD}^{-1/2} \tag{A.56}$$

Appendix E: Reaction volume hamiltonian

In this appendix, we give the derivation of the kinetic and rotational contribution to the reaction volume hamiltonian.

The kinetic energy

From the expressions (5.248–5.250) and

$$\dot{\mathbf{a}}_i = \frac{\partial \mathbf{a}_i}{\partial \rho}\dot{\rho} + \frac{\partial \mathbf{a}_i}{\partial \theta}\dot{\theta} + \frac{\partial \mathbf{a}_i}{\partial \phi}\dot{\phi}$$

$$= \frac{1}{\sqrt{m_i}}\left(\boldsymbol{\ell}_\rho(i)\dot{\rho} + \frac{\rho}{2}\boldsymbol{\ell}_\theta(i)\dot{\theta} + \frac{\rho}{2}\boldsymbol{\ell}_\phi(i)\dot{\phi}\right) \qquad\text{(A.57)}$$

we obtain the kinetic energy (T) relative to the center of mass of the system as:

$$2T = (\dot{\rho}, \dot{\theta}, \dot{\phi}, \{\dot{Q}_k\})\mathbf{A}(\dot{\rho}, \dot{\theta}, \dot{\phi}, \{\dot{Q}_k\})^T \qquad\text{(A.58)}$$

where the matrix $\mathbf{A}$ has the structure:

$$\mathbf{A} = \begin{bmatrix} A_{\rho\rho} & A_{\rho\theta} & A_{\rho\phi} & 0 & 0 & . \\ A_{\theta\rho} & A_{\theta\theta} & A_{\theta\phi} & 0 & 0 & . \\ A_{\phi\rho} & A_{\phi\theta} & A_{\phi\phi} & 0 & 0 & . \\ 0 & 0 & 0 & 1 & 0 & . \\ 0 & 0 & 0 & 0 & 1 & . \\ . & . & . & . & . & . \end{bmatrix} \qquad\text{(A.59)}$$

Inserting the expression (5.256), and retaining only up to linear coupling terms in Q_k, we obtain the following elements for the $\mathbf{A}$ matrix:

$$A_{\rho\rho} = \sum_i \boldsymbol{\ell}_{i\rho} \cdot \boldsymbol{\ell}'_{i\rho} + 2\sum_k Q_k B^\rho_{k\rho} \qquad\text{(A.60)}$$

$$A_{\rho\theta} = \frac{\rho}{2}\sum_i \boldsymbol{\ell}_{i\rho} \cdot \boldsymbol{\ell}_{i\theta} + \sum_k Q_k \left(B^\theta_{k\rho} + \frac{\rho}{2}B^\rho_{k\theta}\right) \qquad\text{(A.61)}$$

$$A_{\rho\phi} = \frac{\rho}{2}\sum_i \boldsymbol{\ell}_{i\rho} \cdot \boldsymbol{\ell}_{i\phi} + \sum_k Q_k \left(B^\phi_{k\rho} + \frac{\rho}{2}B^\rho_{k\phi}\right) \qquad\text{(A.62)}$$

$$A_{\theta\theta} = \frac{\rho^2}{4}\sum_i \boldsymbol{\ell}_{i\theta} \cdot \boldsymbol{\ell}_{i\theta} + \rho\sum_k Q_k B^\theta_{k\theta} \qquad\text{(A.63)}$$

$$A_{\theta\phi} = \frac{\rho^2}{4}\sum_i \boldsymbol{\ell}_{i\theta} \cdot \boldsymbol{\ell}_{i\phi} + \frac{\rho}{2}\sum_k Q_k(B^\phi_{k\theta} + B^\theta_{k\phi}) \qquad\text{(A.64)}$$

$$A_{\phi\phi} = \frac{\rho^2}{4}\sum_i \boldsymbol{\ell}_{i\phi} \cdot \boldsymbol{\ell}_{i\phi} + \rho\sum_k Q_k B^\phi_{k\phi} \qquad\text{(A.65)}$$

where

$$B^\rho_{k\rho} = \sum_i \ell_{i\rho} \cdot \frac{\partial \ell_{ik}}{\partial \rho} \tag{A.66}$$

$$B^\theta_{k\theta} = \sum_i \ell_{i\theta} \cdot \frac{\partial \ell_{ik}}{\partial \theta} \tag{A.67}$$

$$B^\theta_{k\rho} = \sum_i \ell_{i\rho} \cdot \frac{\partial \ell_{ik}}{\partial \theta} \tag{A.68}$$

$$B^\rho_{k\theta} = \sum_i \ell_{i\theta} \cdot \frac{\partial \ell_{ik}}{\partial \rho} \tag{A.69}$$

and so on. We now define the momenta $p_\rho = \partial T/\partial \dot\rho$, and so on. We then obtain the kinetic energy in terms of the momenta as

$$2T_{kin} = \sum_k p_k^2 + (p_\rho p_\theta p_\phi)\mathbf{A}^{-1}(p_\rho p_\theta p_\phi)^T \tag{A.70}$$

The matrix $\mathbf{A}$ can be inverted using, for example, only the linear terms in the coordinates Q_k. The vectors $\ell_{i\rho}$, $\ell_{i\theta}$, and $\ell_{i\phi}$ are constructed such that they are orthonormal for a three-particle system. For $n > 3$, this is only approximately true and depends on the reference vectors $\mathbf{a}_i$ for $(i > 3)$. For some systems, such as H_2+OH [208] or H_2+CN [209], the orthonormality is nearly fulfilled. If this is the case, then simplifications can be introduced. For other systems (large) we may be able to neglect the coupling to the overall rotational motion. Hence, the above expression for the kinetic energy can be used. If the matrix $\mathbf{A}$ is diagonally dominant, the following simplified expression is obtained

$$H = \frac{1}{2}\sum_k (p_k^2 + 2a_k Q_k + \omega_k^2 Q_k^2) + V_0(\rho, \theta, \phi) + \frac{p_\rho^2}{2A_{\rho\rho}} + \frac{p_\theta^2}{2A_{\theta\theta}}$$
$$+ \frac{p_\phi^2}{2A_{\phi\phi}} - \frac{A_{\rho\theta}}{A_{\rho\rho}A_{\theta\theta}}p_\theta p_\rho - \frac{A_{\rho\phi}}{A_{\rho\rho}A_{\phi\phi}}p_\rho p_\phi - \frac{A_{\theta\phi}}{A_{\theta\theta}A_{\phi\phi}}p_\theta p_\phi \tag{A.71}$$

where a_k denotes the linear term

$$a_k = \sum_i \mathbf{g}_i \cdot \ell_{ik} \tag{A.72}$$

where $\mathbf{g}_i = \frac{1}{\sqrt{m_i}}\frac{\partial V}{\partial \mathbf{r}_i}|\mathbf{a}_i$. This hamiltonian can be used for large systems with diagonally dominant $\mathbf{A}$ matrices. If the matrix is not diagonally dominant, we have to use the expression (A.70) above. However, for many systems, such as those mentioned above, it is not expected to be a good approximation to neglect the rotational motion.

The rotational coupling

In the equations above, we have only considered the $3N - 9$ vibrational and the 3 hyper-spherical degrees of freedom. Coupling to the over-all rotational motion is most important for small systems, but it too can be treated within the linear approximation as follows:

The two rotational terms are given as the rotational/vibrational and the rotational contributions to the kinetic energy, that is,

$$2T_{int} = 2\omega \cdot \sum_i m_i \mathbf{r}_i \times \dot{\mathbf{r}}_i \tag{A.73}$$

and

$$2T_{rot} = \sum_i m_i(\omega \times \mathbf{r}_i) \cdot (\omega \times \mathbf{r}_i) \tag{A.74}$$

Introducing the Eckart condition, we get

$$2T_{int} = -\omega_z\rho^2\cos\theta\dot{\phi} + 2\omega_x \sum_k X_k\dot{Q}_k + 2\omega_y \sum_k Y_k\dot{Q}_k + 2\omega_z \sum_k Z_k\dot{Q}_k$$
$$+ \dot{\theta}\rho(\omega_x\bar{X}_\theta + \omega_y\bar{Y}_\theta + \omega_z\bar{Z}_\theta) + \dot{\phi}\rho(\omega_x\bar{X}_\phi + \omega_y\bar{Y}_\phi + \omega_z\bar{Z}_\phi)$$
$$+ 2\dot{\rho}(\omega_x\bar{X}_\rho + \omega_y\bar{Y}_\rho + \omega_z\bar{Z}_\rho) \tag{A.75}$$

where the contribution from the atoms 1, 2, and 3 has been pulled out (the first term) and where

$$X_k = \sum_{k'} Q_{k'}\xi_{k'k}^{(x)} \tag{A.76}$$

$$\bar{X}_\rho = \sum_k Q_k\xi_{k\rho}^{(x)} - \sum_{j=1}^3 \sqrt{m_j} \sum_{k=4}^N \frac{m_k}{M_k}\left(a_{ky}^0\ell_{j\rho}^{(z)} - a_{kz}^0\ell_{j\rho}^{(y)}\right)$$
$$- \sum_{j=1}^3 \sum_{k=4}^N \frac{m_k}{M_k}\left(a_{jy}\frac{\partial a_{kz}^0}{\partial\rho} - a_{jz}\frac{\partial a_{ky}^0}{\partial\rho}\right) + \sum_{j=4}^N \sqrt{m_j}\left(a_{jy}\ell_{j\rho}^{(z)} - a_{jz}\ell_{k\rho}^{(y)}\right)$$
$$\tag{A.77}$$

and similar expressions for $\bar{X}_\theta$ with ρ replaced by θ and so on. The only reason for this partitioning is to pull out the correct hamiltonian for the $N = 3$ case. We have, furthermore, used the notation

$$\xi_{kk'}^{(x)} = \sum_i \left(\ell_{ik}^{(y)}\ell_{ik'}^{(z)} - \ell_{ik}^{(z)}\ell_{ik'}^{(y)}\right) \tag{A.78}$$

$$\xi_{k\theta}^{(x)} = \sum_i \left(\ell_{ik}^{(y)}\ell_{i\theta}^{(z)} - \ell_{ik}^{(z)}\ell_{i\theta}^{(y)}\right) \tag{A.79}$$

$$\xi_{k\phi}^{(x)} = \sum_i \left(\ell_{ik}^{(y)}\ell_{i\phi}^{(z)} - \ell_{ik}^{(z)}\ell_{i\phi}^{(y)}\right) \tag{A.80}$$

$$\xi_{k\rho}^{(x)} = \sum_i \left(\ell_{ik}^{(y)}\ell_{i\rho}^{(z)} - \ell_{ik}^{(z)}\ell_{i\rho}^{(y)}\right) \tag{A.81}$$

The rotational energy can be expressed as

$$2T_{rot} = A\omega_x^2 + B\omega_y^2 + C\omega_z^2 - 2D\omega_x\omega_y - 2E\omega_x\omega_z - 2F\omega_y\omega_z \tag{A.82}$$

where

$$A = \sum_i m_i(r_{iy}^2 + r_{iz}^2) \tag{A.83}$$

$$B = \sum_i m_i(r_{ix}^2 + r_{iz}^2) \tag{A.84}$$

$$C = \sum_i m_i(r_{ix}^2 + r_{iy}^2) \tag{A.85}$$

$$D = \sum_i m_i r_{ix} r_{iy} \tag{A.86}$$

$$E = \sum_i m_i r_{iy} r_{iz} \tag{A.87}$$

$$F = \sum_i m_i r_{ix} r_{iz} \tag{A.88}$$

Introducing the momenta $P_x = \partial T/\partial \omega_x$, and so on, we obtain the following connection between the momenta and the derivatives

$$(p_\rho p_\theta p_\phi P_x P_y P_z)^T = \mathbf{A}(\dot\rho \dot\theta \dot\phi \omega_x \omega_y \omega_z)^T \tag{A.89}$$

where

$$\mathbf{A} = \left\{ \begin{array}{cccccc} A_{\rho\rho} & A_{\rho\theta} & A_{\rho\phi} & \bar{X}_\rho & \bar{Y}_\rho & \bar{Z}_\rho \\[2mm] A_{\theta\rho} & A_{\theta\theta} & A_{\theta\phi} & \frac{\rho}{2}\bar{X}_\theta & \frac{\rho}{2}\bar{Y}_\theta & \frac{\rho}{2}\bar{Z}_\theta \\[2mm] A_{\phi\rho} & A_{\phi\theta} & A_{\phi\phi} & \frac{\rho}{2}\bar{X}_\phi & \frac{\rho}{2}\bar{Y}_\phi & \frac{\rho}{2}(\bar{Z}_\phi - \rho\cos\theta) \\[2mm] \bar{X}_\rho & \frac{\rho}{2}\bar{X}_\theta & \frac{\rho}{2}\bar{X}_\phi & A & -D & -F \\[2mm] \bar{Y}_\rho & \frac{\rho}{2}\bar{Y}_\theta & \frac{\rho}{2}\bar{Y}_\phi & -D & B & -E \\[2mm] \bar{Z}_\phi & \frac{\rho}{2}\bar{Z}_\theta & \frac{\rho}{2}(\bar{Z}_\phi - \rho\cos\theta) & -F & -E & C \end{array} \right\} \tag{A.90}$$

Notice that the derivation is carried to first order in p_k or Q_k. If second-order terms are included, the above coupling matrix will also contain coupling terms to the vibrational modes. The derivation including second-order terms is, in principle, straightforward but the final expression is rather lengthy.

It is, furthermore, convenient to pull out the moment of inertia connected to the three-atomic system. Thus, if we ignore the contribution from the vibrational displacements, we have

$$A^0 = \frac{1}{2}\rho^2(1 - \sin\theta) + \sum_{i=4} m_i(a_{iy}^2 + a_{iz}^2)$$

$$+ M_3\left[\left(\sum_{k=4}^N \frac{m_k}{M_k}a_{ky}^0\right)^2 + \left(\sum_{k=4}^N \frac{m_k}{M_k}a_{kz}^0\right)^0\right] \tag{A.91}$$

$$B^0 = \frac{1}{2}\rho^2(1 + \sin\theta) + \sum_{i=4} m_i(a_{ix}^2 + a_{iz}^2)$$

$$+ M_3 \left[\left(\sum_{k=4}^{N} \frac{m_k}{M_k} a_{kx}^0 \right)^2 + \left(\sum_{k=4}^{N} \frac{m_k}{M_k} a_{kz}^0 \right)^2 \right] \qquad (A.92)$$

$$C^0 = \rho^2 + \sum_{i=4} m_i(a_{ix}^2 + a_{iy}^2) + M_3 \left[\left(\sum_{k=4}^{N} \frac{m_k}{M_k} a_{kx}^0 \right)^2 + \left(\sum_{k=4}^{N} \frac{m_k}{M_k} a_{kz}^0 \right)^2 \right]$$
$$(A.93)$$

We now partition the **A** matrix as

$$\mathbf{A} = \begin{bmatrix} A_{\rho\rho} & 0 & 0 & 0 & 0 & 0 \\ 0 & A_{\theta\theta} & 0 & 0 & 0 & 0 \\ 0 & 0 & A_{\phi\phi} & 0 & 0 & -\frac{\rho^2}{2}\cos\theta \\ 0 & 0 & 0 & A & 0 & 0 \\ 0 & 0 & 0 & 0 & B & 0 \\ 0 & 0 & -\frac{\rho^2}{2}\cos\theta & 0 & 0 & C \end{bmatrix} + \mathbf{\Delta} \qquad (A.94)$$

where $\mathbf{\Delta}$ is defined by the equation. Introducing the notation $\mathbf{A}_0$ for the first matrix, we have

$$\mathbf{A}^{-1} \sim \mathbf{A}_0^{-1} - \mathbf{A}_0^{-1}\mathbf{\Delta}\,\mathbf{A}_0^{-1} \qquad (A.95)$$

where we can use the procedure given in ref. [97] to obtain

$$\mathbf{A}_0^{-1} =$$

$$\begin{bmatrix} A_{\rho\rho}^{-1} & 0 & 0 & 0 & 0 & 0 \\ 0 & A_{\theta\theta}^{-1} & 0 & 0 & 0 & 0 \\ 0 & 0 & A_{\phi\phi}^{-1} + \dfrac{\rho^4\cos^2\theta}{A_{\phi\phi}(4CA_{\phi\phi} - \rho^4\cos^2\theta)} & 0 & 0 & \dfrac{2\rho^2\cos\theta}{4CA_{\phi\phi} - \rho^4\cos^2\theta} \\ 0 & 0 & 0 & A^{-1} & 0 & 0 \\ 0 & 0 & 0 & 0 & B^{-1} & 0 \\ 0 & 0 & \dfrac{2\rho^2\cos\theta}{4CA_{\phi\phi} - \rho^4\cos^2\theta} & 0 & 0 & \dfrac{4A_{\phi\phi}}{4CA_{\phi\phi} - \rho^4\cos^2\theta} \end{bmatrix}$$
$$(A.96)$$

This gives, after some manipulation, the following contribution to the kinetic energy of the hamiltonian

$$2T = \frac{p_\rho^2}{A_{\rho\rho}} + \frac{p_\theta^2}{A_{\theta\theta}} + \frac{p_\phi^2}{A_{\phi\phi}} \left(1 + \frac{\rho^4 \cos^2\theta}{4CA_{\phi\phi} - \rho^4 \cos^2\theta} \right) + \frac{4p_\phi P_z \cos\theta \rho^2}{4CA_{\phi\phi} - \rho^4 \cos^2\theta}$$

$$+ \frac{P_x^2}{A} + \frac{P_y^2}{B} + \frac{4P_z^2 A_{\phi\phi}}{4CA_{\phi\phi} - \rho^4 \cos^2\theta} - \frac{8p_\theta p_\rho A_{\theta\rho}}{\rho^2} - \frac{8p_\phi p_\rho A_{\phi\rho}}{\rho^2 \sin^2\theta}$$

$$- \frac{32p_\theta p_\phi A_{\theta\phi}}{\rho^4 \sin^2\theta} - \frac{2}{A}\bar{X}_\rho p_\rho P_x - \frac{2}{B}\bar{Y}_\rho p_\rho P_y - \frac{2}{C}\bar{Z}_\rho p_\rho P_z - \frac{4}{\rho A}\bar{X}_\theta p_\theta P_x$$

$$- \frac{4}{\rho B}\bar{Y}_\theta p_\theta P_y - \frac{4}{\rho C}\bar{Z}_\theta p_\theta P_z - \frac{4}{\rho A}\frac{\bar{X}_\phi p_\phi P_x}{\sin^2\theta} - \frac{4}{\rho B}\frac{\bar{Y}_\phi p_\phi P_y}{\sin^2\theta}$$

$$- \frac{4\bar{Z}_\phi p_\phi P_z \cos^2\theta}{\rho^3 \sin^4\theta} - \frac{4}{\rho C}\frac{\bar{Z}_\phi p_\phi P_z}{\sin^2\theta} - \frac{2D}{AB}P_x P_y - \frac{2F}{AC}P_x P_z - \frac{2E}{BC}P_y P_z$$

$$\text{(A.97)}$$

For large systems, the overall rotational motion will, as mentioned, be of less importance. In such a case we can use the expression given previously. The rotational motion of the N-body problem is described using the Euler angles α, β, and γ, and the conjugate momenta from which the connection to the variables P_x, P_y, and P_z can be made through the equations

$$P_x = P_\alpha \cos\alpha \cot\beta + P_\beta \sin\alpha - P_\gamma \cos\alpha \csc\beta \tag{A.98}$$

$$P_y = P_\beta \cos\alpha - P_\alpha \sin\alpha \cot\beta + P_\gamma \sin\alpha \csc\beta \tag{A.99}$$

$$P_z = P_\alpha \tag{A.100}$$

Bibliography

[1]   *Encyclopedia of Computational Chemistry*, ed. P. von Rague Schleyer, Wiley, New York, 1998.

[2]   J. B. Delos, W. B. Thorson, and S. K. Knudson, Phys. Rev. **A6**, 709(1972); J. B. Delos and W. B. Thorson, Phys. Rev. **A6**, 720(1972); J. B. Delos, Rev. Mod. Phys. **53**, 287(1981).

[3]   R. P. Feynmann and A. R. Hibbs, *Quantum Mechanics and Path Integrals*, McGraw-Hill, New York, 1965.

[4]   A. P. Penner and R. Wallace, Phys. Rev. **A7**, 1007(1975).

[5]   J. Gauss and E. J. Heller, Comp. Phys. Comp. **63**, 375(1991).

[6]   D. Thirumalai and B. J. Berne, Comp. Phys. Comp. **63**, 415(1991).

[7]   B. J. Berne and D. Thirumalai, Ann. Rev. Phys. Chem. **37**, 401(1986).

[8]   G. Jolicard, J. Chem. Phys. **80**, 2476(1984).

[9]   G. D. Billing, J. Chem. Phys. **86**, 2617(1987).

[10] N. Bohr, Mat. Fys. Medd. Dan. Vidensk. Selsk. **18**, 8(1948).

[11] S. I. Sawada, R. Heather, B. Jackson, and H. Metiu, J. Chem. Phys. **83**, 3009(1985); R. Heather and H. Metiu, J. Chem. Phys. **84**, 3250(1986).

[12] P. A. M. Dirac, Proc. Cambridge Philos. Soc. **26**, 376(1930); J. Frenkel, *Wave Mechanics*, Clarendon, Oxford, 1934.

[13] A. D. McLachlan, Mol. Phys. **8**, 39(1964).

[14] M. F. Trotter, Proc. Am. Math. Soc. **10**, 545(1959).

[15] N. Makri and W. H. Miller, Chem. Phys. Lett. **151**, 1(1981).

[16] M. S. Child, *Semiclassical Mechanics and Molecular Applications*, Clarendon Press, Oxford, 1991.

[17] G. Wentzel, Zeitschrift Physik, **38**, 518(1926).

[18] J. H. van Vleck, Proc. Nat. Acad. Sci. **14**, 178(1928).

[19] F. Y. Hansen, N. E. Henriksen, and G. D. Billing, J. Chem. Phys. **90**, 3060(1989).

[20] R. Heather and H. Metiu, Chem. Phys. Lett. **118**, 558(1985).

[21] N. Makri and W. H. Miller, Chem. Phys. Lett. **139**, 10(1987); B. W. Spath and W. H. Miller, J. Chem. Phys. **104**, 95(1996).

[22] V. S. Filinov, Nucl. Phys. **B271**, 717(1986).

[23] K. G. Kay, J. Chem. Phys. **107**, 2313(1997).

[24] K. G. Kay, Chem. Phys. **137**, 165(1989).

[25] A. Vijay, R. Wyatt, and G. D. Billing, J. Chem. Phys. **111**, 10794(2000).

[26] N. Makri, Comp. Phys. Comm. **63**, 389(1991); J. Math. Phys. **36**, 2430(1995).

[27] M. Winstetter and W. Domcke, Phys. Rev. **A47**, 2838(1993).

[28] Y. Elran and K. G. Kay, J. Chem. Phys. **110**, 8912(1999); ibid. **110**, 3653(1999).

[29] E. J. Heller, J. Chem. Phys. **94**, 2723(1991).

[30] R. E. Wyatt, J. Chem. Phys. **111**, 4406(1999); C. L. Lopreore and R. E. Wyatt, Phys. Rev. Lett. **82**, 5190(1999); R. E. Wyatt, D. J. Kouri, and D. K. Hoffman, J. Chem. Phys. **112**, 10730(2000).

[31] B. K. Day, A. Askar, and H. Rabitz, J. Chem. Phys. **109**, 8770(1998).

[32] J. R. Laing and K. F. Freed, Phys. Rev. Lett. **34**, 849(1975); Chem. Phys. **19**, 91(1977).

[33] M. Ben-Nun, and T. J. Martinez, J. Chem. Phys. **108**, 7244(1998); T. J. Martinez, M. Ben-Nun, and R. D. Levine, J. Phys. Chem. **100**, 7884 (1996); M. Chaija and R. D. Levine, Chem. Phys. Lett. **304**, 385(1999).

[34] M. Ben-Nun and T. J. Martinez, J. Chem. Phys. **112**, 6113(2000).

[35] M. Ben-Nun and T. J. Martinez, J. Chem. Phys. **108**, 7244(1998).

[36] M. F. Herman and E. Kluk, Chem. Phys. **91**, 27(1984).

[37] Z. Kotler, E. Neria, and A. Nitzan, Comp. Phys. Comm. **63**, 243(1991).

[38] N. Makri and W. H. Miller, J. Chem. Phys. **87**, 5781(1987).

[39] R. B. Gerber and M. Ratner, Adv. Chem. Phys. **93**, 4781(1992).

[40] J. Campos-Marinez and R. D. Coalson, J. Chem. Phys. **93**, 4740(1990).

[41] E. Sim and N. Makri, J. Chem. Phys. **102**, 5616(1995).

[42] K. G. Kay, J. Chem. Phys. **101**, 2250(1994).

[43] N. T. Maitra, J. Chem. Phys. **112**, 531(2000).

[44] B. J. Berne and D. Thirumalai, Ann. Phys. Rev. Phys. Chem. **37**, 401(1987).

[45] J. Cao and G. A. Voth, J. Chem. Phys. **104**, 273(1996).

[46] I. Burghardt, H.-D. Meyer, and L. S. Cederbaum, J. Chem. Phys. **111**, 2927(1999).

[47] M. Thoss, W. H. Miller, and G. Stock, J. Chem. Phys. **112**, 10282(2000).

[48] S. A. Lebedeff, Phys. Rev. **A165**, 1399(1968).

[49] E. J. Heller, J. Chem. Phys. **62**, 1544(1975); **64**, 63(1976); **65**, 4879 (1976).

[50] S.-Y. Lee and E. J. Heller, J. Chem. Phys. **76**, 3035(1982); D. Huber and E. J. Heller, J. Chem. Phys. **90**, 7317(1989).

[51] V. Szalay, J. Chem. Phys. **99**, 1978(1993).

[52] R. D. Coalson and M. Karplus, Chem. Phys. Lett. **90**, 301 (1982); J. Chem. Phys. **93**, 3919(1990).

[53] H.-D. Meyer, Chem. Phys. **61**, 365(1981).

[54] K. B. Møller and N. E. Henriksen, J. Chem. Phys. **105**, 5037(1996).

[55] K. G. Kay, J. Chem. Phys. **91**, 170(1989).

[56] Y. Huang, D. J. Kouri, M. Arnold, L. Marchioro II, and D. K. Hoffman, J. Chem. Phys. **99**, 1028(1993).

[57] D. Bohm, Phys. Rev. **85**, 166(1952) ibid. **180**(1952); D. Bohm and B. J. Hiley, *The Undivided Universe*, Routeledge, London, 1993.

[58] A. Messiah, *Quantum Mechanics*, North-Holland, Amsterdam, 1969.

[59] E. R. Bittner and R. E. Wyatt, J. Chem. Phys. **113**, 8888 (2000).

[60] R. E. Wyatt, D. J. Kouri, and D. K. Hoffman, J. Chem. Phys. **112**, 10730(2000).

[61] S. Adhikari and G. D. Billing, Chem. Phys. **238**, 69(1998).

[62] A. M. Arthurs and A. Dalgarno, Proc. R. Soc. London Ser. **A256**, 540(1960).

[63] G. D. Billing, J. Chem. Phys. **99**, 2674(1993).

[64] G. D. Billing, J. Chem. Phys. **99**, 5849(1993).

[65] G. D. Billing, Int. Journ. Thermophys. **18**, 977(1997).

[66] G. D. Billing, J. Chem. Phys. **107**, 4286(1997).

[67] Y. Shimoni and D. Kouri, Chem. Phys. **66**, 675(1977).

[68] J. P. Dahl, Theor. Chim. Acta **81**, 319(1992).

[69] G. A. Worth, H.-D. Meyer, and L. S. Cederbaum, J. Chem. Phys. **109**, 3518(1998).

[70] G. A. Worth, H.-D. Meyer, and L. S. Cederbaum, J. Chem. Phys. **109**, 3518(1998).

[71] I. S. Gradshteyn and I. M. Ryzhik, *Table of Integrals Series and Products*, Academic Press, New York, 1965.

[72] F. A. Gianturco, S. Serna, A. Palma, G. D. Billing, and V. Zenevich, J. Phys. **B26**, 1839(1993).

[73] D. Secrest and B. R. Johnson, J. Chem. Phys. (1966).

[74] S-F. Wu and R. D. Levine, Mol. Phys. **22**, 881(1971); D. G. Truhlar and A. Kupperman, J. Chem. Phys. **56**, 2234(1972).

[75] G. D. Billing and G. Jolicard, Chem. Phys. **65**, 323(1982).

[76] X. Chapuisat and G. Bergeron, Chem. Phys. **36**, 397(1979).

[77] See, for example, J. M. Launay and M. Le Dourneuf, Chem. Phys. Lett. **169**, 473(1990); D. Neuhauser, R. S. Judson, R. L. Jaffe, M. Baer, and

D. J. Kouri, Chem. Phys. Lett. **176**, 546(1991); J. Z. H. Zhang, Chem. Phys. Lett. **181**, 63(1991).

[78] D. Rapp and T. Kassal, Chem. Rev. **69**, 61(1969).

[79] G. D. Billing (Sorensen), J. Chem. Phys. **59**, 6147(1973).

[80] M. R. Flannery, Phys. Rev. J. Chem. Phys. **183**, 321(1969); D. A. Micha, **78**, 7138(1983).

[81] E. P. Wigner, Phys. Rev. **40**, 749(1932).

[82] H. A. Rabitz and R. G. Gordon, J. Chem. Phys. **53**, 1815, 1831(1970); C. Nyeland and G. D. Billing, Chem. Phys. **13**, 417(1976); G. D. Billing, "Nonequilibrium vibrational kinetics," ed. M. Capitelli, *Topics in Current Physics*, vol. 39, Springer-Verlag, Berlin, 1986.

[83] D. M. Brink and G. R. Satchler, *Angular Momentum*, Clarendon, Oxford, 1968.

[84] N. Balakrishnan and G. D. Billing, Chem. Phys. Lett. **233**, 145(1995).

[85] V. A. Zenevich and G. D. Billing, J. Chem. Phys. **111**, 2401(1999); V. A. Zenevich, G. D. Billing, and G. Jolicard, Chem. Phys. Lett. **312**, 530(1999).

[86] G. D. Billing, Chem. Phys. **9**, 359(1975).

[87] G. D. Billing, Chem. Phys. Lett. **50**, 320(1977); Y. Shimoni and D. J. Kouri, J. Chem. Phys. **66**, 675(1977).

[88] G. D. Billing, Comp. Phys. Comm. **44**, 121(1987).

[89] K. P. Lawley and J. Ross, J. Chem. Phys. **43**, 2943(1965).

[90] W. H. Miller and T. F. George, J. Chem. Phys. **56**, 5668(1972).

[91] G. D. Billing, Chem. Phys. **107**, 39(1986).

[92] A. J. Banks, D. C. Clary, and H.-J. Werner, J. Chem. Phys. **84**, 3788(1986).

[93] M.-M. Maricq, E. A. Gregory, C. T. Wickham-Jones, D. C. Cartwright, and C. J. S. M. Simpson, Chem. Phys. **75**, 347(1983).

[94] G. D. Billing, "Vibration-Vibration and Vibration-Translation Energy Transfer, Including Multiquantum Transitions in Atom-Diatom and Diatom-Diatom Collisions," in *Topics in Current Physics*, ed. M. Capitelli, Springer-Verlag, Berlin, 1986; M. Cacciatore and G. D. Billing, J. Phys. Chem. **96**, 217(1992); G. D. Billing and R. E. Kolesnick, Chem. Phys. Lett. **200**, 382(1992); C. Coletti and G. D. Billing, J. Chem. Phys. **113**, 4869(2000).

[95] G. D. Billing and R. E. Kolesnick, Chem. Phys. Lett. **200**, 382(1992).

[96] S. Ormonde, Rev. Mod. Phys. **47**, 193(1975).

[97] B. R. Johnson, J. Chem. Phys. **79**, 1906(1983).

[98] J. T. Muckermann and G. D. Billing, J. Chem. Phys. **91**, 6830(1989).

[99] N. Markovic and G. D. Billing, Chem. Phys. **224**, 53(1997).

[100] J. Z. H. Zhang and W. H. Miller, J. Chem. Phys. **91**, 1528 (1989).

[101] H. Goldstein, "Classical Mechanics," Addison Wesley, Tokyo, 1964. The matrix given in the text is for rotation of the molecule rather than the coordinate system. It is therefore the inverse of what is denoted **A** in Goldstein.

[102] S. Adhikari and G. D. Billing, Chem. Phys. Lett. **305**, 109(1999).

[103] D. Neuhauser and M. Baer, J. Chem. Phys. **90**, 4351(1989).

[104] NIST Chemical Kinetics Database 17-2Q98, NIST Standard Reference Data, Gaithersburg, MD, 1998; A. R. Ravishankara and R. L. Thompson, Chem. Phys. Lett. **99**, 377(1983).

[105] C. Coletti and G. D. Billing, Chem. Phys. Lett. **342**, 65(2001).

[106] K. S. Bradley and G. C. Schatz, J. Chem. Phys. **106**, 8464(1997).

[107] K. G. Kay, J. Chem. Phys. **100**, 4432(1994).

[108] P. Pechukas, Phys. Rev. **181**, 166(1969), 174(1969).

[109] F. J. Webster, P. J. Rossky, and R. A. Friesner, Comp. Phys. Comm. **63**, 494(1991); F. J. Webster, E. T. Tang, P. J. Rossky, and R. A. Friesner, J. Chem. Phys. **100**, 4835(1994).

[110] D. F. Coker and L. Xiao, J. Chem. Phys. **102**, 496(1995).

[111] J. C. Tully, Int. J. Quant. Chem. **25**, 299(1991).

[112] L. Landau, Z. Physik, Sovjetunion **2**, 46(1932); C. Zener, Proc. Roy. Soc. (London) **A137**, 696(1932).

[113] C. Zhu and H. Nakamura, J. Chem. Phys. **101**, 10630(1994); ibid. **102**, 7448(1995).

[114] J. C. Tully and R. K. Preston, J. Chem. Phys. **55**, 562(1972); J. R. Stine and J. T. Muckerman, J. Chem. Phys. **65**, 3975(1976).

[115] A. Bjerre and E. E. Nikitin, Chem. Phys. Lett. **1**, 179(1967).

[116] H. S. W. Massey, Rep. Progr. Phys. **12**, 248(1949).

[117] J. R. Stine and J. T. Muckerman, J. Chem. Phys. **68**, 185(1978).

[118] J. C. Tully, J. Chem. Phys. **93**, 1061(1990).

[119] S. Hammes-Schiffer and J. C. Tully, J. Chem. Phys. **101**, 4657(1994).

[120] O. V. Prezhdo and P. Rossky, J. Chem. Phys. **107**, 825(1997).

[121] Y. L. Volobuev, M. D. Hack, M. S. Topaler, and D. G. Truhlar, J. Chem. Phys. **112**, 9716(2000).

[122] G. D. Billing, Chem. Phys. Lett. **343**, 130(2001).

[123] D. S. Sholl and J. C. Tully, J. Chem. Phys. **109**, 7702 (1998).

[124] A. Bastida, J. Zuniga, A. Requena, N. Halberstadt, and J. A. Beswick, J. Chem. Phys. **109**, 6320(1998); Chem. Phys. **240**, 229 (1999).

[125] S. Adhikari and G. D. Billing, J. Chem. Phys. **111**, 48(1999).

[126] A. Florescu, M. Sizun, and V. Sidis, Chem. Phys. **179**, 53(1994).

[127] L. Wang, J. Chem. Phys. **108**, 7538(1998); L. Wang and D. C. Clary **262**, 284(1996); L. Wang, W. J. Meurer, and A. B. McCoy, J. Chem. Phys. **113**, 10605(2000).

[128] J. D. Kelley, J. Chem. Phys. **56**, 6108(1972).

[129] G. D. Billing, Chem. Phys. **33**, 227(1978).

[130] G. D. Billing, Comp. Phys. Rep. **1**, 237(1984).

[131] P. Pechukas and J. C. Light, J. Chem. Phys. **44**, 3897(1966).

[132] E. Kerner, Can. J. Phys. **36**, 371(1958); C. E. Treanor, J. Chem. Phys. **43**, 532(1965).

[133] G. D. Billing and M. Baer, Chem. Phys. Lett. **48**, 372(1977).

[134] S. Adhikari and G. D. Billing, J. Chem. Phys. **113**, 1409(2000).

[135] G. D. Billing and S. Adhikari, Chem. Phys. Lett. **321**, 197(2000).

[136] G. D. Billing, J. Chem. Phys. **114**, 6641(2001); Chem. Phys. **264**, 71(2001).

[137] T. J. Park and J. C. Light, J. Chem. Phys. **85**, 5870(1986).

[138] G. D. Billing and K. V. Mikkelsen, *Advanced Molecular Dynamics and Chemical Kinetics*, Wiley, New York, 1997.

[139] J. K. Cullum and R. A. Willoughby, *Lanczos Algorithms for Large Symmetric Eigenvalue Computations*, Birkhaüser, Boston, 1985.

[140] M. D. Feit, J. A. Fleck, and A. Steiger, J. Comp. Phys. **47**, 412(1982).

[141] R. Kosloff, Ann. Rev. Phys. Chem. **45**, 145(1994).

[142] G. D. Billing, Chem. Phys. Lett. **339**, 237(2001).

[143] G. D. Billing, Chem. Phys. **46**, 123(1980).

[144] G. D. Billing, Chem. Phys. **61**, 415(1981).

[145] G. D. Billing, Chem. Phys. **70**, 223(1982).

[146] H. Margenau and G. M. Murphy, *The Mathematics of Physics and Chemistry*, D. van Nostrand Co., New York, 1956.

[147] G. D. Billing, Chem. Phys. **51**, 417(1980).

[148] G. D. Billing, *The Dynamics of Molecule Surface Interactions*, Wiley, New York, 2000.

[149] G. D. Billing, Comp. Phys. Rep. **12**, 383(1990).

[150] G. D. Billing and G. Jolicard, Chem. Phys. Lett. **221**, 75(1994).

[151] G. D. Billing, A. Guldberg, N. E. Henriksen, and F. Y. Hansen, Chem. Phys. **147**, 1(1990); G. D. Billing, J. Phys. Chem. **99**, 15378(1995); S. Adhikari and G. D. Billing, J. Chem. Phys. **112**, 3884(2000).

[152] G. D. Billing, J. Phys. Chem. **105**, 2340(2001).

[153] G. D. Billing, J. Chem. Soc. Faraday Trans. **86**, 1663(1990).

[154] W. Miller, Jr. *Lie Theory and Special Functions*, Academic Press, New York, 1968.

[155] K. T. Tang and J. P. Toennies, J. Chem. Phys. **80**, 3786(1984).

[156]   H. H. Nielsen, Rev. Mod. Phys. **23**, 90(1951).

[157]   A. D. Buckingham, Adv. Chem. Phys. **12**, 107(1967).

[158]   G. D. Billing, Chem. Phys. **76**, 315(1983).

[159]   See, for instance, J. R. Maple in *Encyclopedia of Computational Chemistry*, ed. P. von Rague Schleyer, Wiley, New York, 1998: 1015.

[160]   G. D. Billing, Chem. Phys. **79**, 179 (1983).

[161]   G. D. Billing, Chem. Phys. **173**, 167(1993).

[162]   G. D. Billing, Chem. Phys. **104**, 19(1986).

[163]   G. D. Billing, Chem. Phys. **127**, 107(1988).

[164]   G. C. Schatz and M. J. Redmon, Chem. Phys. **58**, 195(1981).

[165]   R. E. Center, J. Chem. Phys. **59**, 3523(1973).

[166]   D. S. Pollock, G. B. I. Scott, and L. F. Phillips, Geophys. Res. Lett. **20**, 727(1993).

[167]   G. D. Billing and D. L. Huestis, Eos, trans. AGU **81**, F931(2000).

[168]   T. N. Truong, W. Duncan, and R. L. Bell, in *Chemical Applications of Density-Functional Theory*, eds. B. B. Laird, R. B. Ross, and T. Ziegler, ACS Symposium Series, Washington D.C. 1996: 85; T. N. Truong and W. Duncan, J. Chem. Phys. **101**, 7408(1994).

[169]   R. A. Marcus, J. Chem. Phys. **45**, 4493(1966); **49**, 2610, 2617(1968).

[170]   G. L. Hofacker, Z. Naturforsch. **18a**, 607(1963); G. L. Hofacker and R. D. Levine, Chem. Phys. Lett. **33**, 2769(1975).

[171]   W. H. Miller, N. C. Handy, and J. E. Adams, J. Chem. Phys. **72**, 99(1980).

[172]   J. T. Hougen, P. R. Bunker, and J. W. C. Johns, J. Mol. Spectry. **34**, 136(1970), P. Jensen, Comp. Phys. Rep. **1**, 1(1983).

[173]   G. D. Billing, Chem. Phys. **89**, 199(1984).

[174]   E. B. Wilson, J. C. Decius, and P. C. Cross, *Molecular Vibrations*, McGraw-Hill, New York, 1955.

[175]   G. D. Billing, Chem. Phys. **146**, 63(1990).

[176]   N. Balakrishnan and G. D. Billing, J. Chem. Phys. **104**, 9482(1996).

[177]   G. D. Billing, Chem. Phys. **159**, 109(1992).

[178]   E. Kraka, "Reaction Path Hamiltonian and Its Use for Investigating Reaction Mechanisms," in *Encyclopedia of Computational Chemistry*, ed. J. W. Schleyer, Wiley, New York 1998: 2463.

[179]   G. D. Billing, Chem. Phys. **135**, 423(1989).

[180]   See, for instance, A. J. C. Varandas and A. A. C. C. Pais, in "Theoretical and computational models for organic chemistry," ed. S. J. Formosinho, I. G. Czisamdia, and L. G. Arnaut, Kluwer, Dordrecht, 1999: 55.

[181]   F. London, Z. Elektrochem. **35**, 552(1929); H. Eyring and M. Polanyi, Z. Physik. Chem. **B12**, 279(1931); S. Sato, J. Chem. Phys. **23**, 592(1955).

[182] G. C. Schatz, in "Reaction and Molecular Dynamics," *Lecture Notes in Chemistry*, eds. A. Lagana and A. Riganelli, Springer-Verlag, Berlin, 2000. See, also, J. N. Murrell, S. Carter, S. C. Farantos, P. Huxley, and A. J. C. Varandas, "Molecular Potential Energy Functions," Wiley, London, 1984; D. M. Hirst, *Potential Energy Surfaces: Molecular Structure and Reaction Dynamics*, Taylor & Francis, London, 1985.

[183] C. A. Mead and D. G. Truhlar, J. Chem. Phys. **70**, 2284(1979).

[184] A. Kuppermann and Y. M. Wu, Chem. Phys. Lett. **205**, 577(1993); Y. M. Wu, B. Lepetit, and A. Kuppermann, **186**, 319(1991).

[185] Y. M. Wu and A. Kuppermann, Chem. Phys. Lett. **201**, 178(1993); B. Kendrick and R. T. Pack, J. Chem. Phys. **104**, 7475(1996); N. Markovic and G. D. Billing, J. Chem. Phys. **101**, 2953(1994); S. Adhikari and G. D. Billing, Chem. Phys. Lett. **289**, 219(1998).

[186] C. Eckart, Phys. Rev. **35**, 1303(1930).

[187] N. Markovic and G. D. Billing, Chem. Phys. **191**, 247(1995); Chem. Phys. Lett. **248**, 420(1996); F. Aguillon, M. Sizun, V. Sidis, G. D. Billing, and N. Markovic, J. Chem. Phys. **104**, 4530(1996).

[188] G. D. Billing, Chem. Phys. **159**, 109(1992).

[189] G. D. Billing, Chem. Phys. **161**, 245(1992).

[190] G. D. Billing and K. V. Mikkelsen, *Introduction to Molecular Dynamics and Chemical Kinetics*, Wiley, New York, 1996.

[191] S. Adhikari and G. D. Billing, Chem. Phys. **259**, 149(2000).

[192] M. Baer, Chem. Phys. Lett. **329**, 450(2000); ibid. **322**, 520(2000).

[193] M. ter Horst, G. C. Schatz, and L. B. Harding, J. Chem. Phys. **105**, 558(1996); T. Takayanagi, M. ter Horst, and G. C. Schatz, J. Chem. Phys. **105**, 2309(1996).

[194] B. Atakan, A. Jacobs, M. Wahl, R. Weller, and J. Wolfrum, Chem. Phys. Lett. **154**, 449(1990).

[195] Q. Sun, D. L. Yang, N. S. Wang, J. M. Bowman, and M. C. Lin, J. Chem. Phys. **93**, 4730(1990).

[196] I. R. Sims and I. W. M. Smith, Chem. Phys. Lett. **149**, 565 (1988).

[197] K. Yamashita, T. Yamabe, and K. Fukui, Theor. Chim. Acta **60**, 523(1982).

[198] W. Quapp, in *The Reaction Path in Chemistry*, ed. D. Heidrich, Kluwer Academic Publishers, Dordrecht, 1995: 95.

[199] See, for instance: R. Elber and M. Karplus, Chem. Phys. Lett. **139**, 375(1987); R. Czerminski and R. Elber, J. Chem. Phys. **92**, 5580(1990).

[200] A. Kuppermann, J. Phys. Chem. **83**, 171(1979).

[201] G. D. Billing, J. Chem. Phys. **107**, 4286(1997); **110**, 5526(1999).

[202] T. Carrington and W. H. Miller, J. Chem. Phys. **84**, 4364(1986).

[203] G. D. Billing, Comp. Phys. Rep. **1**, 237(1984).

[204] G. D. Billing, Chem. Phys. **146**, 63(1990); ibid. **159**, 109(1992).

[205] W. Tsang, J. Phys. Chem. Ref. Data **21**, 753(1992).

[206] B. R. Johnson, J. Chem. Phys. **79**, 1916(1983).

[207] G. D. Billing and N. Markovic, J. Chem. Phys. **99**, 2674(1993).

[208] G. D. Billing, Mol. Phys. **89**, 355(1996).

[209] C. Coletti and G. D. Billing, PCCP **1**, 4141(1999).

[210] A. G. Redfield, Adv. Magn. Res. **1**, 1(1965).

[211] S. M. Anderson, J. I. Zink and D. Neuhauser, Chem. Phys. Lett. **291**, 387(1998).

[212] M. B. Sevryuk and G. D. Billing, Chem. Phys. **185**, 199(1994).

[213] D. W. Oxtoby, Adv. Chem. Phys. **47**, 487(1981).

[214] J. S. Bader and B. J. Berne, J. Chem. Phys. **100**, 8359(1994).

[215] G. D. Billing and K. V. Mikkelsen, Chem. Phys. **182**, 249(1994).

[216] J. L. Skinner, J. Chem. Phys. **107**, 8717(1997).

[217] R. Zwanzig, J. Chem. Phys. **34**, 1931(1961).

[218] G. D. Billing, Chem. Phys. Lett. **76**, 178(1980).

[219] D. W. Chandler and G. E. Ewing, J. Chem. Phys. **73**, 4904 (1980); Chem. Phys. **54**, 241(1981).

[220] J. Cao and G. A. Voth, J. Chem. Phys. **99**, 10070(1993); ibid. **100**, 5106(1994); ibid. **101**, 6168(1994).

[221] G. A. Voth, Adv. Chem. Phys. **93**, 135(1996).

[222] G. D. Billing, *Introduction to the Theory of Inelastic Collisions in Chemical Kinetics*, Copenhagen (1978).

[223] V. S. Filinov, Mol. Phys. **88**, 1517(1996).

[224] S. Jang and G. A. Voth, J. Chem. Phys. **111**, 2357(1999); ibid. **112**, 8747(2000).

[225] U. W. Schmitt and G. A. Voth, J. Chem. Phys. **111**, 9361(1999).

[226] D. Marx, M. Tuckerman, J. Hutter, and M. Parinello, Nature **397**, 601(1999).

[227] D. Zahn and J. Brickmann, Chem. Phys. Lett. **331**, 224(2000).

Index